PRAISE FOR
ISIS: THE STATE OF TERROR

"Jessica Stern and J.M. Berger's new book, *ISIS*, should be required reading for every politician and policymaker.... Their smart, granular analysis is a bracing antidote to both facile dismissals and wild exaggerations.... Stern and Berger offer a nuanced and readable account of the ideological and organizational origins of the group."
—*Washington Post*

"The authors do nimble jobs of turning their copious research and their own expertise on terrorism into coherent, accessible narratives that leave us with an understanding of the Islamic State's history and metastasis, and its modus operandi."
—*New York Times*

"Understanding ISIS, who it appeals to and why, as well as how it sees itself, isn't something we're supposed to do. One purpose of ISIS' savagery is to make us react without thinking, to compel us to view the world as it does, as a stark conflict between good and evil demanding immediate, dramatic action. In that light, consider *ISIS: The State of Terror*, a profound act of counterterrorism."
—*Salon*

"Stern and Berger draw on Internet-based sources, big-brained research on political violence and some of the most acute thinking about the insurgency that is around today."
—*The Literary Review*

"Jessica Stern and J.M. Berger have produced a clear and succinct account of the rise of the fanatics.... This book's achievement is

to demonstrate how ISIS fits within the spectrum of blood-soaked jihadism."
—*The Telegraph* (UK)

"One can only conclude, with the clarity of recent hindsight, that we should have seen it coming—at least when seen through the lens of *ISIS: The State of Terror,* a new history of the threat by US academics Jessica Stern and J.M. Berger. . . . A timely and important history of a movement that now defines the 21st century."
—*The Evening Standard* (UK)

"By far the most important contribution yet to our understanding of an organization that remains cloaked in mystery and misunderstanding. . . . A brisk, readable, and eye-opening account of ISIS's past, present, and future. This is a book every American should read."
—Reza Aslan, author of *No god But God* and *Zealot: The Life and Times of Jesus of Nazareth*

"A timely and urgent book that is essential reading for analysts and policy makers alike. In what is already a cornerstone contribution, Stern and Berger offer the kind of cold-blooded analysis so desperately needed on the poorly understood phenomenon that is the so-called Islamic state."   —John Horgan, author of *The Psychology of Terrorism*

"The first serious book to analyze the rise of ISIS. . . . Stern and Berger write clearly and persuasively and marshal impressive primary research from ISIS's prodigious propaganda to help explain how ISIS became the dominant jihadi group today. It's a terrific and important read."
—Peter Bergen, author of *Manhunt: The Ten-Year Search for Bin Laden from 9/11 to Abbottabad*

ISIS

ALSO BY JESSICA STERN

Denial: A Memoir of Terror

Terror in the Name of God: Why Religious Militants Kill

The Ultimate Terrorists

ALSO BY J.M. BERGER

Jihad Joe: Americans Who Go to War in the Name of Islam

An Imprint of HarperCollinsPublishers

ISIS

THE STATE OF TERROR

JESSICA STERN
and J.M. BERGER

HarperCollins books may be purchased for educational, business,
or sales promotional use. For information please e-mail the
Special Markets Department at SPsales@harpercollins.com.

A hardcover edition of this book was published in 2015
by Ecco, an imprint of HarperCollins Publishers.

FIRST ECCO PAPERBACK EDITION PUBLISHED 2016.

Library of Congress Cataloging-in-Publication Data has been applied for.

ISBN 978-0-06-239555-9

19 20   DIX/LSC   10 9 8 7 6 5

CONTENTS

Abu Sayyaf Group: A jihadist organization in the Philippines founded with funds from al Qaeda. It has pledged loyalty to ISIS.

Ahrar al Sham: The second most significant anti-Assad jihadist group behind Jabhat al Nusra; a member of the Syrian Islamic Front coalition.

al Qaeda, al Qaeda Central (AQ, AQC): A global Salafi Sunni militant jihadi organization founded by Osama bin Laden and others in Afghanistan. It is now run by Ayman al Zawahiri.

al Qaeda in the Arabian Peninsula (AQAP): An al Qaeda affiliate based in Yemen and Saudi Arabia.

al Qaeda in Iraq (AQI): A jihadi group in Iraq founded by Abu Musab al Zarqawi, which would later become the **Islamic State of Iraq** and later still, the **Islamic State of Iraq and al-Sham,** or **ISIS.** It refers to itself now simply as **the Islamic State.**

al Qaeda in the Islamic Maghreb (AQIM): An al Qaeda affiliate that operates in the Sahara and Sahel region of North Africa.

al Shabab: An al Qaeda affiliate in Somalia.

Ansar Bayt al Maqdis (ABM): A jihadist group that arose following the Arab Spring in the Sinai region of Egypt. It has declared its territory in the Sinai to be a province of ISIS.

Ansar al-Islam: A Kurdish separatist and jihadi organization active in Iraq in 2003.

Ansar al-Sharia (AST): A jihadist organization in Tunisia.

Awakening, or **Awakening Movement:** Former Sunni Arab insurgents who joined the fight against jihadi groups in Iraq. Also known as the **Sons of Iraq.**

Ba'ath Party: A political party founded in Syria that merged socialism with anti-imperialism, Arab nationalism, and pan-Arabism. Saddam Hussein and Bashar al Assad were affiliated with the Ba'athist parties in Iraq and Syria, respectively.

Ba'athists: Members of the Ba'ath party.

bayah: A religiously binding oath of loyalty.

Bilad al Sham: Refers to historical Greater Syria that extended into regions of Palestine, Israel, Jordan, Lebanon, Syria, and Iraq; also called **the Levant.**

Boko Haram: A fundamentalist jihadi group in Nigeria.

caliph: Ruler of the Muslim community; a political successor of Muhammad.

caliphate: A political-religious state led by a caliph.

Daesh or **Daash:** A derogatory term for ISIS based on its acronym in Arabic.

Dawla: The Arabic word for "state," often used as a name for ISIS by its supporters.

Eid al-Fitr: The last day of Ramadan, the Islamic month of fasting and religious reflection.

emir: Arabic for commander; literally "prince."

fitna: An Arabic word referring to a period of internal dissent and infighting in Islamic history, also used to refer to similar conflicts in a modern context.

Free Syrian Army (FSA): Originally consisting of Syrian military defectors, the FSA is now an umbrella organization for secular, nationalist anti-Assad fighters.

hadith (plural, ahadith): Stories about Muhammad, his sayings, and historical figures within Islam, which are understood to have varying degrees of authenticity. Many Islamic end times traditions and prophecies are derived from *ahadith*.

Hezb-e-Islami: An Afghan militant group.

hijra: Migration, emigration.

International Security Assistance Force (ISAF): NATO's international security force in Afghanistan. Its role is to support the Afghan National Security Forces as they increase capacity.

Islamic Army of Iraq: A former Iraqi Sunni Arab insurgency group that formed following the 2003 invasion. Following the 2011 withdrawal of American troops it demilitarized and formed a political opposition group.

Islamic Front: A coalition of Islamist rebel groups in Syria, not including the al Qaeda affiliate Jabhat al Nusra.

Islamic State (IS): Name of ISIS after its declaration of a caliphate in June 2014.

Islamic State of Iraq (ISI): The name of the al Qaeda–affiliated insurgent group in Iraq (and its allies) from the death of Zarqawi in 2006 until 2012.

Islamic State of (or "in") Iraq and Syria (ISIS): Also called Islamic State of Iraq and al Sham. The successor group to the **Islamic State of Iraq,** following its expansion from Iraq into neighboring Syria. The acronym **ISIS** is still widely used, despite the fact that the group officially changed its name to the **Islamic State** in June 2014.

Jabhat al Nusra (Nusra): The al Qaeda affiliate in Syria; also known as the Nusra Front.

Jamaat Jaysh Ahl al-Sunnah wa-al-Jamaah: The Army of the Sunni People. A Sunni insurgent group that formed following the 2003 invasion of Iraq. ISIS emir Abu Bakr al-Baghdadi was reportedly a cofounder of this group.

Jemaah Islamiyah: A now-defunct Indonesian jihadi organization that had strong ties to al Qaeda.

jihad: Arabic word meaning "struggle." It has been used to describe a broad range of actions from spiritual struggles to armed conflict.

jihadi Salafism: A branch of Salafism that believes that any government that does not rule though **Shariah** is an illegitimate infidel regime. Jihadi Salafism embraces the use of violence to overthrow these regimes.

Jund al Khalifa: A splinter group of **AQIM** in Algeria that has become part of **ISIS**. It is responsible for the beheading of a French tourist in response to ISIS's call for such actions from supporters.

Khorasan Group: A cell of senior al Qaeda Central operatives dispatched to Syria to plan and coordinate attacks on the West.

kuffar: Infidels; unbelievers.

Kurds: An ethnic group centered in the Middle East, whose ancestral homeland crosses several modern-day borders.

Mahdi: An Islamic end-times figure believed to appear around the time of the Day of Judgment. Sometimes referred to as the Rightly Guided One, or the Hidden Imam.

Mujahid (plural, mujahideen): A Muslim fighter waging military jihad.

Muhajir (plural, muhajireen or **muhajiroun):** Emigrant. Often used to refer to foreign fighters taking part in military jihad. The plural form differs depending on the grammar of a sentence in Arabic.

al Muhajiroun: A radical Islamic organization in Britain led by Anjem Choudary. The organization has been disbanded, but successor social networks remain active.

mujtahidun: Literally "the industrious ones," a term used to refer to very active ISIS supporters on social media.

nasheed (plural, anasheed): An Islamic religious chant.

niqab: A black cloth veil worn by some Muslim women that covers part of the face and the entire body.

nusayri: A derogatory term for people who practice a variant of Shia Islam common among members of the Syrian regime.

peshmerga: Highly trained Kurdish fighters in Iraq; the standing army for the semiautonomous Iraqi Kurdistan region.

political Salafism: A branch of Salafism that pursues the purification of Islam through involvement in politics.

quietist Salafism: A branch of Salafism whose central goal is to purify Islam. They do not identify as political actors nor do they participate in politics.

rafidah: A derogatory term for Shia Muslims.

Salafi: A fundamentalist Sunni Islamic movement that believes in strict adherence to Islam as they believe it was practiced by Muhammad.

Shariah: The Islamic moral code and religious law. There are considerable disagreements among Muslims about how Shariah figures into modern life. **ISIS** and **AQ**-affiliated groups embrace a harsh interpretation, but even they differ over the details.

Shia Islam: A branch of Islam that recognizes Ali, Muhammad's son-in-law, and only his descendants as the rightful leaders of the Muslim community.

shaykh or **sheikh:** An honorific denoting respect for an individual as a leader or influencer within a tribe, clan, country, or Islamic religious group.

shahid: A martyr.

Sons of Iraq: Former Sunni Arab insurgents who joined the fight against jihadi groups in Iraq. More commonly known as the **Awakening Movement**.

sunnah: The recorded traditions of Muhammad.

Sunni Islam: The largest branch of Islam. Frequently referred to as "mainstream" or "orthodox" Islam.

Tablighi Jamaat: An Islamic revivalist movement founded in response to a preserved corruption of moral values. The movement aims to bring Muslims across all social and economic spectra into their understanding of religion by encouraging community service, contemplation, and proselytizing.

takfir: The pronouncement of a Muslim as an apostate. Usually understood by jihadists as a religious authorization to kill the subject.

Taliban: An Islamic fundamentalist organization founded in Pakistan, which later spread to Afghanistan, where it controlled the government from 1996 to 2001. It continues to be a significant insurgent movement.

Tehrik-e-Taliban (TTP): A Pakistani insurgent group linked to al Qaeda, which has splintered in recent years due to a number of internal divisions, including but by no means limited to support for ISIS among some members.

ummah: The worldwide Muslim community.

wilayat: Province. A governing substructure used by ISIS.

Yazidis: A Kurdish-speaking religious and ethnic minority in Iraq; ISIS believes the Yazidis to be devil worshippers who may be killed or enslaved with impunity.

TIMELINE

March 20, 2003	President George W. Bush announces the start of war against Iraq.
April 9, 2003	U.S.-led invasion topples Saddam Hussein's government in Iraq.
May 2003	Zarqawi-led group called the Organization of Monotheism and Jihad begins operations in Iraq.
August 2003	Zarqawi's group bombs United Nations headquarters in Baghdad.
April 2004	Hundreds are reported killed in fighting during the monthlong U.S. military siege of the Sunni Muslim city of Fallujah.
April 2004	Photographic evidence emerges of abuse of Iraqi prisoners by U.S. troops in Abu Ghraib prison near Baghdad.
May 2004	Zarqawi begins videotaped beheadings in Baghdad.
June 2004	United States hands sovereignty to Iraq's interim government headed by Prime Minister Iyad Allawi.
October 2004	Zarqawi swears loyalty to Osama bin Laden and founds al Qaeda in Iraq (AQI).
December 2004	Abu Bakr al-Baghdadi released from United States' Camp Bucca after six months' imprisonment.
January 2005	AQI starts a campaign of public beheadings on the streets of Iraqi cities.

April 2005	AQI becomes a foreign fighter magnet and targets Shi'a, much to the concern of bin Laden's al Qaeda.
May 2005	Surge in car bombings, bomb explosions, and shooting in Iraq.
October 2005	Voters approve a new constitution, which aims to create an Islamic federal democracy in Iraq.
December 2005	Iraqis vote for the first, full-term government and parliament.
February 2006	Bombing of the Shi'a al Askari Mosque in Samarra, Iraq; full sectarian conflict ensues.
April 22, 2006	Newly reelected president Jalal Talabani, a Kurd, asks Shi'a compromise candidate Nouri al Maliki to form a new government in Iraq, ending months of deadlock.
June 2006	Zarqawi killed in U.S. military air strike.
October 2006	Islamic State of Iraq (ISI) is formed; Abu Omar al Baghdadi named new leader.
December 2006	Saddam Hussein is executed by the Iraqis in Camp Justice, a joint Iraqi-American base in a suburb of Baghdad, for crimes against humanity.
January 2007	U.S. military surge and Sunni Awakening begin to greatly diminish ISI.
January 2008	The Iraqi parliament passes legislation allowing former officials from Saddam Hussein's Ba'ath party to return to public life.
March 2008	Prime Minister Maliki orders crackdown on militia in Basra, sparking pitched battles with Moqtada al Sadr's Mehdi Army, a Shi'a militia group.
May 2008	Relentless pressure on ISI and other groups by the U.S. military and government of Iraq results in lowest levels of violence since 2005..
September 2008	U.S. forces hand control of Anbar province, once an insurgent and al Qaeda stronghold, to the Iraqi govern-

ment. This is the first Sunni province to be returned to the Shi'a-led government.

January 2009 Prime Minister Maliki targets Sunni leaders and Awakening groups, increasing sectarian tensions and latent support for ISI in Sunni tribal areas. This lessens the pressure on ISIS, allowing it to stave off disaster.

August 2009 ISI bombs Iraqi ministries of Foreign Affairs and Finance, killing hundreds.

April 2010 ISI leaders Abu Omar al Baghdadi and Abu Ayyub al Masri (aka Abu Hamza al Muhajir) are killed in U.S.-led air strike.

May 2010 Abu Bakr al-Baghdadi named leader of ISI.

March 6, 2011 In the city of Daraa, Syria, near the Jordan border, nearly a dozen boys under the age of fifteen are arrested for anti-regime graffiti. Protests break out in Syria beginning in Daraa, but quickly spreading to neighboring villages.

April 21, 2011 President Assad issues a decree to end Syria's nearly fifty-year-old state of emergency in hopes of quelling the rising protests.

May 2, 2011 Al Qaeda Central leader Osama bin Laden is killed by U.S. special forces in Abbottabad, Pakistan.

May 28, 2011 Hamza al Khatib, a thirteen-year-old boy who was detained during protests in Syria, is delivered to his family as a mutilated corpse, exposing the brutality of the regime.

June 3, 2011 In response to the release of Hamza's body, thousands flood the streets for the "Friday of the Children" protest. The regime responds by blocking access to the Internet from within Syria.

June 14, 2011 The Arab League condemns the Syrian crackdown for the first time.

August 2011	Saudi Arabia, Kuwait, and Bahrain recall their ambassadors to Syria. Leaders from the United States, France, Britain, and Germany call on Assad to resign.
December 2011	The United States concludes its operations in Iraq. The unity government immediately faces disarray, and Maliki issues an arrest warrant for Vice President Tariq Hashimi, a leading Sunni politician. The Sunni bloc boycotts parliament and the cabinet.
January 6, 2012	General Mustafa Ahmad al Sheikh, the highest-ranking person in the Syrian military to defect, joins the Free Syrian Army. He reveals that at least twenty thousand soldiers have already defected.
February 12, 2012	Ayman al Zawahiri calls on all Muslims to help overthrow Assad.
June 16, 2012	The United Nations suspends its monitoring mission in Syria because it is too dangerous to continue operations.
June 2012	ISI releases the first installment in its popular video series, *The Clanging of the Swords*.
July 2012	ISI announces the initiation of "Breaking Down the Walls" campaign, to "refuel" the group by freeing members from Iraqi prisons and by regaining lost ground.
August 2012	President Obama declares, amid rumors of chemical weapons use in Syria, that chemical weapons are a "red line" for action.
September 16, 2012	Iran confirms units of its Revolutionary Guard are helping Assad.
December 2012	Sunni Muslims in Iraq stage mass rallies across the country over several months, protesting perceived marginalization by the Shi'a government.
February 28, 2013	United States promises "nonlethal assistance" to Syrian rebels.
March 2013	Jabhat al Nusra becomes dominant in rebel areas.

March 10, 2013	Islamist groups set up Eastern Council, consolidating control of eastern Syria.
April 2013	The ISI announces that Jabhat al Nusra is its official Syrian offshoot and henceforth the merged group shall be known as the Islamic State of Iraq and Syria/Sham (ISIS). Al Nusra immediately rejects the statement and appeals to al Qaeda Central for judgment.
April 18, 2013	Britain and France claim chemical weapons have been used in Syria.
April 2013	Iraqi troops storm an antigovernment protest camp in Hawija, near Kirkuk, leaving more than fifty dead. This sparks Sunni outrage and the insurgency intensifies. By summer the country has entered full-blown sectarian war.
May 19, 2013	Jabhat al Nusra takes over oil fields and begins selling crude oil.
May 27, 2013	European Union ends arms embargo on Syrian rebels.
June 4, 2013	France and Britain confirm finding evidence of the use of sarin gas in Syria. Within a week the United States also independently confirms that sarin has been used.
July 2013	ISIS announces the initiation of "A Soldier's Harvest" campaign, designed to intimidate/liquidate/assassinate Iraqi security forces, and to establish control over territory. At least five hundred prisoners, mainly al Qaeda members, are freed from Taji and Abu Ghraib prisons.
July 24, 2013	The Israeli director of military intelligence warns that Syria is becoming a "center of global jihad."
August 2013	ISIS begins sustained attacks on Syrian rebel groups such as Liwa al Tawhid and Ahrar al Sham, and then al Nusra in Raqqa and Aleppo. This completely changes the nature of the rebellion in Syria.
August 14, 2013	ISIS pushes Syrian rebels out of Raqqa.

August 31, 2013	President Obama states that the United States has a moral responsibility to act in Syria, but that Congress must approve the use of military force.
October 2013	ISIS creates its first official Twitter account.
November 2013	Several leading Syrian rebel groups form new Islamic Front.
December 2013	Fighting widens between Syrian rebels and ISIS.
January 2014	After serious fighting, ISIS claims complete control over Raqqa, establishing the city as its de facto capital for Syria.
January 2014	Islamist fighters infiltrate Fallujah and Ramadi in Iraq. Iraqi forces recapture Ramadi, but ISIS forces are entrenched in Fallujah.
February 2014	Al Qaeda Central, led by Ayman al Zawahiri, publicly severs ties with ISIS; ISIS responds by saying they represent the spirit of AQ founder Osama bin Laden and not AQ as led by his successor, Zawahiri.
March 2014	ISIS supporters arrested in Switzerland for recruiting fighters and planning a terrorist attack.
April 2014	ISIS launches a Twitter app capable of sending tens of thousands of tweets per day.
May 2014	ISIS releases *The Clanging of the Swords Part 4,* possibly the most popular jihadist propaganda video of all time. The graphic video shows the execution of dozens of unarmed Iraqi soldiers.
May 24, 2014	Returned ISIS fighter Mehdi Nemmouche shoots and kills four people at the Jewish Museum in Brussels, Belgium.
June 2014	ISIS takes control of Mosul, Iraq's second-largest city, and border areas between Iraq and Syria, and claims the borders dating from the Sykes-Picot Agreement of 1916 are void.

June 2014	ISIS spams World Cup hashtags on Twitter with graphic images of executions. Twitter subsequently terminates the ISIS app, reducing the group's ability to broadcast its message.
June 30, 2014	ISIS announces the reestablishment of the caliphate and renames itself "the Islamic State."
July 2014	Abu Bakr al-Baghdadi leads prayer at a mosque in Mosul, his first public appearance. He emphasizes the existence of the caliphate and renames himself Caliph Ibrahim.
July 2014	ISIS releases the first issue of *Dabiq,* an English-language magazine.
August 8, 2014	United States begins air strikes against the Islamic State outside the Kurdish city of Irbil in Iraq.
August 2014	Despite U.S. air strikes and Iraqi, Kurdish, and Iranian forces, the Islamic State maintains control over large areas of Iraq and solidifies its positions in Syria.
August 2014	Twitter bans all official ISIS accounts.
August 25, 2014	The Islamic State releases a video showing the beheading of American journalist James Foley, who had been kidnapped by extremists in Syria in 2012.
September 2, 2014	The Islamic State releases a video showing the beheading of a second American journalist, Steven Sotloff. Obama announces that the United States will take action to "degrade and destroy" ISIS.
September 14, 2014	The Islamic State releases a video showing the beheading of British aid worker David Haines.
September 17, 2014	Australian police break up alleged ISIS plot to behead random people on the streets.
September 2014	Twitter suspends the accounts of hundreds of ISIS supporters.
September 21, 2014	ISIS spokesman Abu Mohammad al Adnani calls on "lone wolves" to attack in the West using whatever tools

are at hand, whether a gun, a knife, or even driving cars into pedestrians.

September 23, 2014   United States and coalition forces begin air strikes in Syria.

September 23, 2014   Australian ISIS supporter stabs two police officers.

October 2014   The Islamic State solidifies its hold in Mosul and in areas of Syria and advances on the vital wheat fields of Kobani, Syria, near the Turkish border.

October 3, 2014   The Islamic State releases the beheading video of Alan Henning, a British cabdriver turned aid worker. His execution causes a widespread campaign of Muslims condemning ISIS.

October 20, 2014   Accused ISIS supporter in province of Quebec hits Canadian soldiers with car, killing one.

October 22, 2014   ISIS supporter shoots and kills a Canadian soldier, then attacks the Parliament Building in Ottawa, where he is killed by police.

November 13, 2014   ISIS announces it is establishing outposts in Egypt, Saudi Arabia, Yemen, Libya, and Algeria.

November 16, 2014   ISIS releases a video confirming the beheading of Abdul-Rahman Kassig, an American aid worker.

November 22, 2014   ISIS supporter shoots a Danish national working in Riyadh, Saudi Arabia.

A NOTE ON SOURCING

Most information described as being derived from jihadist online sources and social media was collected and archived from the primary source at the time of posting, using a variety of tools. The most frequently used tool is a proprietary software package designed by J.M. Berger and coded by Dan Sturtevant and Jonathan Morgan. The software was inspired by the 2012 paper "Who Matters Online," by J.M. Berger and Bill Strathearn, commissioned by Google Ideas.

Data collected using this software is described in endnotes as being "collected by J.M. Berger," or simply as "data collected from Twitter." This description may also be applied to a variety of third-party commercial and open-source tools used from time to time to supplement the software (for instance, to monitor accounts or read tweets). In instances where a third party supplied proprietary metrics, it is cited by name.

Many of these sources are ephemeral, with reference to social media accounts that have already been, or are in constant danger of being, deleted by Internet service providers. Further, the purpose of this book is surely not to facilitate ISIS's efforts to spread propaganda. In most instances, citations of social media accounts will point to secondary sources (when available) for ease of reference and permanence of record.

ISIS

An American is dressed in an orange jumpsuit, apparently intended to echo the garb of al Qaeda insurgents captured and imprisoned by the United States. He kneels next to a man dressed all in black, his face masked, a knife in his hand.

For many, this has become an enduring image of the terrorist and insurgent group known as the Islamic State in Iraq and Syria, ISIS, or simply the Islamic State, as it now calls itself.

In a video posted to the Internet on August 19, 2014, and widely distributed over social media, the American recites a speech, advising President Obama to cease air strikes against the Islamic State. His tormentor speaks, flaunting the British accent that is so central to his performance, warning President Barack Obama that attacks on ISIS would result in the spilling of American blood.

He puts the knife to the American's neck and the camera cuts away to show the victim's severed head, displayed on the back of his lifeless body. Only the beginning of the grisly act is shown. But it is the fear in the American's eyes that is hard to forget.

The dead American was photojournalist James Foley. He was known as a "brave and tireless journalist" who was determined to describe the impact of war on ordinary people's lives.[1] Before he became a journalist, Foley had been a teacher and an aid worker. He had been abducted in November 2012, and had been beaten, starved, and waterboarded for nearly two years before he was finally beheaded.[2] Now the story of this good man had come to a terrible end.

For many people around the world, the methodical, sadistic cruelty of the video was shocking and unbearable, provoking an entirely human desire to avenge Foley's death using any means necessary.

In the Western world, in the twenty-first century, the idea of a beheading was something unreal, archaic, a vaguely understood and little-contemplated relic of a distant past. While there are important exceptions, we have grown used to a less barbaric world, so that when the media bring pictures of terrorists' deliberate savagery to our attention, we recoil.

Other jihadists had used beheadings for this purpose before. Chechen insurgents were known for brutally beheading prisoners. In Bosnia, jihadist fighters once videotaped themselves playing soccer with a decapitated head (Serbs and Palestinians reportedly did the same at different times). But al Qaeda in Iraq—the predecessor to ISIS—made the practice its trademark.

The campaign of horror began with the 2004 beheading of American businessman Nicholas Berg, who had been captured by al Qaeda in Iraq (AQI). It was performed on camera by the group's leader, Abu Musab al Zarqawi, and attracted international attention. Unlike the Foley video, Zarqawi was depicted carrying out the entire beheading with a knife; the camera did not cut away. The act was not swift; it took unbearably long seconds to complete.

The video's impact ensured that more videos would follow, many of which were even more brutal and graphic. The victims included Americans and other foreigners, including British, Russian, Japanese, Bulgarian, Korean, and Filipino citizens.[3]

IT IS DIFFICULT to properly convey the magnitude of the sadistic violence shown in these videos. Some featured multiple beheadings, men and women together, with the later victims forced to watch the first die. In one video, the insurgents drove out into the streets of Iraq cities, piled out of a vehicle, and beheaded a prisoner in full view of pedestrians, capturing the whole thing on video and then driving off scot-free.

The videos were distributed physically on DVDs in Iraq, but they became an Internet phenomenon. Unlabeled online file repositories were linked to by members of jihadist message boards, and the videos were passed around the Web, violence porn with a mission to intimidate and enrage. They succeeded.

It was the birth of a media model that has been transformed, expanded, and refined to a science over the course of years by the group that would eventually spring from the ashes of the American occupation—ISIS, a jihadist army so brutal and out of control that it was officially disavowed by al Qaeda.

ISIS has made its name on the marketing of savagery, evolving its message to sell a strange but potent new blend of utopianism and appalling carnage to a worldwide audience, documenting a carefully manipulated version of its military campaigns, including its bloody 2014 rampage across much of Iraq and Syria. ISIS is using beheadings as a form of marketing, manipulation, and recruitment, determined to bring the public display of savagery into our lives, trying to instill in us a state of terror.

Although some observers followed the rise of ISIS with alarm from late 2013, the Obama administration gave the problem short shrift. In an interview with the *New Yorker* in January 2014,[4] the president himself dismissed concerns about the group and other jihadists fighting in neighboring Syria:

> The analogy we use around here sometimes, and I think is accurate, is if a jayvee team puts on Lakers uniforms that doesn't make them Kobe Bryant. I think there is a distinction between the capacity and reach of a bin Laden and a network that is actively planning major terrorist plots against the homeland versus jihadists who are engaged in various local power struggles and disputes, often sectarian.

The administration continued to downplay the upstart jihadists for months. In June 2014, when ISIS seized control of a substantial chunk of Iraq, in an efficient military campaign marked by the retreat of apparently terrified, U.S.-trained Iraqi soldiers, most within the administration were caught off guard, asking themselves why they hadn't seen the "jayvee team" coming.

Despite the military drama, which sent tremors through regional and Western security services, most Americans and other Westerners were disillusioned and exhausted by more than ten years of a costly War on Terror.

Those who bothered to notice agreed ISIS was a problem. But maybe not our problem, they said. When President Obama authorized air strikes on ISIS positions, depriving them of a fraction of their stolen territory, he quickly moved on to discussions of the economy.

But ISIS would not be ignored. It began by courting American anger specifically, at first with taunting tweets launched over social media, using established marketing and spam tactics to ensure that its invitation to war played not just in Washington, but all over the globe.

For months, ISIS had flooded the Internet with images of hundreds of unnamed Iraqis and Kurds being executed by gun and knife and crucifixion, their heads mounted and displayed on pikes. All of it seemed so far away to those few who even heard about the atrocities, which the media covered sporadically at best.

Then ISIS upped the ante—deliberately re-creating the Nicholas Berg video for a new generation, with a new cast of characters, beginning with the murder of James Foley.

It was perhaps the ending of the video that sealed the incident's place in history. After graphic evidence of the murderous deed had been displayed, there was one final scene: the British jihadist yanked another American up on his knees, by the scruff of his orange jumpsuit—Steven Sotloff, another kidnapped journalist.

"The life of this American citizen, Obama, depends on your next decision," the killer said in a calm, matter-of-fact tone.

This was not a one-off communiqué. It was a promise of more bloodshed to come.

Extensive media coverage highlighted the case, as journalists publicly mourned one of their own, and ISIS spread images of the execution far and wide on social media, even prompting Twitter to intervene in ways it had long scorned, by suspending dozens of ISIS supporters' accounts.

By the time the second execution came, exactly as promised, followed by the addition of a third victim to the queue—this time a British citizen—a slow rumble was spreading through America and the world. ISIS expanded its targeted messaging to include "the allies of America," with special attention to the United Kingdom, and threats to bordering countries such as Turkey and Saudi Arabia.

In corner stores and restaurants, on television and radio broadcasts, over dinner tables and on social media, people began to ask: Why can't the most powerful nations on earth stop these medieval-minded killers? The questions soon transformed into an anger not seen since the days after the September 11, 2001, attacks.

"These guys need to be killed," a middle-aged police officer with a friendly face was heard saying in an even tone to a store owner in Cambridge, Massachusetts—one of the most notoriously liberal cities in the United States—and the sentiment was repeated again and again, around the world, at greater or lesser length, and with greater or lesser intensity.

Who are these men? Where did they come from? What do they want? How are they transforming the nature of terrorism and the war the international community is fighting against it?

What can we do about ISIS? What should we do?

These are the questions that fuel this book.

If journalism is the first draft of history, a book such as this can only be the second draft, and certainly not the final word. It is written at a point in history when ISIS has fully emerged in the world, but before its ultimate fate has become clear.

Regardless of that fate, what ISIS has accomplished so far will have long-term ramifications for jihadist and other extremist movements that may learn from its tactics. A hybrid of terrorism and insurgency, the former al Qaeda affiliate, booted out of that group in part due to its excessive brutality, is rewriting the playbook for extremism. It has inverted many of the dynamics that have applied to violent extremism for a century or longer and changed the rules of engagement on multiple fronts. It is a daring experiment in the power of horror, but also in the marketing of utopia. While most observers view ISIS's "state" as a dystopia, ISIS claims to have formed as a refuge from an impure world, a place where believers can be secure in the knowledge that they are living in accordance with Islam, at least as interpreted by ISIS. And it has documented its attempts at governance with the same attention to detail as its well-publicized atrocities.

There are many dimensions to the rise of ISIS. Some see the problem as explainable only with reference to competition among neighboring states for access to oil, natural gas, and pipelines.[5] Some blame the problem on poor governance and lack of democratic institutions, accusing the U.S. government of evangelism in regard to spreading democracy[6] while paying too little attention to the importance of civil and political rights.[7]

Some view ISIS as a symptom of a kind of "untamed Wahhabism,"[8] deliberately spread by Saudi Arabia and others,[9] or as a prop in a proxy war between Iran and Saudi Arabia, among other states. Still others see it as the public face of the resurgent Ba'athist party, determined to take back what it lost (and more) immediately after the 2003 invasion of Iraq. While researching the book, we heard many points of view.

We are observers of violent extremism, with many years of

experience speaking to terrorists, monitoring their messages, and studying their organizations and beliefs. Therefore our book is an effort to situate ISIS within the global jihadist movement, and within the field of extremism more broadly, so that its true implications can be better understood.

This book is written in the midst of a fast-changing story; in the short period between the book's completion and its publication, ISIS could conceivably double in size or be dealt a massive defeat. Although neither outcome seems probable, ISIS's short history is a series of contradictions and surprises, and we believe that whatever its fate as an organization, it has instituted transformative changes in strategy, messaging, and recruitment that will linger long after its so-called caliphate has crumbled to dust.

Within a short span, ISIS leader Abu Bakr al-Baghdadi and his fanatical followers have sketched out a new model for fringe movements to exploit changing social dynamics and new technologies, exerting an influence over world politics that is wildly disproportionate to its true size and strength.

To cover this ground, we will examine the history of the organization, its innovative propaganda and unprecedented manipulation of social media, and its recruitment of foreign fighters. We also explore the stark contrast it has drawn to the terrorist organization from which it sprang, al Qaeda, as well as a multitude of other extreme ideologies. Finally, although ISIS's evolution is ongoing, we believe some preliminary conclusions can be drawn about how to frame and approach the problem of countering this murderous movement.

There are many other important elements to this phenomenon and the conflict surrounding it, and we look forward to future books that explore some of the issues we could not. Given the fluid nature of the story, updates on ISIS and especially those pertinent to the topics covered in this book will be available at Intelwire.com.

ON NAMES AND DEFINITIONS

Definitions of many of the religious terms used in this book are included in a glossary and an appendix, and readers are encouraged to consult those sections for more information. In addition, we believe it is useful to discuss here the name of the group itself and some terms that are used frequently in relation to the "Islamic State" organization.

The group has renamed and rebranded itself multiple times. It is known as the Islamic State (its most recent self-appellation), but it is also frequently referred to as the Islamic State in Iraq and the Levant (ISIL) or the Islamic State of Iraq and al Sham (ISIS), or as Daesh, a derogatory term extracted from its Arabic acronym.[10]

Differences between ISIS and ISIL stem from issues of technical transliteration and geography. The Obama administration steadfastly referred to the group as ISIL long after most journalists had switched to ISIS (which was also generally the acronym used by the group itself in English communications).[11]

When the Islamic State dropped the -IS or -IL from its name at the end of June 2014, concurrent with its declaration that it was now a caliphate, it seemed this was the end of the naming controversy.[12] But most journalists continued to refer to it as ISIS, while President Obama continued to refer to it as ISIL.[13]

The rationale for the latter, as explained by Matt Apuzzo of the *New York Times*[14] and others, is that referring to the Islamic State by its self-appointed name would legitimize its declaration of an Islamic caliphate.[15]

Extremist groups often adopt a name that reflects their greater ambitions, and as a rule, people refer to them by the names they choose. Does it legitimize the concept of a white-only state to use the name "Aryan Nations"? Ironically, treating the Islamic State differently serves to elevate its claim to legitimacy, making it a special case requiring delicate handling, instead of just another extremist group. The insistence on ISIL also hints at an incorrect presumption that Muslims in general might be inclined to take the extremists seriously, and that the undecided might be swayed by nomenclature.

We prefer Islamic State as the most correct usage, but the vernacular (led by journalists) has embraced ISIS—meaning that for purposes of clarity, ISIS is much more readily associated with the content of the book in the minds of most readers.

On a more mundane level, the acronym *IS* presents challenges in a work of this length. For instance, the contraction "IS's" is unappealing, and the pairing of IS with the verb "is" also leads to the unpalatable "IS is," both of which would recur endlessly in the text.

In concession to these issues, we will generally employ the acronym *ISIS*.

An older semantic debate surrounds the use of the word *jihad*. A more comprehensive definition is included in the appendix, but we will briefly discuss our usage of the term here. The vast majority of the world's Muslims are peaceful people, and many of them object to militants' appropriation of the word and concept of jihad, which they understand to apply to nonviolent activities such as self-improvement or seeking justice.[16] Military jihadists do not make such qualifications when they call their work jihad.

"Whenever jihad is mentioned in the [Quran], it means the obligation to fight. It does not mean to fight with the pen or to write books or articles in the press, or to fight by holding lectures." Those are the words of Abdullah Azzam, the galvanizing force behind the volunteer jihad against the Soviets in Afghanistan.[17] This book will generally follow Azzam's usage. We acknowledge there is a legitimate debate in the public square on this issue, but this book expediently uses the term as jihadists use it.

Another area where definitions are murky involves distinctions among terrorism, insurgency, and war. For purposes of this book, we define terrorism as an act or threat of violence against noncombatants, with the objective of exacting revenge, intimidating, or otherwise influencing an audience. We define terrorists as nonstate actors who engage in violence against noncombatants in order to accomplish a political goal or amplify a message. Two characteristics of terrorism are critical for distinguishing it from other forms of violence. First, it is *aimed* at noncombatants. It is this characteristic of terrorism that distinguishes it from legitimate war-fighting. The laws of war, and both the Islamic and Judeo-Christian just war traditions explicitly forbid deliberately targeting noncombatants.[18] Thus, terrorist acts might usefully be defined as war crimes that are perpetrated by nonstate actors. Second, terrorists use violence for dramatic purpose: instilling fear in the target audience is often more important than the physical result. This deliberate creation of dread is what distinguishes terrorism from simple murder or assault.[19] Terrorists may be supported by states, but they have a fundamental quality of independence—or at least of disavowal and deniability. Thus, the Third Reich would not be considered a terrorist organization, but American and European neo-Nazis would.

The characteristics of terrorism, as we have defined it, raise additional thorny questions. How do we define "noncombatants"?[20] The term is controversial. A soldier on the battlefield is unquestionably a combatant. But what if the country is not at war, and the soldier is sleeping in his barracks, as was the case for the victims of the 1996 Khobar Towers bombing? In our view, noncombatants include civilians, military personnel not engaged in conflict, and political leaders (such as Anwar Sadat). Second, are child soldiers combatants or noncombatant victims? While it is clearly illegal under international law to recruit child soldiers, there is no consensus about the treatment of children who commit war crimes or terrorism.[21] This question is particularly important in regard to ISIS, which, according to the United Nations, "prioritizes children as a vehicle for ensuring long-term loyalty, adherence to their ideology and a cadre of devoted fighters that will see violence as a way of life."[22] (For more on this topic, see Chapter 9).

Will these child-perpetrators of atrocities be treated as victims of ISIS's war, or as terrorists? International law is not yet clear on this issue.[23] A Syrian child, who said that ISIS recruited him by "brainwashing" him with stories about Shi'a soldiers' rape of Sunni women, defected to Iraqi authorities while claiming to his ISIS masters that he planned to carry out a suicide attack.[24] The case highlights the uncertainties regarding how child-perpetrators should and will be treated.

While ISIS claims to be a state, for purposes of this book, we will generally discuss ISIS as a nonstate actor, albeit one at the very edge of the definition, possessing extraordinary infrastructure and expertise, much of it acquired or stolen from state actors, and a will to govern. Similarly, ISIS pushes the boundaries of the definition of insurgency, which is usually defined as an armed rebellion by nonstate actors against a recognized government. At the time of this writing, ISIS was fighting an insurgency against the Iraqi and Syrian governments. It was engaging in acts of terrorism against noncombatants. And it was the de facto authority in parts of both Iraq and Syria. For the time being, we believe ISIS is best defined as a hybrid terrorist and insurgent organization.

ISIS is a movement and an organization that sits at the nexus of a rapidly changing region and world. While it is rooted in history, ISIS has also introduced new elements to our understanding of radical Islamism, terrorism, and extremism writ large. For this reason, it commands a disproportionate share of the world's attention. Into these dark unexplored waters this book intends to wade, in search of understanding.

CHAPTER ONE

THE RISE AND FALL
OF AL QAEDA IN IRAQ

The world awakened to the threat of ISIS in the summer of 2014, but that is not where its story begins.

What we know today as ISIS emerged from the mind of Abu Musab al Zarqawi, a Jordanian thug-turned-terrorist who brought a particularly brutal and sectarian approach to his understanding of jihad.

Many diverse factors contributed to the rise of ISIS, but its roots lie with Zarqawi and the 2003 invasion of Iraq that gave him purpose.

Ahmad Fadhil Nazzal al Kalaylah was born in the industrial town of Zarqa, Jordan, located about fifteen miles from Amman. He was a Bedouin, born into a large, relatively poor family, but part of a powerful tribe. He was a mediocre student who dropped out of school after ninth grade. Like many jihadists, he took on a nom de guerre based on the place he came from, Abu Musab al Zarqawi.

In his hometown, Zarqawi was not known as an especially pious person, but as a heavy drinker, a bully, and a brawler.[1] His biographer reports that those who knew him in Zarqa said he drank like a fish and was covered in tattoos, two practices forbidden by Islam. He was known as the "green man" on account of the tattoos, which he would later try to remove with hydrochloric acid. He was arrested a number of times, for shoplifting, drug dealing, and attacking a man with a knife, among other crimes.[2]

In his early twenties, he joined Tablighi Jamaat, a South Asian Islamic revivalist organization, in part to "cleanse" himself from his life of crime. Tabligh Jamaat aims at creating better Muslims through "spiritual jihad"—good deeds, contemplation, and proselytizing.

According to the historian Barbara Metcalf, Tablighi Jamaat traditionally functioned as a self-help group, much like Alcoholics Anonymous, and most specialists claim that it is no more prone to violence than are the Seventh-Day Adventists, with whom Tablighi Jamaat is frequently compared.[3] But a member of Tablighi Jamaat told coauthor Jessica Stern that jihadi groups were known to openly recruit at the organization's central headquarters in Raiwind, Pakistan.[4]

In 1989, just three months after joining Tablighi Jamaat, Zarqawi joined the insurgency against the Soviet Union's occupation of Afghanistan, by which time the Soviets were already in withdrawal. The war had left him behind.

Zarqawi was not yet a leader, or even a fighter. In Afghanistan and over the border in Pakistan, he spent much of his time working on jihadist newsletters. While it might have seemed a humble start for someone who dreamed of battle, his introduction to jihadi media would later turn out to be useful.

But that was surely not clear at the time. "Zarqawi arrived in Afghanistan as a zero," one of his fellow jihadists told journalist Mary Anne Weaver, "a man with no career, just foundering about."[5]

He later trained and eventually fought in some of the most violent battles to emerge from the post-Soviet chaos in Afghanistan, when Afghan factions began fighting one another for control of the country. He found focus and earned a certain respect in the eyes of his peers. The experience changed him.

"It's not so much what Zarqawi did in the jihad—it's what the jihad did for him," the jihadist said to Weaver.[6]

Perhaps most important were the many relationships he forged during this time. The jihadists he recruited or met during this period would one day form the kernel of an international network. And one new friend turned out to be particularly important to Zarqawi's future—Sheikh Abu Muhammad al Maqdisi, one of the architects of jihadi Salafism, an ideology based on the principle that any government that does not rule through a strict interpretation of Shariah is an infidel regime that must be violently opposed (a fuller description can be found in the appendix).[7]

Maqdisi would become Zarqawi's spiritual father and close friend, despite their very different backgrounds. A trained cleric of Palestinian origin who lived in various Arab countries before settling in Jordan, Maqdisi was the "bookish fatwah monk." Zarqawi would emerge as the man who would test Maqdisi's theories "in real time and in a real war."[8]

Both men returned to Jordan in 1993. They were involved in a series of botched terrorist operations, culminating in their arrest for possessing illegal weapons and belonging to a banned jihadi organization.[9]

Like Afghanistan, prison was transformative for Zarqawi, according to Nir Rosen, who interviewed many of the jihadist's Jordanian peers:

Their time in prison was as important for the movement as their experiences in Afghanistan were, bonding the men who suffered together and giving them time to formulate their ideas. For some, it was educational as well. One experienced jihadi

who knew Zarqawi in Afghanistan told me: "When I heard Zarqawi speak, I didn't believe this is the same Zarqawi. But six
years in jail gave him a good chance to educate himself." [10]

Zarqawi tried to recruit his prison-mates into helping him overthrow the Jordanian leadership. After he was released from prison
in 1999, Zarqawi participated in the foiled "Millennium Plot" timed
for January 1, 2000, a plan to bomb two Christian holy sites, a border
crossing between Jordan and Israel, and the fully booked 400-room
Radisson hotel in Amman.

But he was again thwarted and the plot was disrupted by Jordanian security services.[11] Zarqawi managed to escape, first to Pakistan
and from there to Afghanistan, where he met Osama bin Laden.[12]

By most accounts, the meeting with bin Laden did not go well.
And why would it? The two men were united only by a broad commitment to violent jihad. Bin Laden and his early followers were
mostly members of an intellectual, educated elite, while Zarqawi
was a barely educated ruffian with an attitude.

One version of the meeting, reported by Mary Anne Weaver, described this first encounter as uncomfortable. Bin Laden was put off
by Zarqawi's insistence that all Shi'a Muslims must be killed, an ideological argument accepted by only the most extreme Sunni jihadists, who believe Shi'a are not true Muslims. Zarqawi was reportedly
arrogant and disrespectful of bin Laden. Others in al Qaeda felt the
brash young jihadist was not without his merits, however. He was
eventually allowed to set up his own training camp in Afghanistan,
albeit not officially under al Qaeda's wing. But the differences aired
on the day bin Laden and Zarqawi met would continue to define the
relationship between the two jihadists for years to come.[13]

Over the course of the next five years Zarqawi operated independently from, and yet with the support of, bin Laden and al Qaeda
Central. His training camp in Herat, Afghanistan, was supported
by al Qaeda funds with the consent of Mullah Omar, the leader of

the Taliban. He spent time in Iran, Syria, and Lebanon, where he recruited new fighters and grew his network. He was more focused on jihad in Muslim countries, such as Jordan, while bin Laden at the time was focused on the West, including his long-planned spectacular terrorist attack on the soil of the United States. In the days prior to September 11, bin Laden repeatedly sought *bayah,* a religiously binding oath of allegiance, from Zarqawi, who refused to comply.[14]

Nevertheless, when the Americans invaded Afghanistan after September 11, Zarqawi fought to defend al Qaeda and the Taliban.[15] Wounded in battle, he fled in 2002 to Iran, and from there to Iraqi Kurdistan,[16] where he joined Ansar al-Islam, a Kurdish jihadist group. The Kurds are an ethnic group inhabiting Kurdistan, a region that includes contiguous parts of Iran, Turkey, Syria, and Iraq.

Zarqawi's membership in Ansar al-Islam would later be cited by the United States as evidence that he and al Qaeda were collaborating with Saddam Hussein. But the Kurdish group Zarqawi had joined viewed the Iraqi regime as apostate and aimed to establish a Salafi state governed by Shariah.[17] Ironically, it was the invasion of Iraq that pushed Zarqawi into an alliance with bin Laden and led to al Qaeda's enduring presence in Iraq.[18]

Armed with irrational exuberance and a handful of dubious pretexts for war, the United States and its allies invaded Iraq on March 20, 2003. The invasion had been justified by exaggerated claims that Iraq possessed or was close to possessing weapons of mass destruction, and by the false claim that Saddam Hussein was allied with al Qaeda. While Iraq had a long history of sponsoring terrorist groups, al Qaeda was not one of them.

Zarqawi's name first became widely known in the West when the Bush administration described him as the link between al Qaeda and Saddam Hussein, claiming that Iraq had given safe haven to the terrorists, who now plotted mayhem with impunity inside its borders.

"From his terrorist network in Baghdad, Zarqawi can direct his network in the Middle East and beyond," Secretary of State Colin Powell told the United Nations Security Council.[19] But Zarqawi was neither collaborating with Saddam nor a member of al Qaeda.[20]

In the early days after the invasion, many Iraqis were overjoyed that the brutal dictator had been removed from power. By April 9, Baghdad had fallen and Saddam Hussein had fled. By May, President Bush announced, "Mission Accomplished."

President Bush had spearheaded a strategy of "taking the fight to the terrorists," which he would later repeatedly articulate as "We're taking the fight to the terrorists abroad, so we don't have to face them here at home."[21]

The statement proved half true. Iraq would be a lightning rod for jihadists, who flocked to a country where they had not been able to operate successfully before in order to confront American troops. But the invasion reinforced jihadi claims about America's hegemonic designs on the Middle East, providing a recruiting bonanza at a time when the terrorists needed it most.

Jihadi leaders around the globe described the U.S. occupation as a boon to their movement, which had begun to decline in large measure due to the destruction of al Qaeda's home base in Afghanistan. Abu Musab al Suri, one of the jihad's most prominent strategists, claimed that the war in Iraq almost single-handedly rescued the movement.[22]

As President Bush had claimed, Iraq became a "central front" in the war on terrorism.[23] But it was a front that the United States had created.[24]

Soon after the invasion, terrorism within Iraq's borders rose precipitously.[25] There were 78 terrorist attacks in the first twelve months following the U.S. invasion; in the second twelve months this number nearly quadrupled, to 302 attacks.[26] At the height of the war, in 2007, terrorists claimed 5,425 civilian lives and caused 9,878 injuries.[27] The violence also expanded abroad, as in 2005, when al

Qaeda in Iraq bombed three hotels in Amman, Jordan.[28] The coordinated attack had targeted Western-owned hotels, but the victims were almost all Jordanians, provoking an intense backlash within Jordan and angering many jihadists, who feared the operation would destroy al Qaeda's chances of winning support in the country.[29]

Iraq had erupted into civil war, and the allied mission quickly changed from combat to nation-building. When the mission changed direction, President Bush appointed L. Paul Bremer as head of the Coalition Provisional Authority in Iraq. Bremer's first major decisions would prove critical to the subsequent destabilization of Iraq: disbanding the military, and firing all members of Saddam Hussein's ruling Ba'ath Party from civil service positions.

More than 100,000 Sunni Ba'athists were removed from the government and military, leaving them unemployed, angry, and for the military personnel, armed.[30] Lieutenant General Jay Garner warned that the policy rendered a large number of educated and experienced Iraqis "potential recruits for the nascent insurgency."[31] One particularly important function impacted by the purge was the Iraqi border patrol. The weakened force provided little resistance to the dramatic flow of foreign fighters into the country.[32]

Zarqawi was there to seize the opportunity.

ZARQAWI RISES

Zarqawi's career as a terrorist had been largely marked by failure and frustration, but the American invasion galvanized him to action and created an environment suitable for his brutal tactics and rabid sectarianism.

The Sunni and Shi'a branches of Islam had split soon after the death of Muhammad over the issue of who should succeed the Prophet of Islam as leader of the Muslims, or *caliph*. Sunnis believe that the caliph can be chosen by Muslim authorities. Shi'ites believe that the caliph must be a direct descendant of the Prophet through his son-in-law and cousin Ali.

Over generations, the separation had led to doctrinal differences and, at times, open sectarian conflict or war, although there were equally long periods of peace and cooperation. Today, sectarian tensions are intensely mixed with local and regional politics.

Under the rule of Saddam Hussein, a Sunni Muslim, Iraq's Shi'a majority had been persecuted, massacred by the thousands, and denied political participation. After the 1991 Persian Gulf War, some of the Shi'a had risen up against Hussein, expecting support from the West, only to be crushed by the regime, resulting in tens of thousands of casualties.

"That's when the Hussein regime became far more sectarian and placed a lot of restrictions on Iraqi Shi'a, their religious institutions and leaders, and on Shi'a pilgrims who once came to the country," said Phillip Smyth, who studies Shi'a and Iranian politics and extremism in the region. "Plus, his regime became far more tribally based, meaning he was using Sunnis mainly from the Tikrit area."[33]

The U.S. invasion and subsequent efforts to institute a democratic system in Iraq had elevated the long-suppressed Shi'a into political power, while de-Ba'athification had simultaneously disenfranchised thousands of Sunnis.

The change also represented a significant shift for U.S. interests and relationships in the region. Ever since the Iranian Revolution of 1979, which had installed a Shi'a theocracy fueled by anti-American sentiment, most U.S. allies in the Middle East were ruled by explicitly Sunni regimes.

"The elected officials of Iraq's post-war government are the first Shi'a leaders that the United States has had any direct and meaningful contact with since the Iranian revolution," wrote Vali Nasr, in his book *The Shia Revival,* arguing that America had imagined Sunni democracies would rise in the wake of its intervention and was ill-prepared for the religious politics that followed.

Postwar Iraq was a recipe for sectarian conflict even without Zarqawi to stir the pot, but he wasted little time exploiting the opening.

In August 2003, Zarqawi's men bombed a UN mission center and the Jordanian embassy in Baghdad, setting a rapidly increasing pace of violence. At the end of August, he struck an important Shi'a mosque with a suicide bomber, killing at least ninety-five people, including Zarqawi's primary target, Ayatollah Muhammad Bakr al Hakim, one of the most prominent and beloved Shi'a clerics in Iraq.[34]

Despite the tensions between Zarqawi and bin Laden, in 2004 Zarqawi finally declared *bayah* (allegiance) to bin Laden and announced the creation of a new jihadist movement: Tanzim Qaedat al Jihad fi Bilad al Rafidayn, or al Qaeda in the Land of the Two Rivers, a reference to the fact that the Tigris and the Euphrates converge in Iraq.[35] It became more commonly known in the West simply as al Qaeda in Iraq, or AQI. Aaron Zelin, a leading authority on al Qaeda and ISIS, described the affiliation as a "marriage of convenience,"[36] rather than a meeting of minds.

Over the next few months Zarqawi and his new group continued to sow discord and attract international attention. Suicide bombings became a trademark tactic, leading to a reprimand from his spiritual mentor, Maqdisi.[37]

Foreign fighters also flocked to join AQI in large numbers, many of them passing through established smuggling routes in Syria. Most originated in Saudi Arabia, with significant numbers from Libya, Yemen, elsewhere in North Africa, and Syria. The Syrian regime made a show of trying to crack down on the cross-border trafficking, to little effect. "For every example of co-operation from Syria, there are an equal number of incidents that are not helpful," a U.S. intelligence officer told one reporter.[38]

Zarqawi and AQI also used the Internet to market their cause in a way that al Qaeda Central had never quite mastered. Under pressure from counterterrorism efforts in Afghanistan and Pakistan, al Qaeda Central's media output was devolving into long, boring videos of bin Laden and Ayman al Zawahiri, bin Laden's deputy, lecturing about the jihadi cause; Zarqawi and AQI, in contrast, began to release violent video clips of terrorist attacks and beheadings and distribute them online. (See Chapter 5.)[39]

Despite his pledge of *bayah*, Zarqawi continued to act independently of al Qaeda Central, and he pursued a strategy sometimes at odds with bin Laden's approach. Most central to the dispute were the related issues of *takfir* and the use of extreme savagery as a weapon.

Takfir, the pronouncement of someone as an unbeliever, and therefore no longer a Muslim, is a matter of great gravity in Islam.[40] Among jihadists, such a ruling is understood as a blanket permission to kill the subject or subjects as apostate.

Bin Laden was deeply troubled by jihadi groups that targeted Muslim civilians. Many letters seized from bin Laden's lair in Abbottabad when he was killed in May 2011 emphasize his frustration with regional groups that were undisciplined in their targeting. He urged his subordinates in al Qaeda to avoid domestic attacks that caused Sunni Muslim civilian casualties, pushing them to focus instead on targeting America.[41] Bin Laden was serious about the matter; he had revoked his support of the Armed Islamic Group of Algeria in 1996 because of the group's "worrying ideology."[42]

Zawahiri, who would subsequently take charge of al Qaeda after bin Laden's death, tried to rein in Zarqawi's bloody practices. In a 2005 letter, Zawahiri warned the al Qaeda in Iraq leader that he was far too free in his targeting of Muslim civilians and too prone to display "scenes of slaughter."[43]

Zarqawi reluctantly implemented some of Zawahiri's advice. But what the senior leader saw as weakness and excess, the younger man and his followers understood as design. He found ideological support for his preexisting tendency in an important jihadi text by an ideologue known as Abu Bakr Naji (a pseudonym).

Written in 2004, the 113-page tract in Arabic was titled *Idarat al Tawahhush,* or in English, *The Management of Savagery: The Most Critical Stage Through Which the Ummah Will Pass.* Attributed to an al Qaeda division devoted to research and analysis, it had been posted online to one of the earliest jihadist Internet forums, known as al Ekhlas, which is now defunct. It was translated into English in 2006 by noted scholar of political Islam Will McCants and released by the Combating Terrorism Center at West Point.[44]

The Management of Savagery was a compilation of lessons learned from previous jihadist failures, as well as an advancement in thinking about the movement's future direction. It outlined stages of the jihadist struggle including:

- **Disruption and exhaustion:** In which terrorist attacks damage the economy of enemy powers and demoralize their populations.
- **Management of Savagery:** A phase of violent resistance with an emphasis on carrying out acts of highly visible violence, intended to send a message to both allies and enemies.
- **Empowerment:** The establishment of regions controlled by jihadists which can subsequently grow and unite toward the goal of re-creating the caliphate.[45]

Al Naji recommended drawing the United States into a continual series of conflicts in the Middle East to destroy its image of invincibility, and he also endorsed an embrace and wide broadcast of unvarnished violence as a tool to motivate would-be recruits and demoralize enemies.

Al Naji's tract was widely read and influenced many, perhaps nowhere more than in Iraq. Although al Qaeda in Iraq and its successors were happy to improvise when it suited them, the influence of *Management* could be clearly seen in both its military and media strategies.[46]

The use and depiction of violence are among the most important elements of the strategy:

> Those who have not boldly entered wars during their lifetimes do not understand the role of violence and coarseness against the infidels in combat and media battles. . . . The reality of this role must be understood by explaining it to the youth who want to fight. . . . If we are not violent in our jihad and if softness seizes us, that will be a major factor in the loss of the element of strength, which is one of the pillars of the Umma of the Message.[47]

Zarqawi was also influenced by another jihadi ideologue, Abu Musab al Suri, whose 1,600-page book, *A Call to a Global Islamic Resistance,* famously advocated "leaderless resistance," the use of so-called lone wolf attacks (see Chapter 3).

Less discussed were the book's series of apocalyptic prophecies. Zarqawi published many communiqués that detailed his fulfillment of al Suri's prophecies (see Chapter 10). These included apocalyptic struggles against the "Persians," which can be understood as Shi'a or Iranians.[48]

Zarqawi had long held an implacable hatred for Shi'a Muslims, predating the publications of al Suri's book. The two men may have met during the early 1990s, when they were both taking part in jihadist causes in Afghanistan. Many hours of video featuring al Suri's lectures were distributed widely online.[49]

Even as Zarqawi and AQI were sowing discord with their violent, sectarian attacks, in December 2005, Iraqis voted for their first full-term government and parliament.[50] In 2006, the newly elected president, Jalal Talabani, a Sunni Kurd who was popular among both Sunnis and Shi'a, was pressured to compromise with Shi'a constituencies by appointing Nuri al Maliki prime minister, replacing another Shi'a politician who was perceived as showing favoritism to Sunni Arabs.[51]

At the time, Maliki was also perceived as being the less sectarian option and less beholden to neighboring Iran, which had taken a growing interest in Iraqi politics since the fall of Hussein.[52] Both of these expectations were destined to be met by spectacular disappointment.

ZARQAWI FALLS

Abu Musab al Zarqawi's reign of terror had made an impression in Iraq, igniting a cascade of violence as he continued to focus on sectarian targets, over al Qaeda Central's objections. In February 2006, the al Askari mosque in Samarra was bombed by militants, resulting in severe damage to its structure. AQI did not claim credit for the attack, but a captured member later confessed to orchestrating it. The remains of early Shi'a imams were interred at the mosque, considered a sacred site, and although no one was killed in the attack, it unleashed a wave of back-and-forth sectarian violence. There were dozens of retaliatory attacks on the first day, and thousands killed in the days that followed. The attack was widely seen as precipitating a full-on civil war that threatened the entire nation, portending massive bloodshed to come.[53]

Nada Bakos, the CIA officer charged with taking Zarqawi down, was keenly aware of the missteps that had made her target into a clear and present danger to the ongoing stability of Iraq. Writing in *Foreign Policy,* she said:

The war in Iraq provided al Qaeda *with a new front* for its struggle with the West. . . . The United States didn't "face down" al

Qaeda in Iraq; it inadvertently helped Zarqawi evolve from a lone extremist with a loose network to a charismatic leader of al Qaeda.[54]

In June 2006, the efforts of Bakos and countless others were realized in an air strike that killed Zarqawi. There was optimism that the death of Zarqawi would change the situation on the ground in Iraq. The hope was that by taking out AQI's top leaders—a strategy ironically known as decapitation—the organization would collapse.

In its briefing on the attack, the Defense Department released a photo of Zarqawi's corpse, a miscalculation when dealing with a movement that glorifies martyrdom and has no inhibitions about images of death. Within twenty-four hours, al Qaeda's online supporters were using the photo of Zarqawi's dead body in online banners, videos, and tributes to his martyrdom.[55] (The lesson was taken to heart in 2011 when Osama bin Laden was killed; no photos were released, and the body was secretly buried at sea.)

Zawahiri issued a statement eulogizing Zarqawi, commending him for his commitment to the cause and praising the great services he had done as a servant of al Qaeda.[56] He also used the eulogy as an opportunity to call for AQI to establish an Islamic state. Within a few months, a coalition of jihadist insurgents known as the Mujahideen Shura Council announced the formation of the Islamic State of Iraq (ISI). The council's formation had stemmed in part from AQI's recognition that it could not simply compete against other jihadist factions in its sphere of influence, and that at least some appearance of accommodation was needed.[57] Zarqawi's successor as head of AQI, Abu Hamza al Muhajir, a seasoned Egyptian fighter, pledged loyalty to ISI, and its newly appointed leader, Abu Omar al Baghdadi, about whom little is known.[58]

Brian Fishman, who closely followed al Qaeda in Iraq as a fellow with the New America Foundation, wrote that for a time, Zawahiri's influence took hold. The ISI distanced itself from the sectarian slaughter and committed to the goals Zawahiri had sent to Zarqawi:[59]

> Establish an Islamic authority or emirate, then develop it and support it until it achieves the level of a caliphate—over as much territory as you can to spread its power in Iraq, i.e., in Sunni Arab areas, is [sic] in order to fill the void stemming from the departure of the Americans, immediately upon their exit and before un-Islamic forces attempt to fill this void.[60]

Zawahiri would come to regret some of that advice.

Despite its "clean slate," ISI continued to target civilians, even as violence soared from multiple directions. In December 2006, an average of 53 civilians were killed every twenty-four hours.[61] In response to the growing levels of violence, U.S. General David Petraeus led a "surge" of U.S. troops into Iraq with the goal of securing the Iraqi population against the attacks carried out by ISI and other violent militant groups. This required not only an increase in the number of troops, but an entirely new strategy.

Instead of consolidating U.S. troops on big bases and handing the job as quickly as possible to Iraqi forces, General Petraeus sent the troops into the neighborhoods most affected by jihadi violence. Once the Iraqi people realized the troops were there to protect them they started to tell U.S. forces, by the general's account, "Here, let us tell you where the bad guys are, because we want them out of our neighborhood."[62]

The key to recruiting Sunni Arabs to join the fight against al Qaeda was to reassure them that they would be safe, according to Petraeus. He also had to persuade his own commanders to work with former insurgents who had earlier been targeting U.S. forces.

Over time, tens of thousands of former insurgents joined the fight to secure their communities against violence, as part of the so-called Sunni Awakening, or Sons of Iraq.[63] The Awakening Movement was a critical part of the effort to defeat AQI. Aside from their own revulsion at the al Qaeda affiliate's gory tactics and indiscriminate slaughter, militia members were enticed by the promise that some would be integrated into the Iraqi army and local police forces on a permanent basis. But many were skeptical of these promises, and their skepticism would prove prescient.[64]

The surge achieved its goals, if only temporarily. By 2008, al Qaeda and other violent militants no longer overran the country and the situation stabilized. Relations among the various religious and ethnic communities had greatly improved, as had the economy. In 2008, Maliki surprised observers by sending the Iraqi military against powerful Shi'a militias—which had also opposed the U.S. occupation—in Basra and the Sadr City section of Baghdad, temporarily easing concerns of sectarian favoritism.[65]

"It was a new atmosphere and it was full of promise," wrote Zaid Al-Ali, author of *The Struggle for Iraq's Future: How Corruption, Incompetence and Sectarianism Have Undermined Democracy*.[66] "U.S. officials, state security services, tribal forces, and some armed groups had forged an agreement to work together against the most extreme groups terrorizing Iraq's population. The major roads in those areas were lined with the flags of the Awakening Councils, and local fighters had decided to protect ordinary Iraqis from al Qaeda."[67]

But in 2010, Maliki's State of Law coalition failed to win a clear majority of the seats in parliament, endangering his position as prime minister. A series of political maneuvers ensued, some of which were questionably legal.

Zalmay Khalilzad, U.S. ambassador to Iraq from 2005 to 2009, believes that Maliki turned to Iran for support to keep his position.[68] Phillip Smyth agrees, saying Maliki's coalition was boosted by the addition of Iranian proxies such as the Badr Organization, an outgrowth of Iranian-armed and -funded militias dating back to the 1980s. The organization's reward for supporting Maliki was a voice in government, including the appointment of one of its members as Iraq's minister of the interior.[69]

Khalilzad believes that pressure from Iran is what led Maliki to insist that U.S. forces leave in 2011, a turning point in the sectarian dimension of Iraqi politics. The timing of the exit was initially negotiated by the Bush administration. The Obama administration proposed an extension, but negotiations with the Iraqi government broke down.[70] When the United States withdrew its troops in 2011, it also withdrew its "influence and its interest," according to Ryan Crocker, U.S. ambassador to Iraq from 2007 to 2009.[71]

The administration became politically disengaged. "All at once the regular phone calls among senior-level personnel, senior-level visits, basically ceased," Crocker told PBS *Frontline,* noting there was only one visit to Iraq by a cabinet-level official between the end of 2011 and mid-2014.

"Given that we were hard-wired into their political system, they wouldn't be able to function effectively with each other among [sectarian] communities without us," Crocker said. "I think that [political] disengagement brought them all back to zero-sum thinking."[72]

As Crocker was leaving Iraq, he warned the administration about Maliki's dictatorial and sectarian tendencies, not for the first time. In his view, Maliki was motivated not by the desire to aggrandize himself, but by fear that "sooner or later a coalition of adversaries would overthrow him."[73] Maliki had spent twenty years as a political exile in Syria and Iran, forced by Saddam to flee Baghdad because of his involvement with the underground Shi'a opposition.[74]

Even before U.S. troops left Iraq, Maliki's distrust of Sunni Iraqis led to a crackdown on the leaders of the Awakening Movement, who had been so important in reducing the threat of terrorist violence against civilians. One day after the last U.S. troops left Iraq, Prime Minister Maliki issued an arrest warrant for his Sunni vice president, Tariq Hashimi, on charges of terrorism.[75]

U.S. officials concede that some members of Vice President Hashimi's security forces may have been corrupt or been involved in plots to assassinate Shi'a leaders, but Hashimi had been one of the first Sunni Arabs in Iraq to agree to participate in the political process, at a great personal cost.[76] His removal led Sunni political leaders to boycott the parliament. In addition to costing the Iraqi government the support of the Awakening militias, many disenfranchised Sunni fighters (whose salaries had started to dry up) were now dropped into a boiling cauldron of radicalizing influences.

Rather than attempting to reduce Sunnis' feelings of disenfranchisement, Maliki began to purge the government of prominent Sunnis, further increasing sectarian tensions.[77] He brought terrorism charges against his popular finance minister and a Sunni Arab parliamentarian.[78]

Large protest camps arose in Sunni neighborhoods, including in Ramadi and Hawija, beginning in December 2012.[79] But when al Qaeda's flag rose sporadically in the protests, Maliki panicked. On April 23, 2013, Maliki sent soldiers into Hawija to clear out the "insurgents and extremists."[80]

The Iraqi government reported five civilian deaths, but Human Rights Watch reported much higher numbers.[81] Observers, including the prominent reporter Dexter Filkins, reported seeing hundreds of dead bodies.[82] In December 2013, Maliki again deployed the army against a protest camp in Ramadi, where some 350 Sunnis were protesting abusive antiterrorism laws, reigniting an active insurgency.[83]

According to Amnesty International, several Shi'a militias emerged with the encouragement and support of the Iraqi government, wearing military uniforms, and killing Sunni Arabs with impunity.[84]

Sunni Arabs were left disenfranchised, fearful of their government, and with few options other than supporting insurgency.[85] Patrick Cockburn, a longtime reporter on the Middle East, argues, "Mr. Maliki is not to blame for everything that has gone wrong in

Iraq, but he played a central role in pushing the Sunni community into the arms of ISIS, something it may come to regret." [86]

Conditions eventually deteriorated so far that Iraqi Grand Ayatollah Ali Sistani, leader of the country's Shi'a community, acknowledged that Sunnis had legitimate concerns and that the government had to be more inclusive of Sunni Arab and Kurdish minorities. [87]

CHAPTER TWO

THE RISE OF ISIS

After the death of Zarqawi, the Islamic State in Iraq had been handed setback after setback. When Abu Omar al Baghdadi, head of the ISI, was killed in 2010, it marked a turning point.

ISI's new leader was born Ibrahim Awwad Ibrahim Ali al-Badri al-Samarrai, but he operated under the nom de guerre Abu Bakr al-Baghdadi.

His life story is ambiguous, sparse on details, and few of those uncontested. He was reportedly born in 1971 to a Sunni Arab family in the Iraqi city of Samarra, a city just north of Baghdad. His family was said to be directly descended from the Prophet Muhammad.

According to a disputed but widely distributed biography published under a pseudonym by Turki al Binali, a Bahraini national who joined ISIS, Baghdadi was born into an observant Salafi family and "his brothers and uncles include preachers and teachers." [1]

According to Abu Ali, a neighbor of the family, Baghdadi remained in Samarra until he was eighteen, when he moved to Tobchi, a poor neighborhood on the outskirts of Baghdad. [2] He lived in a run-down apartment attached to the local mosque and reportedly enrolled in the Islamic University of Baghdad, eventually receiving a doctorate in Islamic culture and Shariah law. Abu Ali described him as a "quiet person, and very polite," but also a "conservative practitioner of Islam." He was said to have led prayers at the local mosque from time to time.[3]

During this period, Baghdadi was also a classmate of Ahmed al Dabash, who later became the leader of the Islamic Army of Iraq, a Sunni Arab insurgent group. Dabash remarked that the young Baghdadi "did not show much potential." He described Baghdadi as "quiet, and retiring. He spent time alone. . . . He was insignificant."[4]

Baghdadi reportedly led a quiet life until the United States and its allies invaded Iraq. In 2003, Baghdadi is believed to have begun on the path of jihad.[5]

Jamaat Jaysh Ahl al Sunnah wa-al-Jamaah (the Army of the Sunni People Group) was an insurgent group operating in Samarra, Diyala, and Baghdad. Baghdadi was a cofounder and the head of the group's Shariah committee.[6]

In late 2004 or early 2005, an American-led raid on a home near Fallujah led to the capture of many high-level insurgents and a man who was described as an "apparent hanger-on." The latter was registered at Camp Bucca detention center as Ibrahim Awad Ibrahim al Badri.

There are conflicting accounts of Baghdadi from his time in Camp Bucca. A Pentagon official described him as "a street thug when we picked him up in 2004," a characterization that seems inconsistent with his background.[7]

Andrew Thompson, who served at one of the U.S.-run detention centers in Iraq, wrote an article with Jeremy Suri, a professor at University of Texas at Austin, arguing that the structure of Camp Bucca facilitated further radicalization among the prisoners.

Before their detention, Mr. al-Baghdadi and others were violent radicals, intent on attacking America. Their time in prison deepened their extremism and gave them opportunities to broaden their following. At Camp Bucca, for example, the most radical figures were held alongside less threatening individuals, some of whom were not guilty of any violent crime. Coalition prisons became recruitment centers and training grounds for the terrorists the United States is now fighting. . . .

Small-time criminals, violent terrorists and unknown personalities were separated only along sectarian lines. This provided a space for extremists to spread their message. The detainees who rejected the radicals in their cells faced retribution from other prisoners through "Shariah courts" that infested the facilities. The radicalization of the prison population was evident to anyone who paid attention. Unfortunately, few military leaders did.[8]

In 2007, Major General Douglas Stone became the deputy commanding general of Multi-National Forces in Iraq with responsibility for in-country interrogation and detention. In this capacity, he was responsible for detainees at Camp Cropper, Camp Bucca, and Camp Ashraf. He spent the following year reforming prison conditions and installing innovative deradicalization, rehabilitation, and reintegration techniques, which expedited the release of low-risk prisoners and appeared to reduce recidivism.[9]

Most of the individuals taken into detention did not need to remain for long periods of time, or in many cases should not have been there in the first place, he told us. Many were not jihadists, but were unemployed citizens paid or coerced into joining the resistance. More than 80 percent of the detainees tested illiterate and were largely ignorant about Islam, which made them particularly susceptible to recruitment while in prison.

In interviews for this book, General Stone recounted the reintegration process:

> We studied the detainees: their tribal affiliations, their education level, their employment skills, their purported crimes, their leadership skills, and the extent to which they subscribed to jihadi principles. We decided to separate the hard-core jihadists from the casual insurgents. Our biggest worry was that the real jihadists were using the prison as a terrorist training camp. We wanted to release the individuals who shouldn't have been there, or who could be easily reintegrated into Iraqi society, as quickly as possible. We hired hundreds of teachers to train detainees to read. We hired one hundred and fifty imams from around the globe to preach mainstream Islam. We offered them job training. After a couple of years, we were able to release most of the prisoners, with less than two percent ever returning to the fight. That left only the true problem cases. Only about five thousand were left. The majority were either former regime Baathists, former criminals, or serious *takfiri* ideologues, followers of Zarqawi's extreme beliefs regarding declaring other Muslims to be apostates. Even in American detention these *takfiris* were killing other detainees, cutting their eyes out, and trying to impose a version of Shariah that most Muslims would find quite abhorrent.[10]

Baghdadi's time in detention would only have made him more effective, General Stone said, pointing out that the individuals who spent time in Guantanamo pose a similar problem. Jihadists who get out of U.S. detention develop a kind of aura when reintegrated into their home communities, he said, making it easier for them to recruit others, or to symbolize defiance against a Western power.

Baghdadi was probably systematically organizing while he was in detention. Building up IOUs, getting to know whom to trust.

He must have been plotting while he was incarcerated—he must have planned the whole rollout of the Islamic State. . . .

If you look at how Baghdadi has set up the top leadership of ISIS, you can see how skilled he is. The guys at the top are all very skilled managers. Many of them are former Ba'athists. And to me a most important thing—he's actually designated someone to run ISIS detainee operations. He learned, from being in detention himself, that if you don't manage the prison well, the detainees will just organize themselves against you. And sure enough, his strategy has been to recruit his cadres from the prisons where jihadis were detained. He knows that's where to find hard-core radicals. But even if Baghdadi is ultimately replaced, the ideas that he is promoting will be with us a long time.

Baghdadi left Camp Bucca as an outspoken jihadi and soon joined the ranks of the ISI, then under the leadership of Abu Omar al-Baghdadi.

When a United States–Iraqi joint air strike targeted and killed Zarqawi's successors in April 2010, it wiped out the ISI's senior leadership. With its leadership in disarray and its relevance waning, ISI sought out a leader with both religious authority and a track record of strategic successes.

Abu Bakr al-Baghdadi fit these criteria. His education in Islamic law far exceeded the leaders of al Qaeda. Osama bin Laden studied business in college; his degree was reportedly in public administration.[11] Ayman al Zawahiri was a surgeon.[12] And the strength of Baghdadi's strategies would soon become clear.

In May 2010, he ascended to lead the Islamic State in Iraq (ISI).[13] Baghdadi's first priority after becoming leader was his own personal safety. With ISI in shambles, Baghdadi set out to rebuild the organization, eliminating potential critics and replacing them with trusted allies, many of whom had spent several years with Baghdadi in Camp Bucca.

Among them were several Ba'athist leaders. Although AQI and ISIS are motivated by an ideological commitment to reviving an Islamic state based on their understanding of Shariah, they formed an alliance with the former Ba'athists,[14] who had lost their jobs and status thanks to de-Ba'athification. According to some reports, the "Ba'athification" of ISIS may have been the brainchild of a former colonel in Saddam Hussein's army who spent time with Baghdadi at Camp Bucca.[15]

"In the early days of the alliance, the Ba'athists may have had the upper hand as they brought military and organization skills and a network of experienced bureaucrats that AQI and then ISI lacked," says Richard Barrett of the Soufan Group.[16]

The Ba'athists became a critically important part of ISIS. Baghdadi chose many of them to fill top organizational positions, including Abu Muslim al Turkmani, who became Baghdadi's second in command (until he was reportedly killed in late 2014), and the senior leader of the military council, Abu Ayman al Iraqi. According to Barrett, at least eight of ISI's senior leadership members are former inmates at Camp Bucca.[17]

Learning from past leaders' mistakes, Baghdadi disguised his identity from the earliest days, even in the presence of his closest advisors. Abdul Rahman Hamad, an ISIS fighter who spoke to *Time* magazine, stated, "[He] knew how men can be seduced by money, so he never shared his secrets with anyone."[18] He became known among his men as the "invisible sheikh" or the "Ghost."[19] With between 800 and 1,000 fighters in his ranks,[20] Baghdadi would lead Iraq into its deadliest years since 2008.[21]

Under Baghdadi's leadership, ISI escalated its violence throughout 2010 and 2011, including using coordinated suicide attacks in several locations on the same day. In October 2011, the U.S. Rewards for Justice Program instated a reward of up to $10 million for information leading to the arrest or capture of Baghdadi.[22]

By July 2012, in an atmosphere of growing sectarianism fueled in no small part by the policies of Prime Minister Maliki, Baghdadi had rebuilt the organization so substantially that he apparently felt no qualms about publicly pre-announcing his next move—a campaign called "Breaking Down the Walls," in which Baghdadi promised to liberate Iraqi prisons overflowing with insurgents and jihadists.[23]

Using covert channels to communicate with prisoners in advance, ISI spent the next year making good on Baghdadi's promise. The insurgents attacked eight prisons using improvised explosives. They freed hundreds of prisoners, many of whom were senior leaders of ISI and its predecessors, or experienced fighters who subsequently joined the organization.[24]

During the same one-year period, Baghdadi had courted the wrath of al Qaeda by declaring an expansion of the ISI into neighboring Syria, which was now engulfed in civil war. In defiance of al Qaeda's emir, Ayman al Zawahiri, the Islamic State in Iraq was to be known as the Islamic State in Iraq and Syria, using the now notorious acronym ISIS.

From the ashes of near-total defeat, a new and virulent jihadist idea had emerged, and it aimed to terrorize the world with its brutal ambition.

SYRIA AND THE WAR WITH AL QAEDA

The "Arab Spring" protests began in Tunisia in December 2010, and from there spread throughout the Arab League and beyond. By December 2013, rulers had been replaced in Tunisia, Egypt (twice), Libya, and Yemen; there were uprisings in Bahrain and Syria, and large-scale protests in Algeria, Iraq, Jordan, Kuwait, Saudi Arabia, and beyond.[25]

The protests were fueled by inequality, corruption, and frustration with injustices suffered under long-standing dictatorships.[26]

What began as popular movements turned violent in some countries, but no one had ever seen anything like the civil war that erupted in Syria.

For more than fifty years, the Syrian people have lived under a military dictatorship. A single family has ruled the country since 1970, starting with General Hafez al Assad and his son, Bashar al Assad, who succeeded him in 2000. Speech is extensively censored and those whose words displease the regime are subject to harassment or arrest. Members of the elite have lived very well. But there has been high unemployment, especially among youth.[27]

When popular protests helped unseat the long-standing dictatorships in Egypt and Tunisia in early 2011, young Syrians were inspired to follow suit. For the crime of spray-painting antigovernment graffiti in the town of Daraa, fifteen teenage boys were arrested and brutally tortured. Thousands turned out to protest this vicious act, and the regime responded by opening fire on the assembled crowds.[28, 29]

Soon afterward, a Facebook page called for nationwide protests and thousands flooded the streets to protest the brutality of the Assad regime. In response to these protests, Assad offered concessions, including ending the "state of emergency" that had been in place for nearly fifty years. Still the protests continued to spread. By May of that year, more than a thousand people had been killed by the regime, according to Syrian human rights groups.[30]

On May 28, 2011, the corpse of a thirteen-year old child was delivered back to his family in the town of Daraa, where the protests began.[31] The child's genitalia had been removed, and his corpse was burned and riddled with gunshot wounds. Some fifty thousand protesters gathered outside Daraa. The Syrian government responded by again firing on the protesters and disconnecting the Internet.[32]

Western governments called on Assad to step down, and the Arab League condemned the crackdown. According to Human Rights Watch, the Syrian government has taken tens of thousands of detainees into custody, solely on the basis of their peaceful opposition to the regime.

Many of the detainees were brutally tortured. Even the hospital staff treating wounded protesters were arrested and tortured.[33] Human Rights Watch and others have reported that Syrian security forces were using rape systematically to torture men, women, and children, some as young as twelve years old.[34]

If the sectarian clashes in Iraq provided an opening for ISI to regroup, the violence in Syria gave Baghdadi a pretext to expand. The border between Syria and Iraq had long been porous. Long-standing smuggling routes that were used to move fighters and supplies from Syria during the war in Iraq were now reversed to bring fighters and supplies back into Syria.

In support of this effort, Baghdadi sent a number of operatives into Syria with the task of setting up a new jihadist organization to operate there. Among them was Abu Mohammed al Jawlani, a Syrian-born member of al Qaeda in Iraq who had spent time in Camp Bucca with Baghdadi and had more recently served as the regional leader of ISI in Mosul.[35] Jawlani quickly established himself as leader of a group that came to be known as Jabhat al Nusra, which at first positioned itself as an independent entity with no ties to either al Qaeda Central or the ISI.[36]

Within a year, al Nusra was a recognized leader among insurgent groups in Syria.[37] Moderate opposition groups gradually found themselves struggling to acquire funding and weapons, while al Nusra and other Islamist groups were funded externally by donations and internally by the seizure of equipment and resources on the battlefield. Islamist groups soon had the upper hand over the secular opposition.[38]

For the first six months after the announcement of its creation, Nusra engaged in the same kinds of brutal attacks that had been the favorites of AQI and ISI: it bombed urban areas, killing civilians by the dozen, and targeted alleged government sympathizers and coop-erators.[39] These tactics alienated both the civilian population and the local Syrian revolutionaries.

In late summer 2012, al Nusra changed its approach. It started to cooperate with Syrian nationalist groups such as the Free Syrian Army, but it also reached out to forge relationships with groups with widely divergent ideologies, as long as they shared Nusra's commit-ment to ousting the Assad regime.[40]

The new strategy worked. By late 2012, Aaron Zelin described Nusra as "one of the opposition's best fighting forces, and locals viewed its members as fair arbiters when dealing with corruption and social services."[41]

At the same time, Baghdadi and ISI remained busy in Iraq. The two groups were expanding in different countries, but via markedly different strategies. Each was also growing in influence, setting the stage for the rivalry and confrontation that would ultimately end in al Qaeda Central's disavowal of ISIS.

On April 9, 2013, Baghdadi announced a merger of ISI and al Nusra, calling the new group the Islamic State of Iraq and the Levant (ISIS). In effect Baghdadi was unilaterally establishing himself as the leader of both organizations (ISI and al Nusra), now merged into one. The announcement surprised both Zawahiri and Jawlani. Neither of them had signed off on the decision, and neither was enthusiastic about it. Al Nusra immediately announced its allegiance to Zawahiri and al Qaeda Central, placing al Nusra and ISIS in direct confrontation.[42]

Zawahiri scrambled to solve the crisis between the groups and to assert AQC's dominance over its affiliates. In a private letter that leaked to the press, he declared the merger null and void, ruling that Baghdadi would continue to run operations in Iraq and Jawlani would continue in Syria.[43]

But Baghdadi rejected the ruling in a defiant and very public audio statement released through jihadi media outlets: "When it comes to the letter of Sheikh Ayman al-Zawahiri—may God protect him—we have many legal and methodological reservations," he said. Baghdadi said he would continue to pursue a united Islamic state crossing the border between the two countries.[44]

Unsurprisingly, relations between ISIS, al Nusra, and al Qaeda Central continued to deteriorate as ISIS peeled off fighters from al Nusra and sent reinforcements from Iraq. Unlike al Nusra, which had forged alliances and won respect from other rebel factions, ISIS took an unyielding approach, refusing to share power in areas where it operated. Starting in mid-2013, these tensions evolved into violence, and by early 2014, a war within a war was being fought across northern Syria, with ISIS battling a number of other rebel factions, including al Nusra.[45]

On February 2, 2014, al Qaeda formally disavowed ISIS in a written statement: "ISIS is not a branch of the [al Qaeda] group, we have no organizational relationship with it, and [al Qaeda] is not responsible for its actions."[46]

Ever since the days of Zarqawi and bin Laden, al Qaeda's Iraqi affiliate had been troublesome, but the differences over tactics and ideology had been fought out in private and papered over in public. Baghdadi's outright defiance and his escalating violence against other jihadists in Syria had forced Zawahiri's hand.

If the emir of al Qaeda expected contrition, he was gravely mistaken. ISIS responded swiftly and with characteristic violence. On February 23, 2014, a suicide bomber assassinated Abu Khaled al Suri, a longtime al Qaeda member believed to be Zawahiri's personal emissary in Syria, who had been charged with seeking a resolution to the dispute. There was little doubt who was responsible.[47]

In May, ISIS spokesman Abu Muhammad al Adnani issued a scathing speech addressing Zawahiri, sarcastically titled, "Sorry, Emir of al Qaeda," in which he mockingly apologized for ISIS's failure to follow Zawahiri's weak example.

"Sorry for this frank report," he said, but members of al Nusra had been heard saying that the 63-year-old Zawahiri was "senile."

"Sorry, emir of al Qaeda," he said, but Zawahiri had made a "laughingstock" of al Qaeda. "Sorry," he said, but ISIS had questions about why it should continue to follow al Qaeda's losing example. "We await your wise reply." [48]

ISIS had successes to back up its swagger. In a sustained campaign throughout 2014, it seized and consolidated control of Raqqa, Syria, and most of the surrounding area, driving out both the regime and other rebels. It established Raqqa as its capital in Syria, populating it with hordes of foreign fighters and implementing ISIS's harsh interpretation of Shariah law. [49] It also won significant control of Syrian city Deir ez Zour from al Nusra and other opposition forces, shifting considerable resources from al Nusra to ISIS and providing a crucial political and logistical way station near the border with Iraq. [50]

A CALIPHATE CLAIMED

ISIS continued to make steady gains in both Iraq and Syria, controlling ever larger swaths of territory and aggressively governing in the areas where it could consolidate control. It captured Fallujah in January and kept on going. [51]

To accomplish this feat, ISIS crafted a series of complex alliances with Sunni Arab tribes in Iraq, even with tribes that did not necessarily share ISIS's extreme ideology. Many Sunni Arabs were fed up with the Maliki regime, which had continued to describe the Sunni Arab uprising against his sectarian policies as terrorism. Members of the Awakening Movement (who had sided with the U.S. military in the 2007 surge) felt particularly betrayed. Maliki had agreed to offer them a role in the military and police forces, but had not fulfilled his promise. Some angry members joined ISIS, while others chose to sit out the battle. [52]

Tensions were exacerbated by the regime's reliance on Shia militias to fight ISIS in Anbar province and other areas. Many of these groups were Iranian proxies, owing more allegiance to Tehran than

Baghdad, and some had returned to Iraq after fighting ISIS in Syria.[53] For Iran, the growing chaos presented an opportunity to solidify its influence over Iraq and its prime minister.

More than eighty Sunni tribes reportedly fought alongside ISIS, and at times it was difficult to know who was in control of any specific area.[54] But ISIS was content to take the credit, and no one else stepped up to speak for the insurgency. The coalition seemed legitimately shaky on its face, and reports of the internal tensions led many to speculate that it could tear itself apart at any moment. But somehow, it kept hanging on.[55]

In early June 2014, ISIS captured Mosul, a city of 1.5 million people and the site of Iraq's largest dam.[56] Because it was so dangerous for journalists and other noncombatants to operate in areas afflicted with insurgency, the victory seemed to come out of nowhere. Certainly Western governments seemed to be caught flat-footed.

In addition to the unusually thick fog of war, however, there was a truly unexpected development. The United States had invested $25 billion in training and equipping the Iraqi army over the course of eight years.[57] That investment evaporated in the blink of an eye as Iraqi soldiers turned tail and fled in the face of ISIS's assault on Mosul.

According to the *Los Angeles Times,* which interviewed some of the soldiers who had served in Mosul, the senior commanders fled when they saw ISIS's now-infamous black flags moving into the city. Corruption and sectarian tensions with the army itself may also have played a role; the regime had systematically driven Sunnis out of senior military positions, often in favor of less experienced Shi'a officers who had important friends.[58] In a Reuters investigative report, Iraqi military commanders also detailed the breakdown and said the government had declined offers of help from powerful Kurdish fighting forces.[59]

Reports circulated on social media that ISIS had looted the banks in Mosul, which were later denied by the Iraqi government, but the denials—sourced to Iraqi bankers and officials whose own businesses rested on their ability to secure funds and the country's economy—were not any more credible than the original reports.[60]

It hardly mattered. No one disputed that ISIS had become the richest terrorist organization in the world, and was getting richer by the day. Most agreed its cash reserves ran into hundreds of millions of dollars, perhaps even a billion, and by November, some estimated it was generating $1 million to $3 million per day, although a large number of unknowns plagued such questions.[61] Unlike al Qaeda and many other terrorist groups, which rely on external sources of funding, including "charitable" donations, much of ISIS's revenue was generated internally, from taxes on local populations, looting, the sale of antiquities, and oil smuggling, with the latter seen as one of the most important sources.[62] ISIS tapped into "long-standing and deeply rooted" black markets and smuggling routes, making traditional instruments for fighting terrorist financing far less useful.[63] It also raised millions by ransoming Western hostages.[64] While the United States and the United Kingdom have government policies that forbid paying ransoms, many other countries, including some in Europe, have paid to have hostages released.[65]

Tikrit, the hometown of Saddam Hussein, fell soon after Mosul. At many stops along its march, ISIS captured U.S.-supplied military equipment from fleeing Iraqi soldiers, which they trumpeted with photos on social media.[66]

On June 29, ISIS made a move in the world of ideas that was as bold as its military blitzkrieg on the ground. In an audio recording from its chief spokesman, Abu Muhammad al Adnani, ISIS declared that it was reconstituting the caliphate, a historical Islamic empire with vast resonance for Muslims around the world, but especially for Salafi jihadists, whose efforts were all nominally in the service of that goal.

ISIS emir Abu Bakr al-Baghdadi was announced as the new "Caliph Ibrahim," and he showed his face in public for the first time a few days later, delivering a sermon at a Mosul mosque. The new caliphate would simply be known as the Islamic State, the announcement said, dropping "Iraq and Syria" from the organization's name to reflect its global claim of dominion.[67] Neverthless, many outside observers (and even some supporters) continued to use the acronym ISIS to refer to the group.

The announcement (discussed at more length in Chapter 5) demanded the loyalty of all Muslims around the world (a laughable concept) and specifically from other jihadist groups. It was met by wild enthusiasm from ISIS supporters and a mix of hostility and incredulity from almost everyone else.

The jihadists continued pushing south into territory controlled by ethnic Kurds under Iraq's federal system. The Kurdish militia, known as the *peshmerga,* was no match for the heavily armed ISIS fighters. While they put up a better fight than the Iraqi forces, they too were forced to retreat.[68]

The advance created a humanitarian crisis. The area that ISIS had captured had a large population of religious and ethnic minorities, including an estimated 35,000 to 50,000 Yazidis, who practice an ancient, complex religion mixing beliefs from a number of sources. ISIS views them as devil worshippers and constructed a religious justification to kill all the men and enslave the women and children (see Chapter 9).[69] The Yazidis were now defenseless against ISIS's genocidal intentions, and ISIS hunted and then surrounded them as they fled to Iraq's Mount Sinjar with no food and no water. The United States, the United Kingdom, and France made emergency airdrops of food and water to the Yazidi refugees to forestall what the UN referred to as a threatened genocide.[70]

But still, the siege continued. On August 7, President Obama announced that the United States would take military action against ISIS to help secure the safety of the refugees and American personnel in Iraq.

"I know that many of you are rightly concerned about any

American military action in Iraq, even limited strikes like these," the president said in an address. "I understand that. I ran for this office in part to end our war in Iraq and welcome our troops home, and that's what we've done. As commander in chief, I will not allow the United States to be dragged into fighting another war in Iraq."

U.S. air strikes, combined with air support from the Iraqis and ground support from the *peshmerga* and the Kurdish militant groups Kurdistan Workers' Party (PKK) and its Syrian offshoot, the People's Protection Units (YPG), allowed tens of thousands of Yazidis to escape the newly rechristened Islamic State, but the continuing expansion of the insurgency put thousands more in harm's way, leading to mass killing of the men and the institutionalized slavery of women and children, including horrific ongoing sexual abuse of captured women.[71]

Faced with U.S. air strikes, the group began implementing a strategy from *The Management of Savagery* called "paying the price," in which it responded to any hint of aggression with extreme violence. In September, ISIS began to release videos online featuring the execution by beheading of Western hostages, which continued into the winter (see Chapter 5).

Nevertheless, ISIS continued its aggressive military campaign, even as the world slowly awakened to its depredations. Everywhere they controlled territory, ISIS instituted a harsh theocratic rule, which included at least skeletal governance, with a functioning economy and civil institutions. The initial wave of strikes in Iraq slowed ISIS's advance but did not significantly reduce its dominion.[72]

The effects of U.S. engagement in Iraq rippled over into Syria. On September 9, an explosion massacred the senior leadership of Ahrar al Sham, perhaps the most important jihadist group fighting the Assad regime after the al Qaeda–linked Jabhat al Nusra, and several other leaders within the Islamic Front, a broad coalition of Islamist rebel groups. The bombing targeted a meeting in which top leaders of the group were hashing out an internal dispute over its recent alliance with a coalition that included all of the remaining U.S.-supported rebels, and the question of whether to pursue a more inclusive strategy in Syria. It was unclear whether the attack originated with the regime or with ISIS, but the dramatic assault threw the alliance of Syrian fighters not aligned with ISIS into deep turmoil.[73]

Despite his promise of a limited role for the United States in Iraq, President Obama faced mounting pressure to do something about the group. In an address on September 10, he announced the goal of U.S. intervention had expanded.

"Our objective is clear: We will degrade, and ultimately destroy, [ISIS] through a comprehensive and sustained counterterrorism strategy," he said, despite the fact that ISIS was far more significant as an insurgency than as a terrorist group. As part of this objective, the president said, an international coalition would strike ISIS in Syria as well as in Iraq.[74]

Soon afterward, the participants in the coalition grew to include the United Kingdom, France, Australia, Canada, Germany, the Netherlands and—significantly—Bahrain, Jordan, Saudi Arabia, Turkey, Qatar, and the United Arab Emirates, Sunni-majority countries with the most to lose from ISIS's imperial ambitions and efforts to recruit in the region.[75] The United Arab Emirates sent a female fighter pilot to lead one of its missions.

This significant expansion of the rules of engagement with ISIS became much more complicated with the first coalition strikes in Syria. During the first raid on September 22, 2014, American planes bombed not just ISIS targets but Jabhat al Nusra, which had broken away from ISIS months earlier and established itself as a leading force in the rebel alliance to overthrow Bashar al Assad.

According to the administration, and backed up to some extent by open-source reports out of the Syrian civil war, the strikes were aimed at the "Khorasan Group," a virtually unheard-of cell of senior al Qaeda Central operatives that had been dispatched to Syria to plot attacks against the West.[76]

Information about the Khorasan Group was sketchy and conflicted, but the impact of the strikes was clear. Jabhat al Nusra responded by taking the offensive to the few remaining "moderate" rebels supported by the United States, dealing them a devastating blow. The future of the secular rebellion in Syria teetered on the brink of annihilation as ISIS continued to fight.[77] Charles Lister of the Brookings Institution, one of the most insightful followers of jihadist movements in Syria, wrote in early December:

> . . . while surprising to outsiders, the Al-Qaeda affiliate Jabhat al-Nusra is still to this day perceived by many as an invaluable actor in the fight against Damascus and as such, the strikes on its positions are seen by many as evidence of U.S. interests being contrary to the revolution [against Assad].[78]

However, the situation is fluid, Lister noted in an email weeks later, and al Nusra's expanding conflict with other rebel factions may be starting to undermine its position. In late December, as this book was going to press, the largest Islamist factions in Syria announced a new coalition that excluded both al Nusra and ISIS.[79]

As of this writing, the advance of ISIS on the ground had been slowed by coalition air strikes and other action. While it continued to cling to the vast majority of its territory, there were signs that the coalition campaign was having some effect, for instance a protracted battle for the town of Kobane, defended by Kurdish *peshmerga*. The fate of Kobane was still undecided as this book went to press, but the contrast to ISIS's swift seizure of Mosul was stark.[80] The group faced other setbacks, including repeated strikes by both the coalition and the Assad regime on its strongholds in the Raqqa region (the latter killing large numbers of civilians), but it also showed signs that it was adapting to coalition strikes by hiding operatives.[81]

Lister wrote in November that ISIS was fielding approximately 25,000 fighters, including terrorist and insurgent divisions, as well as a force more resembling a traditional army's infantry. According to Lister, ISIS controls territory from the Aleppo region of Syria to the Salah ad Din province in Iraq,[82] an area larger than the United Kingdom.[83]

It rules using a structure of *wilayat* or "provinces," each with its own governor, and local governments beneath them, as well as a series of administrative units, in many ways replicating a typical government bureaucracy. Its military force is primarily dominated by Iraqis, while many of its civil institutions are staffed by foreigners (see Chapter 4).[84] The structure is designed to survive the death of Baghdadi, and while the symbolic impact of killing the so-called caliph could be destabilizing in a number of ways, it is by no means certain that removing ISIS's leadership would cripple the organization.

ISIS's strength on the ground is an important part of the story, but only a part. Through a media strategy as aggressive as its military tactics, ISIS seeks to extend its influence around the world.

It has set its sights on winning support from members of the global al Qaeda network and it has created remotely directed outposts, *wilayat* as far away as Algeria and Libya.

ISIS intends not just to "remain" in Iraq and Syria, but to "expand" around the world, in the words of Baghdadi and other top leaders. In order to achieve this goal, it has projected its influence to potential recruits and hoped-for allies around the world using methods unlike any other extremist group. To understand how this projection works is to open a window on ISIS's goals, beliefs, and its ultimate fate.

CHAPTER THREE

FROM VANGUARD
TO SMART MOB

It was 1988, and the Soviet occupation of Afghanistan was entering its final days. International agreements had been signed and sealed, and the enemy forces slowly but inexorably withdrew. For ordinary Afghans, this prospect must have been a relief, a hopeful moment. Perhaps the long and costly war might finally end and some semblance of ordinary life finally return.

For the interlopers, it was a hopeful moment as well, but their desires were different. Foreign fighters, subscribed to a jihadist ideology, had flocked to the country by the thousands. They believed, not without some merit, that they had defeated one of the world's two superpowers. But for their leaders, the end of fighting provided no relief. Their passions and hopes were stoked by the prospect that this war would not only continue but expand to encompass the world.

For their plan to work, secrecy was required. Although thousands had come to fight the Soviets in the first stage, part two would be different. Through August and September, small meetings of two to fifteen leaders of the "Arab Afghans" were convened in Peshawar, Pakistan, to lay down plans for the next generation of violent jihad.[1]

The new organization would consist of two groups, one with limited scope and wide membership, and one with more ambitious scope and limited membership. The broad group would consist of would-be foreign fighters and Islamic radicals from around the world. These would be trained in insurgent and terrorist tactics in Afghanistan, then sent forth into the world to pursue their own agendas—always remembering the relationships they had forged.

From this large pool, which would eventually sprawl into the tens of thousands, the "best brothers" would be invited into a more exclusive circle and indoctrinated into the overarching conspiracy to change the path of history. These men would form a small and tightly cohesive organization of elites, which they referred to as the military base, and later simply as the base—in Arabic the word was *al qaeda*.

At its inception, al Qaeda numbered just over three hundred men, and while the ranks would fluctuate over time, they rarely exceeded several hundred. In addition to those few hundred, its employees and allies numbered in the thousands. Members of the core group had to swear complete obedience (*bayah*) to the emir (Arabic for "prince") of al Qaeda, Osama bin Laden. One of the terror group's founding memos listed four requirements for becoming an al Qaeda member, in bullet-point format—two of them were obedience. (The other two were a personal referral from a trusted member of the inner circle and "good manners.")

Al Qaeda was exclusive, but not isolationist. With a substantial sum of money drawn from Osama bin Laden's deep pockets, it began to send tendrils around the world, financing and providing technical support to everything from a Muslim insurgency in the Philippines to the first World Trade Center bombing to the full-on war in Bosnia. Key al Qaeda members moved in and out of these activities. They played a critical role but were rarely the prime drivers of events. Al Qaeda guided and it supported, but it did not claim credit and it did not advertise its name.

Instead, al Qaeda was a vanguard movement, a cabal that saw itself as the elite intellectual leaders of a global ideological revolution that it would assist and manipulate. Al Qaeda would set the stage for a global Muslim revolution by priming the pump.

It trained skilled fighters and terrorists using a network of training camps, some that it owned directly and others that it financed or supplied.[2] It funded the spread of propaganda and ideology, often relying on the work of high-profile clerics and scholars who were not obviously cogs of the core organization, such as the late Abdullah Azzam and Omar Abdel Rahman, the "blind sheikh."[3] It facilitated and eventually directly committed terrorist attacks in order to teach the global community of Muslims, known in Arabic as the *ummah,* that it could fight back.

But like many other terrorist organizations, al Qaeda imagined the revolution would be a spontaneous happening. The function of terrorism was to awaken the sleeping masses and point them in the right direction.[4] The masses would then rise up and more or less take matters into their own hands.

Through the 1990s, al Qaeda grew into a corporation, with a payroll and benefits department, and operatives who traveled around the world inserting themselves into local conflicts, either to assist radical movements on the ground or profit from them, as when it laundered money through Bosnian relief charities or trained members of a jihadist cell in the United States that carried out the 1993 World Trade Center bombing and tried to bomb New York City landmarks just weeks later.[5]

During this phase, it increasingly devoted assets to committing its own terrorist attacks, instead of acting through proxies. Its simultaneous truck bomb attacks on U.S. embassies in Kenya and Tanzania in 1998 represented its most important move into this arena. By the end of the 1990s, bin Laden's deep pockets were starting to show the strain of all this activity (helped along by some catastrophic business developments), and the organization regrouped in Afghanistan in the late 1990s, where its resources turned more and more toward spectacular terrorist attacks, culminating in terrible fashion on September 11, 2001.

Throughout this, al Qaeda remained the vanguard, the elite. It laid plans but did not broadcast them. After the embassy bombings and the bombing of the USS *Cole* in 2000, it came out of the closet with a feature-length propaganda video that showcased many of its key leaders and its very basic message of armed resistance.[6]

But the video's simple problem/solution formulation did not offer al Qaeda as a political force, only as a paramilitary force multiplier for the hypothetical Muslim silent majority waiting to be mobilized.

Al Qaeda was the spark. The existence of gasoline was assumed.

The hoped-for spontaneous Muslim revolution did not emerge in the days and weeks that followed 9/11, but the attack thrust al Qaeda into its own sort of revolution—it would no longer lurk in the shadows, pulling strings from a remote enclave in Afghanistan. The terrorist organization was now one side in a full-fledged war, and with war came the necessity of politics.

Al Qaeda was slow to adapt, and it never fully assimilated the implications of the change in its role. Weeks turned into months, and no claim of responsibility for the 9/11 attacks or taunting challenge was forthcoming[7] (a partially completed but unreleased video was later found on an al Qaeda hard drive).[8] It released only a smattering of uninformative and uncompelling press statements and video clips.

It was as if bin Laden believed al Qaeda could somehow continue to act as the hidden hand after killing thousands of Americans in a single unforgettable spectacle (although the invasion of Afghanistan may have derailed possible plans to claim the attack). It took years for al Qaeda to begin fully exploiting the media-ready elements of September 11, although the response from Western news outlets helped fill the void.

Between its failure to plan and a failure to anticipate the fury of America's response, al Qaeda was so slow off the mark with its messaging that the CIA beat it to the punch, airing an intercepted video featuring bin Laden discussing the planning for the attack before al Qaeda could even attempt to claim it.[9]

As the full force of the U.S. military descended on al Qaeda in Afghanistan, its ability to keep operations centralized began to decay almost immediately.

Al Qaeda had previously maintained operatives under its core organization in Yemen and Saudi Arabia, who were subjected to a severe crackdown in the wake of 9/11.[10] While the group was semi-independent and intermingled with other jihadist communities dating back to the Soviet days, few policy makers and analysts saw reason to delineate between the Yemeni and Saudi branches and their parent. Each branch fell under attack, and both suffered serious losses.[11]

Beyond the Gulf, the core al Qaeda had resources and loose alliances around the world, most of which came under greater or lesser amounts of pressure in the months that followed. Al Qaeda adapted to this new reality, but it did not assimilate its implications.

RISE OF THE AFFILIATES

American media and scholarship tend to treat the affiliate system as if it were a robust, well-defined structure, rooted in history. In fact, it is barely a decade old, with much of its activity weighted toward the second half of that span, and its history is one of fractiousness from the start.

When America turned its full attention to al Qaeda in the wake of September 11, the result was like a fist smashing down on a ball of clay.

The terrorist organization was flattened and thinned, bent out of shape, and spread out over a wider area, as key personnel fled the onslaught in Afghanistan for points abroad, and operatives who were already in the field found themselves increasingly isolated.

Tight lines of control became attenuated. Orders had to travel more slowly, over longer and more exposed routes, to get from the central command to those who carried out the kinetic work of terrorism. Secondary nodes sprang up to mitigate the dragging response times, in which directives from on high could take weeks or months to arrive via courier, thanks to al Qaeda's elaborate security precautions.

After the United States invaded Iraq in 2003, an existing group of Jordanian-influenced jihadists led by Abu Musab al Zarqawi, one of the terror organization's informal allies, directly began fighting United States forces (as discussed in Chapter 1). In 2004, Zarqawi pledged loyalty to Osama bin Laden and renamed the group al Qaeda in Iraq, the first formal AQ affiliate under the franchise model.[12]

In 2007, the Salafist Group for Preaching and Combat announced it was joining al Qaeda and would henceforth be known as al Qaeda in the Islamic Maghreb (AQIM).[13] Other affiliates soon followed. In 2009, the survivors of the Yemen and Saudi branches announced they were merging to form al Qaeda in the Arabian Peninsula (AQAP).[14] And in 2012, after years of being rebuffed by Osama bin Laden, Somalia's al Shabab was accepted into the fold by Zawahiri.[15]

In April 2013, Jabhat al Nusra split from the remnants of al Qaeda in Iraq to become al Qaeda's Syrian affiliate,[16] and in September 2014, Zawahiri announced a new affiliate, al Qaeda in the Indian Subcontinent, whose membership is still unclear but whose domain extends over geographical territory once considered the stomping ground of the core al Qaeda. A flurry of terrorist attacks soon followed in the new affiliate's name.[17]

FROM TERRORISM TO INSURGENCY

The affiliate structure immediately began to shift al Qaeda's focus away from global terrorism toward local insurgencies.

Under bin Laden, the terrorist group certainly had its hands in the insurgency business. For example, it had bankrolled, trained, and organized Muslim separatists in the Philippines into a fighting force that used terrorist tactics alongside open war in an effort to carve out an extreme Islamist political space in the island nation.[18]

Al Qaeda training camps had long focused on teaching military tactics, including many lifted from the U.S. Army. After training in Afghanistan, fighters might return to their home countries or be deployed to another conflict. In Bosnia, al Qaeda supported Egyptian radical networks in creating a division of foreign mujahideen to fight the Serbs, and it played a direct role in recruiting U.S. military veterans to serve both as trainers and soldiers.[19]

But these efforts, and others, were local conflicts with local combatants, and while al Qaeda's role was important, it was also in some ways peripheral and in all cases covert. Mujahideen in Bosnia and the Caucasus and Kashmir and other hot spots did not fight under the name of al Qaeda, and if their leaders owed bin Laden great respect and deference, they did not owe him obedience. Al Qaeda was neither attributed as the cause of these conflicts, nor was it responsible for their outcomes.

Each new affiliate that joined al Qaeda after 9/11 was, to a greater or lesser extent, mounting an insurgency in its home region, and each was allocating far greater resources to such battles than to striking international targets with the elaborately planned terrorist plots that had become AQ's trademark. When they ventured out with bombings and other civilian massacres, they often struck at geographic neighbors—the near enemy—rather than al Qaeda's preferred symbolic targets in the West.

Of all the affiliates, AQAP was most directly controlled by al

Qaeda Central, and it was the most active in plotting against the United States homeland. But while it quickly earned a reputation among policy makers and terrorism analysts as the "most dangerous" of the affiliates,[20] perhaps even more than al Qaeda itself,[21] the resources it allocated to terrorism were meager.

One would-be suicide bomber on a U.S.-bound plane succeeded in injuring only his private parts when his "underwear bomb" caught fire but did not detonate.[22] An intercepted cargo plane bombing was done so cheaply—$4,200—that its frugality became the cover story in the branch's English-language propaganda magazine, *Inspire*, rather than its lethality (casualties: zero).[23]

As the affiliate structure snapped into place, so too did an ideological current (bolstered by practical necessities) that further fractionalized the parent. Abu Musab al Suri, one of the most influential modern jihadist ideologues, outlined the case for decentralization in a 2005 book, *A Call to a Global Islamic Resistance*. One of the movement's most important elites, he laid out a blueprint for al Qaeda's obsolescence—leaderless resistance, in which the jihadi revolution would be carried forward by small cells who answered to no central authority.[24]

Leaderless resistance, essentially an optimistic idea that radicals would self-organize into independent cells and take violent action without direction, was not especially new. It was appealing to weak movements that faced significant external pressures without enjoying popular support. Leaderless resistance was urged on the white nationalist movement during the 1980s and 1990s by some of its most prominent ideologues, including Louis Beam and Tom Metzger,[25] ultimately hastening that movement's irrelevance as precious few volunteers stepped forward to risk prison without financial, technical, or even verbal support from a leadership figure.

For al Qaeda Central, this shift took the wind out of more than a decade of active, if selective, recruitment. Leaderless resistance is a tactic adopted when operational security concerns outweigh an organization's desire for a steady influx of new blood and spectacular, highly sophisticated attacks.

While al Qaeda Central contracted under heavy pressure from drones, raids and military strikes, the affiliates grew, attracting new recruits to take part in an increasingly militarized environment, one that al Qaeda was unsuited to lead. Supplying armies is very different from commanding them. Osama bin Laden had only minimal military experience; Ayman al Zawahiri had even less. As secure communication between the slow-moving core and its fast-moving satellites grew ever more difficult, centrifugal force began to degrade al Qaeda's identity as a cohesive whole.

Al Qaeda was no longer a vanguard leadership movement, playing chess on the world stage with a variety of resources at its disposal. While still representing a radical fringe and a tiny minority of the world's Muslims, the affiliates were gaining recruits and dragging al Qaeda, painfully, into the turbulent waters of populism.

THE AGE OF *FITNA*

With all these moving parts, it did not take long for the rumblings of *fitna*—an Arabic word referring to a period of internal dissent and infighting in Islamic history—to surface.

The new affiliate structure was inherently a field of land mines. Three of the affiliates—AQIM, al Shabab, and al Nusra—had their origins as splinter groups, the products of earlier waves of *fitna*, while AQAP was a merger between two badly damaged organizations. Although all four had some measure of al Qaeda influence in the DNA of their predecessor organizations (including shared personnel and Osama bin Laden's money), their histories hardly recommended them as islands of stability.

Furthermore, the affiliates had served mostly local interests prior to joining al Qaeda. While each leader made an oath of loyalty to Osama bin Laden, and after his death to Zawahiri, membership in the world's elite jihadist network had not visibly resulted in a substantial change to their priorities.

The very first affiliate, al Qaeda in Iraq, was a disaster almost from the start. Its leader, Zarqawi, was bullheaded and brutal, favoring the ideological approach known as *takfirism*, which refers to the practice of deeming someone to be a nonbeliever in Islam based on specific actions or practices. The concept had been around for a long time in various forms, but Zarqawi took the practice to new heights (in terms of who might be targeted) and new lows (in terms of requiring evidence of guilt). In his mind, there were any number of reasons one might be deemed to have left the fold—such as following the Shi'a branch of Islam, or by inconveniencing AQI in almost any way—and he used it as a pretext for wanton murder. (A more complete discussion of *takfir* may be found in the appendix.)[26]

Al Qaeda's efforts to rein in Zarqawi's excesses,[27] which it felt were hurting the image of jihadists everywhere, were only partially successful, winning some grudging concessions but forming the foundation of a deeper frustration that would linger after Zarqawi's death in 2006.

As more and more affiliates entered the system, al Qaeda faced new and different organizational challenges. AQAP was better behaved, but its successful English-language media operations threatened to overshadow its operations such that at one point, a plan was floated to make its highly visible English-speaking provocateur, Anwar Awlaki, an American citizen, into the affiliate's actual leader. The proposal was nixed by bin Laden.[28]

AQIM had significant internal tensions, in part thanks to the popular Mokhtar Belmokhtar, a legendary but fiery figure, who balked at the chain of command that placed him under the affiliate's leadership. Belmokhtar broke with AQIM and eventually made his own pledge directly to Zawahiri, although his organization has not been recognized as an affiliate to date.[29]

The Islamic State in Iraq, the successor group to al Qaeda in

Iraq, was similarly beset with strife and a long grudge over differing tactics that had festered since the Zarqawi days. In late 2011, it had sought to expand its reach by sending operatives to Syria, who then formed Jabhat al Nusra, a new fighting organization that soon took on a life of its own. In 2013 it changed its name to the Islamic State in Iraq and Syria (ISIS) and tried to reestablish its dominance over al Nusra, only to be rebuked by the latter's leadership and later by al Qaeda Central.[30]

The rift would set the stage for the worst crisis in al Qaeda's history, the rise of ISIS, but the problem would be dramatically foreshadowed in Somalia, where social media pulled back the veil on internal strife within al Shabab and pointed toward a revolution in jihadi culture.

AN AMERICAN HERALD OF CHANGE

Al Shabab was a splinter from a Somali Islamist group, the Islamic Courts Union, and it thrived after its parent perished, in large part because of the charisma and brutality of its emir, Ahmed Godane.[31]

Al Shabab quickly earned a reputation for attracting foreign fighters, especially Westerners. Many of these were from Somali diaspora communities in Minnesota, but a young Syrian-Irish-American from Alabama had become the insurgent group's public face. Omar Hammami had catapulted to fame of a sort in an al Shabab propaganda video titled "Ambush at Bardale," in which he and other American recruits rapped in English about jihad to the delight of radicals.[32]

In March 2012, Hammami posted a video to YouTube claiming that al Shabab wanted to kill him and asking for help from "the Muslims," a plea essentially directed at the leadership of al Qaeda Central. The dispute, it later emerged, had several dimensions, including Hammami's objections to corruption within Godane's regime, poor treatment of foreign fighters, and the American's quixotic view that jihadist groups should immediately declare a caliphate then fight to defend it, a perspective parallel to the thinking of the leaders in the Iraqi affiliate of al Qaeda. He also audaciously accused al Shabab of assassinating al Qaeda emissaries and allies in Somalia, charges that were later supported by other evidence.[33]

To promote his "help me" video, Hammami took to social media. Although he opened a number of accounts, some private and others public, he was most successful on Twitter. Using the handle @abumamerican and posing as his own "PR rep" at first, he tried to engage Western terrorism analysts, with an eye toward drawing media attention to his plight.[34]

Hammami's rebellion was part of a broader *fitna* within al Shabab that involved leaders with long-standing ties to al Qaeda. But none of them had been able to appeal to Zawahiri and receive any sort of response, a command-and-control logjam that grew more and more conspicuous as weeks of infighting dragged into months. Hammami's theory was that news coverage would eventually find its way back to al Qaeda Central and prompt a reply. He waited, but only silence followed.

Hammami's presence on Twitter had a cascading effect, drawing out al Shabab loyalists who proceeded to attack and threaten him over the social media service. In a second wave, Hammami's supporters within Somalia signed up and set about to discredit and expose Godane's maneuvers against the dissidents in daily Twitter fights. Infighting was nothing new to al Qaeda and its progeny, but the public spectacle was unprecedented, and it further stoked the flames of discontent. Much of the sniping took place in Hammami's absence. He spent his time hiding in the forests of Somalia, occasionally emerging to recharge his phone and fire off a new volley of provocations.

While Hammami was an important catalyst, he was not the only jihadi taking his grievances to social media. In the years since September 11, al Qaeda had taken to the Internet, in part to offset its lagging communications from senior leadership, but mostly because everyone else was using the Internet.

The terrorist group had generally kept up with the technology of the day, but in the realm of social media, it was slightly slower to adopt the latest trends. The center of gravity for jihadist extremists online had settled onto password-protected message boards, highly structured discussion forums that were carefully moderated by activists who were members of al Qaeda, or very closely aligned with such (see Chapter 6).[35]

The arrangement had numerous advantages, mostly revolving around control. Because the forums were moderated by people with legitimate terrorist connections, they were an important vehicle for authenticating official statements from al Qaeda and its affiliates, making false claims almost unheard-of.

The moderators could also clamp down on anyone who was sowing dissent and even ban them altogether if they could not be brought to heel. At the time, this seemed like a secondary benefit, but it soon became apparent that it was a crucial control mechanism for the post-9/11 al Qaeda.

During the terrorist heyday of the 1990s, al Qaeda was able to indoctrinate and manage recruits within its training camps and by virtue of the secrecy of its operations. Insiders were compartmentalized, and casual supporters were kept at a distance (except when it was time to pass the collection plate). When the U.S. invasion of Afghanistan fractured this infrastructure, the forums offered a method for achieving a similar effect remotely, albeit much less effectively.

Before it renamed itself ISIS, the Islamic State in Iraq had experimented with the idea of trying to launch viral content from the forums, including soliciting online *bayah* and seeking popular affirmation for its leadership, but the efforts fell flat, mainly because of the moderated format and the strength of al Qaeda Central's control.[36]

Hammami had tried his appeal in the forums but was categorically rejected. The moderators, fearing the effects of infighting might discredit al Shabab, censored any attempt to distribute his messages and grievances, and later suppressed similar attempts by more senior Shabab allies with more established reputations.[37]

But in early 2013, the Islamic State of Iraq announced that it was reabsorbing the Syrian al Qaeda affiliate, al Nusra, into a new controlling entity, the Islamic State of Iraq and al Sham, or ISIS. Al Nusra was having none of it and invoked the authority of Zawahiri in rejecting the bold power play.

Chaos broke out on the jihadist forums, and people started to take sides. One very prominent forum member, a widely admired jihadi analyst known as Abdullah bin Mohammed, began to criticize ISIS on the forums, accusing it of committing crimes against other jihadi groups and insinuating that it had been infiltrated by external evildoers who were now steering it down a dark path.[38]

After much drama, bin Mohammed was banished from the forums. But he found a new home on Twitter, where he quickly amassed tens and eventually hundreds of thousands of followers, including many who followed him from the forums.[39] In this new wilderness, no moderator held the power to silence dissent and his audience was vastly bigger, including people with a casual interest in jihadism as well as hard-core operatives. Other forum celebrities soon followed his example. Dirty laundry was aired, debates were held right out in the open, and support could be quantified in follower counts and retweets. It was very nearly democratic.

All this while, Hammami held court for terrorism analysts and traded jibes with Somali haters, occasionally going silent for long periods as he fled for his life. He began to achieve celebrity, or at least notoriety, both in the West and among jihadis. At one point, he even exchanged private messages with bin Mohammed over Twitter.[40]

Al Shabab was forced to fire back at Hammami's allegations over its official Twitter account, in addition to its many proxies who never wearied of attacking the American, even as he documented more scandals and the growing violence by al Shabab against its own ranks. The *fitna* began to spill back onto the forums, discrediting their legitimacy compared to the free expression available on Twitter.[41]

Al Jahad, a second-tier forum, began to take up Hammami's cause. It published an open letter to Zawahiri from a senior Shabab foreign fighter with long-standing al Qaeda ties, Ibrahim al Afghani, who begged Zawahiri to exert his authority over Godane.[42] If Zawahiri tried, it never became public—another pitfall of the communications breakdown. Godane could simply ignore private communiqués or pretend he never received them, confident that al Qaeda Central would be unable to do anything about it.

One of al Jahad's administrators, identifying himself as Sa'eed ibn Jubayr, took to Twitter in defense of Hammami but also stepped forward as an unlikely champion for a particularly Western value.

"Maybe jihadis are adopting freedom of speech," he tweeted at one point in response to a comment by one of the authors. "And I don't see anything wrong or messy with jihadis accepting open criticism from within."[43]

Hammami's quest ended with his apparent execution at the hands of al Shabab in September 2013. Even that news broke and disseminated over social media, confirmed by both pro- and anti-Godane factions.[44]

Many online jihadis honored him as a martyr and adopted his picture as a Twitter avatar in protest of his slaying. Hammami had lost the battle, and his life, but he had helped inaugurate a new era for the jihadist movement. This new paradigm was not democratic, but it was a feedback loop, in which jihadist supporters and even fighters found themselves with a new voice and a bully pulpit.

ENTER THE ISLAMIC STATE

ISIS had been the victim of social media criticism when Abdullah bin Mohammed turned his pariah status on the forums into Twitter celebrity, but it was quick to turn the tool to its advantage. As 2013 rolled into 2014, more and more jihadist fighters from every Syrian faction signed up for Twitter as their platform of choice.

Many factors came into play. Aside from the question of freedom of speech, the *fitna* in the forums had turned ugly. The two most important arenas were al Fidaa, al Qaeda Central's official forum, and al Shamukh, a designated forum for authenticated al Qaeda official releases.[45]

The administrators of the forums had tried hard to suppress the *fitna* plague in their online realms, but now they themselves had been infected. Users were thrown out for expressing pro- or anti-ISIS views, and eventually the administrators—including some of al Qaeda's inner circle of media operatives—turned on each other. Scores took to Twitter as the forums blew up. Shamukh defected entirely to ISIS, only to be wrested back in a coup by al Qaeda–loyal admins who controlled the message board's technical features. (By September, it was swinging back toward ISIS.)[46]

Tensions began to mount between ISIS and other Syrian mujahideen. After ISIS's attempted power grab in early 2013, Zawahiri had ordered it to stay in Iraq and leave Syria to al Nusra. ISIS ignored his commands and fighting broke out between ISIS and al Nusra, later expanding to include a number of other Syria-based mujahideen groups.[47]

In February 2014, Zawahri was backed into a corner. He had no leverage over ISIS, which, like many of the affiliates, was largely self-sufficient in terms of cash flow, weaponry, and terrorist expertise, and he apparently lacked either the will or the operational capacity to have ISIS's recalcitrant emir, Abu Bakr al-Baghdadi, assassinated. His emissaries, including the veteran mujahid Abu Khaled al Suri, tried to mediate a settlement to stanch the flow of bad blood, to no avail.

Zawahiri finally played his only remaining card, issuing a statement in February 2014 that publicly disavowed ISIS, essentially firing it from the al Qaeda affiliate network. ISIS responded quickly, assassinating al Suri before the month had ended. It was not just a divorce, ISIS meant to wage war, and it soon began fighting al Nusra and several other Islamic rebel factions within Syria.[48]

The fighting was not confined to the battlefields. ISIS also mounted a systematic and devastating campaign for hearts and minds on social media, most visibly and noisily on Twitter. This propaganda program (discussed more fully in Chapters 5 and 7) had multiple purposes and multiple fronts, but its most immediate effect was to project strength and highlight al Nusra's weakness, a perception that became increasingly concrete as ISIS gained ground against its fellow rebels over the next few months.

But the information war was not limited to al Nusra. In March 2014, ISIS launched a Twitter hashtag campaign, with its supporters seemingly rising up as a populist mass to tweet, "We demand Sheikh Al Baghdady declare the caliphate." In fact, the campaign, like many that would follow, was an orchestrated social media marketing effort, but as such, it was a rousing success. For some jihadi sympathizers, the idea of reviving the historical Muslim empire was exciting, and many rallied around the demand. Others were horrified, tweeting their angry objections about an idea they found heretical.[49]

The caliphate trial balloon (see Chapter 7) was the first of many shots across the bow of al Qaeda Central, which had for years been playing a long game, an incremental strategy whose goal of a global caliphate was constantly off on the vanishing horizon. Al Qaeda's affiliates, born-again insurgents, had been toying with seizing territory and attempting to govern for some time, but none had the audacity to claim the mantle of the caliphate, a concept freighted with huge religious and historical significance.

Just days later, ISIS leaked an al Qaeda Central video featuring Adam Gadahn, an American believed to be close to Zawahiri, who had guided and professionalized the parent's media operations in the post-9/11 era (see Chapter 5). The leak was almost certainly a direct result of the *fitna* on the forums, where most of al Qaeda's media operatives were members, including some who had defected to side with ISIS.[50]

In the video—which al Qaeda never officially released, perhaps intending it for an internal audience—Gadahn slammed ISIS as "extreme" and "radical" and intimated it was responsible for the "sinful attack" on al Suri. Gadahn went on to honor a number of "martyrs"—notably including Omar Hammami and Ibrahim al Afghani, who had both died at the hands of al Shabab.[51] ISIS used the video's harsh criticisms as a bludgeon against its critics and tried to make it a wedge between al Qaeda Central and al Shabab—unsuccessfully, at least in public. The few surviving Shabab dissenters were buoyed by the apparent support from Gadahn, backhanded or not.[52]

The March "caliphate" hashtag was a broad clue to a plan that ISIS fully intended to implement, although many were still shocked when it came to fruition.

The official announcement came in late June 2014, at the start of Ramadan. The announcement included an official announcement of al Baghdadi's real name and lineage, and video of his appearance in public and unmasked to deliver the Friday *khutba* (sermon) at a mosque in Mosul.[53]

Each of these details conveniently undermined objections to al Baghdadi's ascension that had been raised during the trial balloon in March.

The age of terrorist focus-group testing had arrived. Instead of the jihadi elite living (sometimes literally) on the mountaintop, reading the *New York Times* and watching Al Jazeera to gauge the mood of the Muslim masses, the newly rechristened Islamic State had ad-opted a feedback loop model, polling its constituents and making shrewd calls about when to listen and who could safely be ignored.

Offline, ISIS followed the model of a functional—if limited—government. Online, it played a different game. It amassed and em-powered a "smart mob" of supporters—thousands of individuals who shared its ideology and cheered its success, all the while orga-nizing themselves into a powerful tool to deploy against the world, harassing its enemies and enticing new recruits.

The concept was defined by Howard Rheingold, a technologist who has written extensively about how virtual communities affect human behavior, in his 2002 book, *Smart Mobs:*

> Smart mobs consist of people who are able to act in concert even if they don't know each other. The people who make up smart mobs cooperate in ways never before possible because they carry devices that possess both communication and com-puting abilities. Their mobile devices connect them with other information devices in the environment as well as other people's telephones.[54]

The smart mob paradigm kicks in when a large group of people spontaneously begin to act in synchronized ways due to the density of connections in their technology-assisted social network, where it is possible to connect with more people at different levels of inti-macy than allowed by simple physical proximity.

Although ISIS methodically shaped and manipulated its social media networks, it also benefited from this sort of self-organization. Small blended groups of ISIS members and supporters would take jobs upon themselves, including translating communiqués and pro-paganda into multiple languages and crafting armies of Twitter "bots"—scraps of code that mindlessly distributed its content and amplified its reach. In some ways, it was the realization of al Suri's leaderless jihad, except that activity that appeared spontaneous could often be traced back to the organization's social media team, which in turn coordinated with its leadership.

Meanwhile, back on the ground, ISIS had routed Iraqi government forces and seized a significant swath of northern Iraq,[55] and it pushed its message out on social media at the same time and with similar aggression.

By mid-2014, its messaging machine was well oiled and effective. The differentiation from al Qaeda was sharp. Despite the occasional dud, the overall storytelling and production quality of ISIS video was often incredible, the likes of which had been rarely seen in propaganda of any kind, and certainly leaps and bounds ahead of its predecessor's often-sophisticated attempts.

ISIS benefited from the constant, lethal pressure on al Qaeda Central, which was forced to abandon its more ambitious media efforts in favor of sporadic talking-head releases. Ayman al Zawahiri was not a particularly strong orator to start with. His charms were not enhanced by a format that boiled down to him lecturing tediously while staring straight at a fixed camera for forty-five minutes.

But the change in content was even more striking. ISIS was offering something novel, dispensing with religious argumentation and generalized exhortation and emphasizing two seemingly disparate themes—ultraviolence and civil society. They were unexpectedly potent when combined and alternated.[56]

The ultraviolence served multiple purposes. In addition to intimidating its enemies on the ground (Iraqi troops who fled before the IS advance had reportedly been terrified by footage of mass execution of prisoners),[57] ultraviolence sold well with the target demographic for foreign fighters—angry, maladjusted young men whose blood stirred at images of grisly beheadings and the crucifixion of so-called apostates.

But the emphasis on civil society, in videos and print productions, provided a valuable counterpoint and validation of the violence, off-setting its repulsion. ISIS would not shy away from whatever needed to be done, but its goal was to create a Muslim society with all the trappings—food aplenty, industry, banks, schools, health care, social services, pothole repair—even a nursing home with the insurgents' unmistakable black flag draped over the walls.[58]

The narrative tracks ultimately advanced the same message—come to the Islamic State and be part of something.

Throughout its long history, al Qaeda never put forward such an open invitation. Following the model of a secret society, al Qaeda had created significant obstacles for would-be members, from the difficulty of even finding it to months of religious training that preceded battle. The ISIS message was exactly the opposite—you have a place here, if you want it, and we'll put you to work on this exciting project just as soon as you show up (although in reality, some less radical recruits were quietly subjected to indoctrination anyway).

It was yet another lesson from Abu Bakr al Naji's *The Management of Savagery*. The media campaign's "specific target is to (motivate) crowds drawn from the masses to fly to the regions which we manage," Al Naji wrote, as well as to demotivate or create apathy and inertia among who might oppose the establishment of the self-styled Islamic State.[59]

The vanguard was dead. The idea of a popular revolution had begun.

In the end, al Qaeda's failure was the ironic failure of all vanguard movements—an assumption that the masses, once awakened, will not require close supervision, specific guidance, and a vision that extends beyond fighting.

Al Qaeda's vision is—often explicitly—nihilistic.[60] ISIS, for all its barbarity, is both more pragmatic and more utopian. Hand in hand with its tremendous capacity for destruction, it also seeks to build.

Most vanguard extremist movements paradoxically believe that ordinary people are afflicted with deep ignorance, yet such movements also expect that once their eyes have been opened, the masses will instinctively know what to do next.

ISIS does not take the masses for granted; its chain of influence extends beyond the elite, beyond its strategists and loyal fighting force, out into the world. Its propaganda is not simply a call to arms, it is also a call for noncombatants, men and women alike, to build a nation-state alongside the warriors, with a role for engineers, doctors, filmmakers, sysadmins, and even traffic cops.

It's the opening act on a brave new world. It's too soon to know how the invitation to the masses will be received, or even if ISIS will last long enough to find out. But win or lose, extremism will likely never be the same again.

CHAPTER FOUR

THE FOREIGN FIGHTERS

In August 2014, ISIS marked Eid al-Fitr, the end of Ramadan, with a twenty-minute, high-definition video offering its greetings to the Muslim world.[1]

Gauzy images of smiling worshippers embracing at a mosque cut to children passing out sweets to break the Ramadan fast. Scenes of laughing children on the streets were interspersed with scenes of the *muhajireen* (Arabic for "emigrants")—British, Finnish, Indonesian, Moroccan, Belgian, American, and South African—each repeating a variation on the same message.

"I'm calling on all the Muslims living in the West, America, Europe, and everywhere else, to come, to make *hijra* with your families to the land of Khilafah," said a Finnish fighter of Somali descent. "Here, you go for fighting and afterwards you come back to your families. And if you get killed, then . . . you'll enter heaven, God willing, and Allah will take care of those you've left behind. So here, the caliphate will take care of you."

Hijra is an Arabic word meaning "emigration," evoking the

Prophet Muhammad's historic escape from Mecca, where assassins were plotting to kill him, to Medina. Abdullah Azzam, the father of the modern jihadist movement, defined *hijra* as departing from a land of fear to a land of safety, a definition he later amplified to include the act of leaving one's land and family to take up jihad in the name of establishing an Islamic state. For most Islamic extremists today, the concepts of *hijra* and jihad are intimately linked. (See the appendix for a fuller discussion.)[2]

As the video continued, an Islamic religious chant known as a *nasheed* played over and over again, its chanted lyrics emphasizing the video's message.

> *Our state was established upon Islam,*
> *and although it wages jihad against the enemies,*
> *it governs the affairs of the people.*
> *It looks after its flock with love and patience.*
> *It does so carefully, and thereby does not receive any censure.*
> *The Shariah of our Lord is light, by it we rise over the stars.*
> *By it, we live without humiliation, a life of peace and security.*

As the verse about peace and security played for the first of several times in the video, the camera focused on a child holding a realistic-looking submachine gun.[3]

A few months later, the Eid video's sidelong references to fighting and jihad were placed in a much starker contrast, in a release that again focused on ISIS's substantial foreign fighter contingent.

In a procession were a long line of foreign fighters, each guiding with his left hand a prisoner identified as a Syrian soldier. They walked up to a bin containing serrated daggers, each fighter taking one with his right hand. There were at least seventeen fighters and as many prisoners. Many of the fighters, emphasized by the camera angles, were white-skinned Europeans. Only one wore a mask, the British fighter known as "Jihadi John," who had executed James Foley and other American and European hostages.

The camera lingered on the knives and the terrified prisoners for long, long seconds before the fighters began to hack through the necks of their victims. The video was intensely graphic, showing parts of the executions in slow motion and lingering over each horrific detail.

After, the camera played over the faces of the executioners, ensuring that the foreign fighters were clearly visible and sparking a rush to identify them. Media reports identified the perpetrators as French, German, British, Danish, and Australian citizens, although some of these claims were tentative.[4]

The contrast between these two scenes could not be more stark, and it highlights the two most important elements of ISIS's aggressive campaign to recruit fighters and supporters from around the world.

ISIS propaganda and messaging is disproportionately slanted toward foreign fighters, both in its content and its target audience. Important ISIS messages are commonly released simultaneously in English, French, and German, then later translated into other languages, such as Russian, Indonesian, and Urdu.

"Foreign fighters are overrepresented, it seems, among the perpetrators of the Islamic State's worst acts," said Thomas Hegghammer, a leading scholar of jihadist history, in an interview with BillMoyers.com. "So they help kind of radicalize the conflict—make it more brutal. They probably also make the conflict more intractable, because the people who come as foreign fighters are, on average, more ideological than the typical Syrian rebel."[5]

Of course, Syrian and Iraqi allies of ISIS, often initially motivated by pragmatic local concerns, may be equally vulnerable to radicalization in such a volatile environment, and local participants are also represented in ISIS's ultraviolent propaganda. But because of ISIS's outsize emphasis on publicizing foreign fighters while restricting the flow of information from independent sources, clear evidence is less abundant.

HOW MANY FIGHTERS?

One of the most important questions about the threat presented by ISIS, and the conflict in Syria and Iraq in general, is numerical: How many foreign fighters are there, where do they come from, and what will they do after fighting?

Unfortunately, the question is nearly impossible to answer with any kind of specificity, due to the dangers that ISIS presents for journalists and intelligence operatives on the ground. It's difficult enough to accurately assess the total size of ISIS's fighting force, let alone break it down into demographic components.

In the open-source world, there are only estimates, and the situation does not appear to be much better in the world of secret intelligence. While anecdotal information on foreign fighters exists in abundance, no one claims to be able to see the whole picture.

In October 2013, Radio Free Europe/Radio Liberty published a compilation of data on all foreign fighters in Iraq and Syria, drawn from multiple sources.[6] The data broke down according to country of origin, and included both high and low estimates from the various sources. The fighters counted came from all jihadi groups in the region, not just ISIS.

REF/RL found between 17,000 and 19,000 fighters, with about 32 percent originating in Europe (including Turkey). The majority of fighters identified in the data originated in the Middle East and North Africa, with the greatest numbers coming from Tunisia and Saudi Arabia. The remainder came in smaller numbers from other places around the world, including former Soviet republics, the Americas, and Australia. This figure is continuing to grow.

The nature of the data set provides multiple challenges in creating a clear picture of the foreign fighter phenomenon. Three-quarters of the country estimates came from studies by the International Centre for the Study of Radicalisation and Political Violence, based at King's College London, without reference to when the estimates were compiled. The majority of the remaining country estimates were taken from a mix of journalistic reports and government estimates by source countries, involving different methodologies.

Many of the estimates that were available for the RFE/RL report are likely too low. For instance, the report cited "3,000 plus" for Saudi Arabia and Tunisia, the two largest contributors from the Middle East and North Africa. In an interview with Al Arabiya, a source based in the ISIS foreign-fighter stronghold of Raqqa said fighters from both countries received preferential treatment and leadership positions. Chechen fighters, renowned for their viciousness and military skills, were also highly valued.[7]

More problematic, numbers were unavailable for several countries known to have provided fighters, including Azerbaijan, Indonesia, the Philippines, and Somalia.

In general, foreign fighter estimates from both government sources and news reports are often unclear as to whether fighters were affiliated with ISIS and whether the estimates pertain only to Syria or to Iraq and Syria.

Government-provided estimates are especially problematic, given the closed nature of intelligence reporting and the political considerations accompanying disclosure.

In October 2014, FBI director James Comey told CBS News' *60 Minutes* that an estimated "dozen or so" Americans had joined ISIS. In November 2014, a government official speaking off the record told us that more than one hundred Americans had traveled to Syria to fight over the entire course of the conflict, including those who had returned, matching a previous estimate that we believe to be considerably too low.

The day after that interview, Comey told reporters that 150

Americans had traveled to Syria "in recent months." Earlier news reports citing unnamed government sources said there were "several hundred American passport holders running around with ISIS."[8] In earlier interviews, Comey also suggested that whatever number he provided was likely too low. These wild inconsistencies lead us to question the usefulness of any such official estimates.[9]

Based on both social network analysis and anecdotal observation of comments by foreign fighters on social media, we believe that as of this writing, a minimum of 30 to 40 Americans are *currently* affiliated with jihadists in Syria and Iraq, in both fighting and noncombat capacities, and we estimate that well over a dozen are currently affiliated with ISIS. This figure represents what we can confidently assess from open sources, meaning the real figure is certainly higher, possibly by a wide margin.[10]

For the United Kingdom, similar disclaimers apply, but the range of estimates is much higher, especially on a per capita basis. In August, the United Kingdom estimated to reporters that 500 British citizens were affiliated with ISIS in Syria and Iraq, with another 250 who may have returned. It is unclear whether the returnees are still affiliated with ISIS, but reports indicate it is difficult to simply leave the organization.[11] Dramatically higher estimates began to circulate toward the end of 2014.[12] British ISIS members were significantly more numerous and visible than Americans on social media platforms, in our observations.[13]

French- and German-speaking fighters have also been observed in large numbers on social media, and low-end estimates point to more than 550 fighters from Germany, and more than 1,000 from France. From the West, significant numbers of Canadian fighters also made their presence known on social media, although like Americans, many of them kept a lower profile.[14]

A typical jihadi foreign fighter is a male between 18 and 29 years old, according to a study by the Soufan Group, although there are many exceptions. Some are well over 30, and it is not uncommon to see fighters between 15 and 17.

Beyond age and gender, there are few consistent patterns and no reliable profile of who is likely to become a foreign fighter, but among Western recruits, a disproportionate number of converts can typically be found. (Converts are often especially vulnerable to fundamentalist ideas, often combining wild enthusiasm with a lack of knowledge about their new religion, making them susceptible to recruiters.) This approximate profile has endured for decades, through multiple jihadist conflicts.[15]

WHY JOIN?

Why do individuals travel abroad to take part in somebody else's violent conflict, a markedly different behavior from taking part in a conflict that involves one's home community?

There is no single pathway, no common socioeconomic background, not even a common religious upbringing among individuals attracted to foreign fighting in general or jihadist fighting in particular.

"Four decades of psychological research on who becomes a terrorist and why hasn't yet produced any profile," according to John Horgan, director of the Center for Terrorism and Security Studies at the University of Massachusetts Lowell, who has studied the subject intensively. While efforts to generalize the problem have failed, he says, it is possible to understand some pathways for individuals.[16]

A variety of studies, using different frameworks and concepts, have approached the question of why people join violent extremist groups. Many of these boil down to a distinction between external and internal motives.

External motives have to do with an individual's perception of large-scale events in the world. While many analysts and policy makers have pointed to factors such as weak states, education, and social and economic disadvantage as external motivating factors, among those who study extremism in depth there is little consensus and much dispute on the importance of these factors.

More often than not, the external factors cited by extremists themselves point toward the importance of much more specific situations, for instance, a military conflict or genocidal campaign, usually but not always involving victims from a potential recruit's identity group.

Jihadist propaganda has often relied on exactly these flashpoints, such as the Soviet invasion of Afghanistan or the genocide in Bosnia, using them as a point of entry to leverage narratives about the event, characterizing participation as not only a reasonable choice, but an obvious moral obligation. Indeed, jihadi ideologues often focus on the obligation of individual jihad when some or all of the *ummah*, or the Muslim nation, is under threat.[17]

But these flashpoints do not necessarily provide adequate motivation on their own merits. They offer outlets, either for social pressures in a fighter's native land or for his own internal struggles and dilemmas.

Internal motives stem from what an individual wants or needs for himself, in terms of the perceived benefits of membership in an extremist group, such as a feeling of belonging, escape into a new identity, adventure, or money. Foreign fighters have personal needs that are met by joining an organization, and those personal needs may become more important over time.

"They want to find something meaningful for their life," in the words of John Horgan. "Some are thrill-seeking, some are seeking redemption."[18]

According to Scott Atran, Western volunteers are often in transitional stages in their lives. They are often "immigrants, students, between jobs or girlfriends . . . looking for new families of friends and fellow travelers. For the most part they have no traditional religious education and are 'born again' into a radical religious vocation through the appeal of militant jihad." [19]

Social acceptance and reinforcement is also an important factor. Atran's research found that three out of four foreign fighters in Syria traveled together with others, a figure consistent with previous studies on the subject.[20]

Traditionally, jihadist fighters have found internal motivation in the promise of perceived religious rewards such as entry into heaven and the benefits that promise includes, such as the much-discussed seventy-two virgins (the role of religion is emphasized in Atran's research).

But for many, perhaps most, jihadists, religious motivations are necessary but not sufficient to explain the leap to violent action. Some mix of political sentiment, religious belief, and personal circumstance is required. Parsimonious explanations, which focus only on single external factors, whether religious or political, cannot explain why one sibling becomes a jihadist and another a doctor. Clearly, something happens that makes an individual willing to risk his or her life for a cause.

During the course of the civil war in Syria, the balance of internal and external factors has shifted over time. At the start of the conflict, a diverse coalition of imported religious fighters and secular Syrian rebels united loosely around the goal of overthrowing the oppressive Assad regime. For the jihadists, a longer-term goal was the establishment of a state governed by Islamic law, but the initial focus for most combatants was on fighting the regime. In the wake of ISIS's rise, according to research by Peter Neumann, Scott Atran, and others, that goal has shifted noticeably to establishing Shariah law and supporting the institution of the caliphate, regardless of the wishes of the local Syrian population.[21]

With the emergence of large numbers of foreign fighters on social media, providing a conversational and continual commentary on the conflict, internal motivations soon came to the fore, and they went beyond the promise of heaven. While few would dispute the importance of religious allure in attracting fighters to the field, the conversation online frequently turned to the theme of fun and adventure.

British fighters, for instance, often posted pictures and stories about their day-to-day experiences. One of them, twenty-three-year-old Ifthekar Jaman, coined the phrase "five-star jihad" to describe the fun he was having fighting in Syria, which caught on as a rallying cry to his countrymen, who showed up in ever-increasing numbers. (Jaman was killed in December 2013.)[22]

A number of "celebrity" fighters upped the ante. One of the most popular was a former Dutch soldier named Yilmaz, who helped train mujahideen fighters with various factions in Syria. He documented his Syrian experience with a wealth of photographs, posted on Instagram under the name "chechclear," a reference to a gruesome video of Chechen insurgents beheading a Russian soldier in the 1990s.

As chechclear, he documented the war itself, posting pictures of battles and fighters, but also images of the people of Syria, including children, and seemingly incongruous snapshots of jihadists cuddling with cats, all of the photos enhanced by the photographic filters that helped make Instagram so popular.[23]

Yilmaz and other fighters also took to sites such as Ask.fm, a social media platform oriented around answering questions from other users. Questioners often asked how to donate to fighting groups or how they could get to Syria themselves, which fighters answered with greater or lesser amounts of specificity.

"I will personally assist you insha'Allah," Ifthekar Jaman told one questioner on Ask.fm. "But know this, if you are a spy, when you are caught, your punishment will be with little or no mercy."[24]

Others asked what to expect if they joined, querying everything from food choices to bathroom facilities to what sort of gear they should pack.

"Cargo pants (combat trousers), 511 brand is good," wrote Abu

Turab, a twenty-five-year-old American who had drifted among fighting groups. "I have Old Navy, lol, but water-resistant stuff is the best. Don't hesitate to buy expensive stuff, for you're spending as [an act of worship]. Jackets and boots, try to buy GORE-TEX."[25]

The rise of violent infighting among jihadist factions in early 2013 and the subsequent disavowal of ISIS by al Qaeda put a significant damper on the five-star jihad. On social media, an explosion of discontent emerged as the focus of the conflict irrevocably shifted from fighting the Assad regime to a battle for supremacy among the mujahideen. Although combat with the regime continued, the infighting among the rebels racked the conscience of many participants.

"Have you forgotten your enemies who have destroyed a part of the Ummah?" one fund-raiser tweeted. "They are the people [you're] fighting, the KAFIRS [unbelievers], not MUSLIMS."

Others were alarmed at the effect this would have on potential recruits.

"Many will avoid hijra because of what just happened," one tweeted mournfully.

An Indonesian fighter was at a loss to answer a potential recruit who privately messaged him to say he feared he would be killed by his fellow Muslims.

"I don't know how to answer this Muslim brother from Morocco who planned to join ISIS with me," he tweeted plaintively.[26]

But ISIS was already moving to provide a new answer to the question: "Why join?" With the rollout of its plans for a caliphate in mid-2014, the focus shifted to promoting a sense of inclusion, belonging, and purpose in its demented utopia.

FOREIGNERS IN ISIS

With the declaration of its "caliphate" in July 2014, ISIS began to enhance and amplify themes relating to the society it wanted to create.

While these ideas had already been present in its propaganda, the declaration of the caliphate had a dimension that went beyond simply showing ISIS in its best light. The new focus reflected a mandate given by Abu Bakr al-Baghdadi in his first speech as putative caliph:

"O Muslims everywhere, whoever is capable of performing *hijrah* (emigration) to the Islamic State, then let him do so, because *hijrah* to the land of Islam is obligatory," Baghdadi said. "We make a special call to the scholars, [Islamic legal experts] and callers, especially the judges, as well as people with military, administrative, and service expertise, and medical doctors and engineers of all different specializations and fields." For these professionals, as well as for fighters, emigration was a religious obligation, he said.

In July 2014, ISIS's Al Hayat Media Center released an eleven-minute video that drove this point home. Titled "The Chosen Few of Different Lands," the video showed a Canadian fighter named Andre Poulin, a white convert known to his comrades as Abu Muslim. It was a masterpiece of extremist propaganda.[27]

The video opened with stunning high-definition stock footage of Canada (or a reasonable facsimile) as Poulin described his life back home.

"I was like your everyday regular Canadian before Islam," he said. "I had money, I had family. I had good friends."

The barbaric nature of ISIS can lead observers to conclude its adherents are simplistic, violent, and stupid. "The Chosen Few" displayed a keen self-awareness of this perception and actively argued against it, with Poulin as its telegenic exemplar.

"It wasn't like I was some social outcast," Poulin said. "It wasn't like I was some anarchist, or somebody who just wants to destroy the world and kill everybody. No, I was a very good person, and you know, mujahideen are regular people too. . . . We have lives, just like any other soldier in any other army. We have lives outside of our job."

Life had been good in Canada, Poulin said, but he realized he could not live in an infidel state, paying taxes that were used "to wage war on Islam."

In reality, Poulin was not quite the model of social integration that he portrayed on film. He developed an interest in explosives early and had dabbled in Communism and anarchism before settling on radical Islam as an outlet for his interests. He had been arrested at least twice for threatening violence against the husband of a man whose wife he was sleeping with. These facts were conveniently omitted from his hagiography.[28]

In the video, Poulin said ISIS needed more than just fighters.

"We need engineers, we need doctors, we need professionals," he said. "We need volunteers, we need fund-raisers." They needed people who could build houses and work with technology. "There is a role for everybody."

A narrator gave a brief account of Poulin's life, with pictures, which concluded with an action sequence showing him taking part in an attack on a Syrian military air base in Minnigh. Shot in high definition, the footage was remarkable, depicting Poulin rushing toward the enemy, highlighted among his fellow combatants using sophisticated digital techniques. Poulin was clearly visible in action, running out in front of his comrades until he was struck down in a massive explosion. After, his dead body was shown sprawled on the ground and later being prepared for burial.

"He answered the call of his Lord and surrendered his soul without hesitation, leaving the world behind him," said a narrator in perfect, unaccented English. "Not out of despair and hopelessness, but rather with certainty of Allah's promise."

At the end, Poulin spoke again, his visage filtered in a gauzy light.

"Put Allah before everything," he said.

The "whole society" pitch had been presaged for some months. ISIS supporters on social media tweeted Photoshopped images of an "Islamic State" passport, for instance. Their enthusiasm for these tokens of future legitimacy was, at times, reminiscent of a child trying on his father's shoes, pretending to be grown up.

But as ISIS cemented its control of territory in Iraq and Syria, such images took on an increasingly material reality, albeit presented through carefully filtered glimpses. Each of ISIS's provinces issued a steady stream of images showing the infrastructure of government taking form—police cars and uniforms emblazoned with the black flag, markets overflowing with food.

ISIS selectively amplified its nation-building efforts, but it did not entirely fabricate them. While some of its outreach involved active image management, some parts were pragmatic, such as its offer of handsome salaries for engineers able to maintain the oil fields on which ISIS relies for black-market income.[29]

In November 2014, ISIS announced it would mint its own currency in keeping with the "prophetic method," posting images of the new coins to Twitter. ISIS military uniforms in Mosul sported black patches with white writing in Arabic citing its adherence to the "prophetic method." As Will McCants of the Brookings Institution wrote, this "nightmarish bureaucracy" was intended to invoke echoes of Islamic prophecies related to the end times (see Chapter 10).[30]

But all of this also provided important markers of stability and substance. The stark black flag, which had come to be emblematic of ISIS's fighting force, was not just a symbol of war, the images argued wordlessly. It was the symbol of a society; no distant dream, but a living, breathing institution waiting to be populated by the believers.

In an intelligence environment where credible estimates were unavailable on critical issues such as the size of ISIS's fighting force, or even just that force's foreign component, information on noncombatant emigration was sparser still. But one element of that campaign was sensational enough to grab the headlines—ISIS's recruitment of women.

THE WOMEN'S BRIGADES

Many of ISIS's most vocal and visible supporters online are women. Analysis of social networks linked to ISIS on Twitter found hundreds of users identifying themselves as women and actively spreading the organization's message.[31]

The leader of this online recruiting effort was a veteran of online agitation using variations on the online username "al Khansa'a." The name corresponded to a female poet who was among the earliest converts to Islam in the days of the Prophet, known for ordering her sons into battle on behalf of Islam. All four died. "I feel proud to be the mother of martyrs," she is famously reputed to have said.[32]

Al Khansa'a had been active on al Qaeda–linked forums well before ISIS's rise. Among members of the forum community, she was an early adopter of social media, opening a Twitter account under the handle @al_khansaa2 in September 2012, as well as establishing a presence on Facebook and other channels.

She was not only an influential figure; she was also well-connected to other al Qaeda users, actively participating in networks connected to AQAP and al Shabab, with a special interest in connecting other female jihadist supporters to each other and to the broader al Qaeda network.[33]

Al Khansa'a was also ahead of the curve with her allegiances, defecting to ISIS at the outbreak of the *fitna* (infighting) with al Qaeda. At first, she was heavily engaged in the heated battles that fired up between top jihadist forum members, but as the weeks passed, she transitioned into a new role—leading an online "brigade" that shared her name and was devoted to recruiting women to join ISIS.[34]

Aqsa Mahmood is another of the many women now tirelessly working to recruit foreigners to join ISIS. As a teenager growing up in Glasgow, Scotland, she turned away from a typical, seemingly happy life spent consuming young adult novels and rock music and toward an increasingly militant outlook on the world and on her Muslim heritage, a sharp break from her family's views.

Mahmood documented her transformation with all the enthusiasm a teenager can bring to bear on her Tumblr blog, describing a swift transition from a mainly secular lifestyle into radicalism, noting her family's disapproval along the way and sometimes laughing it off.

"My parents genuinely think I'm extremist," she wrote.

Instead, she wrote in March 2013, her online friends—steeped in Salafist interpretations of Islam and the horror of the emerging Syrian civil war—were "the new family."[35] She immersed herself in ever-more radical content from YouTube, Tumblr, and other online sources, citing al Qaeda–linked clerics such as Abu Muhammad al Maqdisi and Abu Yahya al Libi as "my men of *haqq* (truth)."[36]

"I just want to make *hijrah* ok," she typed.[37]

Throughout 2013, her content turned more and more to openly jihadist ruminations and the growing obligation she felt to be involved in the struggle in Syria. In November, now nineteen, she abruptly bid her horrified family farewell.

"I will see you on the day of judgment. I will take you to heaven, I will hold your hand," her father recounted her saying. "I want to become a martyr."

She traveled to Turkey and from there to Syria, where she joined ISIS and married a Tunisian fighter.[38] From Aleppo, she kept up her online activities, using Twitter and Tumblr to encourage others to follow her example.

"And to those who are able and can still make your way, please [fear Allah] and don't delay anymore, hasten hasten hasten to our lands and live in [honor]," she tweeted.[39]

Uncounted other young women like Mahmood were lured to join ISIS in Syria and Iraq, including hundreds of Westerners and many more from Arabic-speaking countries.

"Most foreign girls will be married off to foreign fighters upon their arrival," wrote Mia Bloom, a leading expert in the role of women in jihadist movements. "In fact, many are offered up as a form of compensation to the men fighting for al Baghdadi."[40]

Two teenage girls from Vienna, Austria, ages fifteen and sixteen, discovered this reality immediately after they left home to travel to Raqqa, Syria, where they were promptly married off to Chechen fighters. They reportedly became pregnant almost immediately and wrote to their families to say they wanted to come home, but there was no escape for them. Austrian police sources quoted in a British tabloid said the girls' social media accounts were taken over by other ISIS members, who sent a stream of happy messages encouraging others like them to make *hijra*.[41]

For some, on their arrival in Syria, the virtual al Khansa'a Brigade transformed into a physical reality. The bricks-and-mortar al Khansa'a Brigade was a grim counterpoint to the illusion that its namesake sold online, according to one Syrian woman who defected from ISIS. In an interview with CNN, she described joining the brigade in Raqqa, Syria, where many ISIS foreign fighters were concentrated.

The defector, referred to as Khadija to protect her identity, told a jarring story of a women's squad of morality police, who whipped women seen on the streets wearing anything that did not measure up to ISIS's rigid ideal of female modesty.

The punishments were meted out by a woman Khadija knew as Umm Hamza (*umm* is Arabic for "mother of," and is used as a *kunya*, a form of alias, by female jihadists in a manner similar to how *abu*, Arabic for "father of," is used by males).

"She's not a normal female. She's huge, she has an AK, a pistol, a whip, a dagger and she wears the niqab," Khadija told CNN.[42]

Khadija was initially seduced by the power of her position, but over time the grinding horror of life under ISIS's rule began to take a toll. She witnessed crucifixions and brutal beheadings. As the commander of her brigade tried to push her toward marriage, she was increasingly alarmed by the domestic and sexual violence she saw ISIS wives endure.

"I started to get scared, scared of my situation," she said. "I even started to be afraid of myself." She was smuggled to Turkey before she could be given to a husband.

ISIS's bid to build a society didn't stop at the recruitment of women, however. Foreigners were encouraged to bring their whole families to Iraq and Syria to "live under the shade of the caliphate."

In November 2014, ISIS released a video introducing "some of our newest brothers from Kazakhstan," who had "responded to the crusader aggression with their *hijra* and raced to prepare themselves and their children." The video showed dozens of smiling boys, the sons of a unit of Kazakh fighters, clambering into a bus and going to a schoolroom described as "the ultimate base for raising tomorrow's mujahideen." [43]

"We spent our childhood far away from this blessing," their Kazakh teacher explained. "We were raised on the methodology of atheism. . . . The *kuffar* (unbelievers) poisoned our minds. . . . Our children are happy. They're living in the shade of the Quran and Sunnah."

Another teacher was shown supervising a class of pre-teenage boys in uniforms.

"They've completed lessons in Quran, [proper recitation of the Quran], and the Arabic language," he said. "They will move on to do physical and military training."

The scene shifted to show a Kazakh boy of perhaps nine combat-stripping an assault rifle, then training with others in its use. The physical training included hand-to-hand combat and calisthenics. At the end of the day, a member of ISIS's media team questioned one of the students.

"What will you be in the future, if God wills it?" the interviewer asked.

"I will be the one who slaughters you, oh *kuffar*," the boy responded with a grin pointed at the camera. "I will be a mujahid, if God wills it." One ten-year-old boy from the video was depicted in a subsequent release as executing two prisoners.

Such videos and images are far from rare. ISIS members on social media routinely post images on social media of children holding severed heads and playing on streets where dismembered bodies are splayed carelessly on the sidewalk. One image posted to Twitter showed a child playacting the beheading of American hostage James Foley using a doll.[44]

A UN report on war crimes in Syria pointed to the indoctrination of children as a "vehicle for ensuring long-term loyalty" and creating a "cadre of fighters that will see violence as a way of life." While children have often been victims of such manipulation in war zones, ISIS approached their "education" as it did almost everything else—systematically.

"This is not a marginal phenomenon. This is something that is being observed and seems to be part of the strategy of the group," Leila Zerrougui, the UN special representative for children and armed conflict, told the Associated Press.[45]

For many families, of course, the reality of life under the Islamic State does not match the idyllic picture painted in ISIS propaganda. Some of the uncounted families who have moved to ISIS territory in Syria reported conditions deteriorating throughout 2014 as the organization came under increasing external pressure. In the most important cities under ISIS control, Raqqa in Syria and Mosul in Iraq, electricity is reportedly limited, with garbage lying in the streets for days. In Mosul, a shortage of chlorine has rendered the water dangerously undrinkable, and ISIS has cut off most communications to the outside world in its effort to suppress news about the reality on the ground.[46]

LEFT BEHIND

The potent projection of ISIS's "caliphate" exerted a gravitational pull on vulnerable people around the world, but not all of these individuals entered its orbit. Some were unable to travel to the Middle East, thwarted by personal circumstance, external obstacles, or lack of imagination. Denied participation in the ISIS project abroad, some chose to participate at home, through acts of violence.

ISIS had been born out of al Qaeda, a traditional terrorist group, transforming itself into a formidable insurgency with substantial territory under its control. But its apocalyptic plan had always included a confrontation with the West, and it had stretched its influence out both virtually and physically in preparation for a new phase of war.

The threat took a variety of forms. In some cases, individuals living in the West acted on their own initiative. In others, ISIS operatives guided their actions, either remotely over social media or in person, using returned foreign fighters and other operatives abroad.

By March 2014, when few in the West were even contemplating an intervention in Iraq or Syria, ISIS already had operatives working on mayhem. In Switzerland, authorities disrupted a terrorist cell, led by three ISIS recruiters, which was in the midst of plotting a terrorist attack using explosives and poison gas. The arrests were kept quiet for months as Swiss authorities searched for additional conspirators.[47]

In May, a French citizen of Algerian descent named Mehdi Nemmouche shot and killed four people at the Jewish Museum of Belgium before fleeing the scene. When he was arrested, in a railway station in France days later, police found in his luggage a video featuring the ISIS flag and claiming responsibility for the attack. Further investigation revealed that Nemmouche was a returned foreign fighter. A French hostage who had been imprisoned with James Foley and Steven Sotloff subsequently identified Nemmouche as one of his jailers.[48]

In Malaysia, nineteen alleged ISIS supporters were arrested between April and June 2014, accused of planning to bomb places where alcohol was served or brewed.[49]

In June 2014, President Obama announced the United States would increase its troop presence in Iraq to protect U.S. personnel, and on August 7 he informed the world that he had ordered air strikes against ISIS targets to slow its military advances and protect the beleaguered Yazidi minority in Iraq, which faced an imminent genocide. The pace of ISIS's "external operations"—terrorist plots and attacks—picked up significantly.

The incidents took a number of forms. In mid-August, a nineteen-year-old British citizen was arrested on a London street carrying a knife, a hammer, and the flag of ISIS. He was charged with preparing a terrorist act.[50] In France, two teenage girls—ages fifteen and seventeen—were arrested for planning to bomb a synagogue in Lyon, part of a network of Islamic radicals online, although reports did not specify ISIS.[51] In September, Australian police arrested fifteen people in a series of police raids to prevent a plot to randomly behead Australian citizens and wrap their bodies in the ISIS flag for public display. The plan was directed over the phone by an Australian ISIS recruiter based in Syria.[52]

On September 21, ISIS's chief spokesman, Abu Muhammad al Adnani, called for supporters around the world to rise up and respond to Western-led air strikes by carrying out attacks against any citizen of a country that belonged to the coalition against ISIS.

Do not let this battle pass you by wherever you may be. You must strike the soldiers, patrons, and troops of the [unbelievers]. Strike their police, security, and intelligence members, as well as their treacherous agents. Destroy their beds. Embitter their lives for them and busy them with themselves. If you can kill a disbelieving American or European—especially the spiteful and filthy French—or an Australian, or a Canadian, or any other disbeliever from the disbelievers waging war, including the citizens of the countries that entered into a coalition against the Islamic State, then rely upon Allah, and kill him in any manner or way however it may be. Do not ask for anyone's advice and do not seek anyone's verdict. Kill the disbeliever whether he is civilian or military, for they have the same ruling. Both of them are disbelievers. . . .

If you are not able to find an IED or a bullet, then single out the disbelieving American, Frenchman, or any of their allies. Smash his head with a rock, or slaughter him with a knife, or run him over with your car, or throw him down from a high place, or choke him, or poison him. . . . If you are unable to do so, then burn his home, car, or business. Or destroy his crops. If you are unable to do so, then spit in his face.[53]

The same day that the speech was released, Algerian terror group Jund al Khilafah, which had split from AQIM and thrown its support behind ISIS months earlier, kidnapped a French hiker and immediately issued a video threatening to behead him if the French government continued to support Western air strikes against ISIS. On September 24, it issued a second video, fulfilling its threat on camera.[54]

Short days later, an eighteen-year-old stabbed two Australian police officers he was scheduled to meet with after his passport had been suspended. The officers survived. The teen's Facebook page was filled with ISIS material.[55]

On October 20, twenty-five-year-old Martin Couture-Rouleau drove a car into two Canadian soldiers in a parking lot in St.-Jean-sur-Richelieu, Quebec, then jumped out of the vehicle with a large knife. Police killed him before any more mayhem could occur. Couture-Rouleau had tried to leave Canada to go to Syria, but his passport was suspended because he had come to the attention of authorities. The Quebecois's social network on Twitter was filled with French-speaking ISIS members and supporters.[56]

Two days later, thirty-two-year-old Michael Zehaf-Bibeau shot and killed a soldier at a war memorial in Ottawa, Ontario, adjacent to the Canadian Parliament, then stormed the legislature, making his way into the building before being shot and killed by police. He had made a video condemning Canada's foreign policy. He had applied for a passport to travel to Syria, but his application was under investigation at the time of the attack. ISIS supporters online obtained a distributed image of the killer and celebrated the attack.[57]

And just one day after that, a thirty-two-year old American, Zale Thompson, attacked two New York City policemen on patrol with a hatchet. Thompson, who was killed by police, reportedly consumed jihadist content online, although other reports suggested a scattered fixation on a wider range of issues.[58]

It was a remarkable string of so-called lone wolf attacks.[59] For years, al Qaeda had been encouraging such attacks with only rare successes, spread out over months and years. Al Qaeda in the Arabian Peninsula was especially persistent in promoting such attacks in its English-language magazine, *Inspire,* widely distributed online, but it had racked up only a handful of debatable successes over its four years of publication, along with a somewhat larger number of failed attempts.[60] In contrast, ISIS had inspired three successful attacks within a span of days. In November, ISIS later took credit for all three, as well as the earlier incident in Australia.

"All these attacks were the direct result of [Adnani's] call to action, and they highlight what a deadly tinderbox is fizzing just beneath the surface of every western country, waiting to explode into violent action at any moment given the right conditions," stated an article published under the name of a British prisoner of ISIS who had been co-opted into the role of spokesman (see Chapter 5).[61] More attacks in the name of ISIS soon followed in December, including a hostage situation in Sydney, Australia.[62]

In addition to the "lone wolf" threat, the question of returning fighters loomed large in the minds of Western security services. Returning fighters, like Nemmouche, were arrested in countries from Norway to Luxembourg to Indonesia, with many being detected in Europe, and certainly more still who escaped detection.[63]

In 2015, the terror threat in Europe began to heat up. A series of lone-wolf attacks inspired by ISIS in France in December and January (including stabbings and hit-and-runs) had been capped by an al Qaeda in the Arabian Peninsula assault on the French magazine *Charlie Hebdo,* which published cartoons of the Prophet Mohammed. An ISIS supporter associated with the attackers also jumped in to attack police while they searched for the first team. More than a dozen people were killed, and European governments began a massive crackdown, rolling up returned ISIS fighters and other jihadists in a sweep that was ongoing as this book went to press.[64]

These cases broke down into two distinct challenges. First, there were unrepentant fighters who returned either of their own accord or at the direction of ISIS, presenting a very high risk that they would carry out terrorist attacks on behalf of the group. For intelligence and law enforcement agencies, it was imperative to detect and interdict such active operatives.

The second challenge was more confounding. As the conflict wore on, reports began to grow about foreign fighters who had become disenchanted with the conflict and wanted to return home.[65] It was in the interest of Western governments to see radicals disengage with their extremist causes, but it was impossible to know for certain who was sincere and who presented a risk of future terrorism.

Some fighters might be lured by an offer of a deal for cooperation, but these almost always involved significant prison time. And while a fighter might be disillusioned with the cause or the experience, he might still dislike Western policies and be disinclined to turn on his former friends. Denmark launched a deradicalization initiative for former fighters, and other countries were considering similar programs, but such efforts were plagued by broad, unanswered questions about their effectiveness and the risks that they incurred.[66]

Additionally, there was a difficult question of accountability. Justice demanded that there be consequences for crimes, particularly the horrific war crimes and atrocities carried out under the banner of the so-called Islamic State. To incentivize defections, was it necessary to allow some crimes to go unanswered? Western policy makers were paralyzed by the complexity of these issues and a dearth of research on disengagement and deradicalization.[67]

And even on topics where research was available, such as the risk of terrorism among former foreign fighters, it was unclear whether past trends would continue in light of the new dynamics of ISIS.

A 2013 study by Thomas Hegghammer found that relatively few Western jihadist fighters had taken up terrorism upon leaving the battlefield, over the history of the movement.[68] But the percentages were still significant enough to make foreign fighting one of the few reliable indicators of future terrorism risk, at least compared to any other criteria. And, Hegghammer found, the presence of former fighters in a terrorist plot increased the chance a plot would be successful and significantly increased the lethality of a terrorist attack.

The percentages, combined with the soaring numbers of foreign fighters in Syria generally, and in ISIS specifically, pointed to an increased risk of terrorism that could linger for years.

Another important variable raised the question of whether historical jihadist conflicts such as the 1990s war in Bosnia could serve as a barometer of future events. The 2014 surge in the number of jihadist foreign fighters and inspired lone-wolf attacks was attributable, at least in part, to a revolution in the style and content of messaging that ISIS had deliberately pursued.

ISIS was rewriting the rules of jihadist extremism using sophisticated tactics of manipulation and distribution. It was not just a splinter from al Qaeda, it was an evolution. ISIS was reinventing al Qaeda's model of terrorism and radicalization, and its new ideas were sending shock waves around the world.

CHAPTER FIVE

THE MESSAGE

Jihadists have been making "slick" propaganda for decades, but for a long time, these productions catered to an exclusive audience of potential recruits, never making the evening news or creeping into the collective consciousness of the West.

Since the war against the Soviet Union in Afghanistan during the 1980s, jihadist organizations have used video and print media in sophisticated ways. From full-color magazines to audio lectures on cassette and TV-style talk shows, the genre is overstuffed with decades of material that flew under the radar of Western media.

In a crowded field, there are some standouts. *The Martyrs of Bosnia* is a sweeping feature-length video documentary released in English and Arabic versions that comprehensively describes the arc of the 1990s Balkans war from the perspective of the jihadist foreign fighters who took part.[1]

In 2001, al Qaeda released *The State of the Ummah*, nearly two hours of lavishly produced propaganda that came to define the group for footage-hungry Western media after September 11, providing now-ubiquitous images such as masked al Qaeda terrorists endlessly advancing along a set of monkey bars.

But *The State of the Ummah* was much more than simply B-roll for twenty-four-hour news networks. An ideological incitement, it served to define al Qaeda to potential recruits and apologists in the Muslim world.[2]

The movie is broken into parts, which boil down to "The Problem" and "The Solution." The problem, described at length, was the political weakness of Muslims and the corruption of Arab regimes, who were supported by the United States.

"This tape that you are viewing now are real-life scenes that portray, with blood and tears, the sorry state of the Muslim nation," said an unidentified narrator.

"The wounds of the Muslims are deep, very deep, in every place," Osama bin Laden reiterated a few minutes later.

The video continues in this vein for nearly forty-five minutes.

The solution was, of course, al Qaeda. Although the terrorist group is not named in the video, its chief leaders and ideologues are featured at length, discussing the need for Muslims to violently resist the conspiracies of the West and Israel.

"The Solution" was a carefully stage-managed affair. As the cameras rolled, often at interesting angles, a series of masked men went through a pantomime of military training in a desert backdrop identified as the al Farouq training camp, including running, jumping, diving, swimming, shooting, demolitions, motorcycle gymnastics, and, of course, monkey-barring.

The visuals were memorable and effective, yet they were notably contrived. The overall effect, likely intentional, made al Qaeda look like an adventure camp for young men.

"So it is incumbent on the Muslims, especially those in leadership positions from among the faithful scholars, honest businessmen, and heads of the tribes to migrate for the cause of Allah and find a place where they can raise the banner of jihad and revitalize the *ummah* to safeguard their religion and life," bin Laden intoned professorially near the end of the video. "Otherwise they will lose everything."

The State of the Ummah was the last major release by al Qaeda prior to September 11. After the United States invaded Afghanistan and rousted the organization from its fixed bases, it took time for the media arm to regroup. It began to recover in 2002 and 2003, with the help of Adam Gadahn, a California native, mentioned earlier, also known as Azzam the American.[3]

Gadahn helped modernize the media operation. He embraced digital recording and editing, and online methods of distribution. With his involvement, al Qaeda produced a documentary/dramatization of the planning for the 9/11 attacks called *The 19 Martyrs* and a faux news program, *Voice of the Caliphate,* which lasted only one episode.[4]

Gadahn had a knack for what the television industry refers to as "high concept" ideas—a two-sentence pitch that sounds novel and exciting—but his execution was not especially memorable. Most of his overproduced videos disappeared like rocks thrown into a pond, their ripples fading quickly. The lack of traction and the toll of avoiding ongoing counterterrorism activity in the Afghanistan-Pakistan border region resulted in a steady decline in the quality of al Qaeda's propaganda releases.

By the time Osama bin Laden was killed in 2011, al Qaeda's media output largely consisted of tedious position papers delivered by a succession of ideologues staring straight into a camera, sometimes for an hour or more. Sometimes the media branch provided high-tech computer-generated backdrops in a desperate bid to add some visual interest. Raw videos captured when bin Laden was killed showed him delivering speeches in front of a closet in his house, which was later digitally replaced with a neutral backdrop.[5]

In the early days, al Qaeda in Iraq hewed closely to traditional jihadist propaganda, but it did not take long to distinguish itself. Echoing videos from Soviet-era Afghanistan and later Bosnia, AQI put a premium on combat scenes, releasing clip after clip of IED explosions, mortar fire, and sniper attacks. The quality of the footage was frequently terrible, much of it shot on cheap handheld digital cameras. But the quantity was remarkable.

The combat clips were distributed individually, then collected by technically proficient online supporters, who strung them together into lengthy montages with a sound track of *anasheed* (Islamic a capella religious songs; simply *nasheed* when referring to just one), bookended by computer animated title sequences. Sometimes they added clips from the news to frame the mayhem.

Soon there were more ambitious efforts, such as "The Expedition of Shaykh Umar Hadid." In these early productions, AQI's media department had found the germ of an idea—storytelling. Although the videos were still often bloated with exposition and rambling religious lectures, more examples began to emerge with self-contained narratives that fit within the broader story of the war.

But most of all, al Qaeda in Iraq differentiated itself with graphic violence. Starting in 2004, with the videotaped execution of American contractor Nicholas Berg, AQI released a seemingly unending series of videos showing the execution of hostages and prisoners, often by decapitation (or near decapitation). At least eighty such videos were released during the AQI era, many featuring multiple victims. They came in a remarkable variety, from nearly anonymous snuff films to at least three videos showing public executions in front of sometimes-cheering spectators on Iraqi city streets.[6]

The pace and quality of these productions ebbed and flowed with the strength of al Qaeda in Iraq. Although its output was voluminous, the quality was spotty and with a few rare exceptions, most of its videos would have been forgettable if not for the shocking brutality, which came to define the group so completely that even the leaders of al Qaeda Central objected.

"Among the things which the feelings of the Muslim populace

who love and support you will never find palatable . . . are the scenes of slaughtering the hostages," wrote Ayman al Zawahiri, then al Qaeda's second in command, to AQI's emir, Abu Musab Zarqawi, in 2005. "You shouldn't be deceived by the praise of some of the zealous young men and their description of you as the shaykh of the slaughterers. . . . We are in a battle, and more than half of this battle is taking place in the battlefield of the media."[7]

Al Qaeda in Iraq tempered, but did not stop, its documentation of atrocities. But its successor group would eventually take Zawahiri's last point to heart.

In 2010, as the rechristened Islamic State in Iraq was reaching new lows, control of the organization passed to Abu Bakr al-Baghdadi. While terrorist groups are often shrouded in secrecy, Baghdadi took his anonymity to extraordinary heights, forgoing al Qaeda–style communiqués and functioning largely in the shadows.

The burden of communicating ISI's message and agenda fell instead on the group's spokesman, Abu Muhammad al Adnani. Born in Syria under the name Taha Sobhi Falaha, Adnani articulated ISI's talking points in a series of audio statements.

In 2011, one of his speeches proved to be a defining moment. Titled "The Islamic State Will Remain," it acknowledged the group's setbacks but set a defiant tone. Predicting a return to the glory days, he vowed that ISI would fight on despite any setback.[8]

"How powerful you are!" Adnani told ISI's supporters, over and over again. "How good you are!" And to its enemies, his message was also clear: "The Islamic State will remain," he said, using the Arabic word baqiyyah, which can also be read as "survive."

"The Islamic State will survive despite your sects, alliances, armies, and weapons," he proclaimed. "It will survive despite your plots and conspiracies."

The contrast between Adnani's speech and the besieged attitude emanating from al Qaeda Central could hardly be starker. AQC's messaging felt increasingly disconnected from the battle. One of Osama bin Laden's final speeches focused on climate change. Zawahiri, his successor, had launched into an hours-long series of commentaries on the Arab Spring that seemed to emerge in slow motion, sometimes referencing events months past, in which he seemed to grope for relevance.[9]

Adnani's words were electrifying to the supporters of ISI, who transformed *baqiyyah* into a slogan and battle cry. In subsequent videos, fighters shouted it defiantly. On jihadist Internet forums and social media outlets, they adopted it as a marker of loyalty. It also set a tone that ISIS would, over time, refine and propagate throughout its messaging.

The prodigious propaganda output of ISIS in all its incarnations could fill a book by itself, spanning books, lectures, magazines, audio, video, tweets, and Facebook posts (for more on social media, see Chapters 6 and 7).

THE CLANGING OF THE SWORDS

One series perhaps best illustrates the dramatic transformation that made the nearly extinguished al Qaeda affiliate into a powerful independent force.

Salil as-Sawarim (*The Clanging/Clash of the Swords*) launched in June 2012.[10] Its opening installment was billed as the first in an ongoing series, a tactic increasingly favored by the group's propagandists. Those watching *The Clanging of the Swords Part 1* would have seen few clues about what was to come.

A little more than an hour long, it was a fairly typical piece of jihadi agitprop for the late 2000s, if slightly more violent. Its author mashed together static clips of jihadi ideologues lecturing, talking-head segments lifted from Middle Eastern news broadcasts, and a number of unevenly filmed guerrilla-style combat scenes. Most of the malice in the video was directed against Iraqi Shi'a politicians, whom it described (not entirely inaccurately) as being under the influence of Iran. The video waxed on about the atrocities Muslims were suffering at the hands of the Shi'a in Iraq.

There is a well-known saying of unknown origin, "War is long periods of boredom punctuated by moments of sheer terror." *The Clanging of the Swords* consisted of long periods of boredom punctuated by distant explosions and images of dead bodies.

In July, ISI's notoriously anonymous emir Abu Bakr al-Baghdadi released his first audio lecture, a defiant speech full of fire and seemingly irrational optimism. The thirty-three-minute speech used some variation of the word "victory" twenty-one times. "Allah promised us victory, triumph, and power," he said. "Allah will keep His promise at all times." He also announced a new initiative, "Breaking Down the Walls," an ambitious strategy to free the many jihadist prisoners languishing in Iraqi jails.[11]

A few weeks later, in August, *The Clanging of the Swords Part 2* was released.[12] It was significantly different than its predecessor. Gone were the polemics. Instead, the new video consisted almost entirely of combat footage. But that too was different. The quality of the video and the camera work were significantly better. In places, the quality was comparable to a professional television program, telling a story in narrative form.

It's possible that the creator of Part 1 was a quick study, but the huge improvement suggests Part 2 was the work of a different filmmaker.

Where the first film had strung together many small combat clips with little context, the forty-nine-minute sequel followed a single operation, an assault on Haditha, Iraq, from training to the death of two fighters in a friendly fire incident, all presented in a cinema verité style.

In the video, ISI fighters attack checkpoints outside the city, then storm the homes of men identified as the local counterterrorism officials. At least eight prisoners are taken during the operation and executed.

While past jihadi videos had followed specific operations in some detail, *The Clanging of the Swords Part 2* was a remarkable leap forward, thanks to its combination of tight editing, technical quality, attention to detail, and graphic violence.

It had also subtly dropped a key element of al Qaeda propaganda.

In many ways, al Qaeda's ideology and strategy were explicitly predicated on assumptions of weakness.

In its worldview and favored ideological justifications, jihad was an act of defense, or at least that was the line they sold to the world. Self-defense was easier to rationalize—and sell—than an improbable vision of global domination. So al Qaeda's recruitment materials and fund-raising activities brimmed over with talk of "the plight of Muslims," steeped in pathos.[13] According to al Qaeda's ideologues, this urgent and existential danger was the entire reason for the organization's existence.

The concept of weakness also figured heavily in strategy. Tactically, weakness justified asymmetrical warfare in the form of terrorist attacks on soft civilian targets, on the premise that al Qaeda was too weak to militarily confront its enemies.[14]

Weakness also factored into the choice of enemy. Over time, al Qaeda had adopted the view that "apostate" Arab regimes—al Qaeda's real enemy—dominated the Middle East thanks to American military and economic support. As bin Laden famously put it, the United States was "the head of the snake," which must be cut off before the day could be won.

Jihadists commonly characterized this as a distinction between the "near enemy" (Middle East regimes) and the "far enemy" (Western governments). Because of the far enemy's support, direct opposition to the near enemy was believed to be impossible.

The tumult of the Arab Spring, along with the growth of al Qaeda's affiliate system (Chapters 3 and 8), had already begun to undercut this concept, and ISI was poised to directly challenge it.

The Clanging of the Swords Part 2 sent a clear message, video proof that the near enemy was vulnerable. It wasted no time on justifications and dropped the theme of persecution and oppression that had been present in the first installment just weeks earlier.

Instead, the sequel depicted ISI as a strong force meting out rough justice against deserving enemies. Although there had been examples of jihadist propaganda before that combined many of these elements, *The Clanging of the Swords Part 2* had a special power, thanks to the combination of technical prowess and aggressive tone.

Part 3 was released in January 2013. The new release was a documentary about "Breaking Down the Walls," delivering on the campaign Baghdadi had promised in July.[15]

With much of the action recorded in high definition, *Clanging 3* showed distinct signs of being filmed with professional video equipment by experienced cameramen. Some scenes were shot with multiple cameras, allowing the action to unfold from different angles. Others continued the verité style of Part 2, with handheld footage of live combat operations.

Several operations were labeled "Breaking Down the Walls." The filmmakers also filmed discussions among masked ISI fighters and interviewed prisoners who had been freed by ISI or escaped of their own accord.

At one point, the video displayed a sly and unexpected sense of humor, showing ISI members' efforts to rescue a camel that had fallen into a pit. A caption described it as an operation to "liberate a prisoner in the desert."

It was an even grander affair than the previous installment, but less dramatic and effective, clocking in at an overstuffed eighty minutes. Although Part 3 was another step forward in ambition and technical execution, it was a step backward in terms of focus and storytelling, to some extent lapsing into the earlier model of ISI propaganda, which resembled a laundry list of armed confrontations.

The through-line of "Breaking Down the Walls" was not strong enough to hold the video together as a unified narrative. Even the greatly escalated body count could not compensate for the repetitive nature of the footage.

The propagandists were still learning.

In May 2014, *The Clanging of the Swords Part 4* premiered on the Internet. The release marked a graduation of sorts. The members of ISIS's media team could no longer be considered students; they were now fully professional.

The sixty-two-minute video opened with aerial footage of Fallujah filmed by a drone. The ISIS drone was little more than a hobbyist's toy, a flying camera remote controlled by radio, but the symbolism was powerful and clear: The enemy's most feared and hated weapon was now part of ISIS's arsenal.

What followed was an untrammeled show of strength. As the narrator boasted of the vast area controlled by ISIS, masked jihadis paraded in armored columns through the streets, with apparently admiring throngs gathered to watch. After a rousing speech, a *nasheed* played over gripping scenes of car-to-car combat, incongruously framed by a Native American dream catcher ornament swinging from the driver's rearview mirror.

Captions claimed the victims were Shi'a soldiers on their way to join Iraqi military units, but to all appearances, the ISIS fighters were driving around shooting at whatever random cars they passed and even pedestrians. When the camera panned over the dead occupants of one beat-up old vehicle, the victims were young men dressed in shorts and T-shirts. Most of their targets were visibly unarmed. Only the captions differentiated the scene from an indiscriminate massacre.

"The clash of swords is the song of the defiant," singers chanted in Arabic over the slaughter, "and the path of fighting is the path of life."

Following this brutal carnage, the tone changed. In a public meeting, ISIS fighters offered clemency to anyone who had fought them in the past if they would only renounce the errors of their ways. One man after another stepped up, publicly recanted, and received warm embraces.

A considerable amount of combat followed, this time against visibly armed, military targets, followed by a suicide bombing and a checkpoint operation. Foreign fighters were shown burning their passports and renouncing the citizenship of their native lands. Unlike the previous installments in the series, the clips were shorter and punchier. The shift between fighting scenes, executions, and noncombat events helped elevate the ultraviolent video, giving a sense that while ISIS was unapologetically brutal, it had more to offer than just violence.

"Oh our people, *Ahlus Sunna* [adherents to the traditions of Islam], indeed the Islamic State exists only to defend you, and protect your rights, and stand in the face of your enemies," a narrator said, using the name the group had not yet formally adopted. "Indeed, the Islamic State is your one true hope, after Allah."

About halfway into the hourlong video, the executions of prisoners began, followed by scenes of sniper killings. The body count at this point reached into the dozens, and ISIS wasn't finished yet. At the thirty-seven-minute mark, a cameraman interviewed captured Iraqi soldiers who were being forced to dig their own graves. More combat and ambushes followed, periodically interspersed with scenes of ISIS's mercy toward those who would disavow their previous opposition.

"We don't want you to come to this place and repent out of fear of us, because if you fear us, there's no good in you," a masked speaker told one gathering. "We want your repentance and return to be due to the fear of Allah."

"Oh my *ummah*, a new dawn has emerged, so witness the clear victory," the singers chanted. "The Islamic State has been established by the blood of the truthful. No one will ever stand between the mujahideen and their people in Iraq after this day."

"The Islamic State has attacked, and surrounded the tyrants," they sang.

Over the final scene of a mujahid slowly walking, carrying the black ISIS flag, a narrator closed out the film with reference to an apocalyptic prophecy.

"And so the flame was started in Iraq, and its heat will increase by the will of Allah until it burns the crusaders in Dabiq," a town in Syria that ISIS adherents believed would be the location of a decisive battle with the "Crusaders." (See Chapter 10.)

The Clanging of the Swords Part 4 was wildly successful. It racked up millions of views on video-sharing platforms, although the numbers were almost certainly inflated (perhaps exponentially) by ISIS's deceptive social media techniques. Regardless of the total number of viewers, the video created vast excitement among those who followed ISIS online and many who were vulnerable to its message. The overarching theme of ISIS propaganda had been condensed and purified, and the message was "We are strong, and we are winning."

RACE TO THE CALIPHATE

While the quality of ISIS video releases would continue to fluctuate, overall the media team improved steadily over time, even as the quantity of its output increased. The growing focus on the packaging of the message corresponded to a new emphasis on its content. While ISIS made gains on the ground in Iraq, it was also expanding the definition of both the war and the organization itself. The media efforts fertilized the ground where ISIS would plant its next bold claim to religious authority—the declaration of the caliphate.

The precise composition of the ISIS media team was unknown

(or more accurately, it was the subject of conflicting reports with uncertain sourcing), but some elements became clear over time. Many regional hubs where ISIS operated had their own media departments, including Raqqa and Deir Ez-zoor in Syria, and Diyala, Saladin, Mosul, and Kirkuk in Iraq. Their Twitter accounts routinely published photos, videos, and text updates about ISIS activities, creating a remarkably robust (if carefully manipulated) record of ISIS's activities.[16]

A number of Westerners were involved in the media project. In May 2014, ISIS debuted an outlet dedicated to disseminating material in English and European languages. The Al Hayat (Arabic for "Life") Media Center ramped up at a critical time for ISIS, just weeks before the dramatic military offensive and caliphate proclamation that would put it on the front pages. Al Hayat translated ISIS's Arabic propaganda into English, including *The Clanging of the Swords Part 4,* but it also produced original content that revealed the complexity of the organization's media strategy.[17]

In May and June, Al Hayat rolled out multiple English-language magazines, some of which recycled content from social media, and others that included original reporting from areas ISIS controlled. The stories included coverage of battles but also devoted many pages to ISIS's efforts to govern, such as the execution of a "sorcerer" and religious training for imams. One issue spotlighted ISIS's consumer protection bureau in Raqqa, which held merchants responsible for the quality of goods they sold.[18, 19]

More issues of the magazines came out in quick succession, seven issues by mid-June. After their initial release in English, most of the issues were also distributed in French and German editions.

The publications continued to present the society that ISIS was building, including reports on agriculture and the ISIS police force. One issue was devoted to the dramatic capture of Mosul in early June. Concurrently, another spotlighted the violent side of ISIS, with page after page of graphic images showing the execution of criminals and prisoners, some with their brains splattered on the ground, others cut to pieces.[20]

The strange dichotomy of ultraviolence and civil order was echoed throughout ISIS's many streams of propaganda. Although the image was to some extent contrived, the overall package represented something new and different in the world of jihadism. ISIS was projecting its vision of a comprehensive society that went beyond the nihilistic destruction associated with the jihadist movement. This society, ISIS argued, existed in the here and now, and the organization approached the project with clear enthusiasm.[21]

The concept of governing had been circulating through al Qaeda for years, and its affiliates in Mali and Yemen had both made efforts to seize territory and build out social services. But neither had been able to hold its ground for long. Furthermore, they seemed uninterested in the work based on its own merits, acting instead out of a cynically manipulative impulse.

"Try to win them over through the conveniences of life and by taking care of their daily needs like food, electricity and water," the emir of al Qaeda in the Arabian Peninsula (in Yemen) wrote to the emir of al Qaeda in the Islamic Maghreb (in North Africa). "Providing these necessities will have a great effect on people, and will make them sympathize with us and feel that their fate is tied to ours."[22]

Unlike its counterparts in Yemen and North Africa, ISIS seemed to relish providing services, rather than simply seeing it as a PR strategy (although the sustainability of these efforts was an open question). When it took control of an area, ISIS wasted no time outfitting police cars, ambulances, and bureaucracies with its ubiquitous black flag emblem. ISIS put traffic cops at intersections; in addition to its law enforcement and consumer protection bureau, it opened a complaints desk and nursing homes. Its members radiated enthusiasm for these projects.[23]

AQAP had also advised AQIM to refrain from immediately instituting the jihadists' harsh interpretation of Islamic law. "You can't beat people for drinking alcohol when they don't even know the basics of how to pray," one letter stated.

ISIS had other ideas. Not only did it implement a draconian regime of crime and punishment, which its members believed to be divinely ordained, but it celebrated and painstakingly documented the process in its propaganda, publicizing everything from the destruction of cigarettes and drug stashes to the amputation of thieves' hands "under the supervision of trained doctors" to the genocidal extermination and enslavement of Iraqi minorities.[24]

In many ways, the combination of elements was unprecedented. Nazi Germany, whose parallels in propaganda and brutality often invited comparisons to ISIS, had produced masterful propaganda while carrying out a painstakingly documented program of genocide, but these were separate efforts. Its propaganda did not celebrate the genocide; rather it served to justify an imperative to act in the name of national and racial purity without sharing the gruesome reality. The Nazis did not broadcast their atrocities to the world.

In stark contrast, ISIS presented its vision of a demented utopia in which children played with severed heads and ran laughing down streets lined with mangled bodies instead of trees. A seemingly endless procession of atrocities was captured in photographs and videos, and distributed through both official and unofficial channels on social media.[25]

To some extent, the shocking violence seen in these messages owed a debt to *The Management of Savagery*, the jihadist tract that heavily influenced ISIS's strategy across multiple fronts.[26] Al Naji wrote of the necessity of violence, in all its "crudeness and coarseness," in order to awaken potential recruits to the reality of the jihadis' war and to intimidate enemies by showing the price they would pay for their involvement. But, he wrote, "we find that every stage of our battle needs methods that are soft and the like in order to counterbalance that (violence) so that the situation will be in good order."[27]

While much of the propaganda was intended for a Western audience, it also served audiences in Syria and Iraq, where for many sectarian hatred equaled or trumped dreams of caliphate building.

In its publications and in countless videos, ISIS extolled the virtues of killing the *rafidah* (a derogatory term for Shi'a Muslims) and the *nusayri* (a derogatory term for Alawites, members of a sect of Shi'a Islam practiced by members of the Syrian regime). ISIS videos documented the grisly killing of unarmed Shi'a prisoners by the hundreds, compared to the relative handful of Westerners who captured the attention of the media. Away from the cameras, the blood flowed even more freely, with reports of thousands of sectarian killings, often of unarmed prisoners.

The flood of propaganda in May and June was a deliberate prequel to the June 30 proclamation that ISIS had reestablished the "caliphate" and renamed itself simply "The Islamic State," dropping the limiting geographic identifiers of Iraq and Syria. ISIS had been telegraphing the audacious move for months, and the flurry of new publications in the weeks before the announcement were branded simply with the "Islamic State" name. Although many users still referred to it by the acronym ISIS, the shortened name had been heard in numerous propaganda videos for months.[28]

The announcement came on June 29, the start of Ramadan, in the form of an audio message from ISIS spokesman al Adnani, titled "This Is the Promise of Allah." In addition to the Arabic audio, translations of the statement were released in English, French, German, and Russian.[29]

In the speech, Adnani argued that ISIS was obliged to declare the return of the caliphate, and that Muslims everywhere were obliged to pledge loyalty to the new caliph, Ibrahim, formerly known as Abu Bakr al-Baghdadi. In addition, he said, "all emirates, groups, states, and organizations" were now null and void.

This specifically included all other jihadist groups, Adnani explained. "We do not find any (Islamic legal) excuse for you justifying holding back from supporting this state," he said, adding ominously, "And if you forsake the State or wage war against it, you will not harm it. You will only harm yourselves."

Adnani urged Muslims from around the world to come to the Islamic State, again dispensing with the narrative of the weakness of the Muslim world and reinforcing months of messaging about the organization's strength and purpose, using the word "victory" fifteen times in the course of thirty-four minutes.

"We fight for an ummah to which Allah has given honor, esteem, and leadership, promising it with empowerment and strength on the earth," he said. "Come O Muslims to your honor, to your victory. By Allah, if you disbelieve in democracy, secularism, nationalism, as well as all the other garbage and ideas from the West, and rush to your religion and creed, then by Allah, you will own the earth, and the east and west will submit to you. This is the promise of Allah to you. This is the promise of Allah to you."

The refrain of victory again reflected the advice of the jihadist tract *The Management of Savagery*. However ISIS took a page from the playbook of the enemy, at least as it was understood by Abu Bakr al Naji, the tract's author. Al Naji opined that the world's superpowers had created a "deceptive media halo which portrays these powers as non-coercive and world-encompassing," projecting an "aura of invincibility."[30]

As ISIS took full form, refining its media output carefully at each stage, it adopted its own halo. Victory was not only near, it was here. Regardless of how tenuous or risky its actions might appear to an objective observer, ISIS put a halo on its actions at every step, co-opting the very approach al Naji attributed to enemy powers.

Although ISIS was continually honing its messaging machine, the reaction to the announcement was mixed at best, exciting those who were already fully in ISIS's camp but leaving other jihadists incredulous. ISIS adherents who expected a groundswell of support from ordinary Muslims everywhere were destined to be sorely disappointed.

ISIS's online supporters rushed to celebrate the few pledges that trickled in during the early days, which came almost exclusively from small groups that had previously pledged allegiance to ISIS. Many supporters on social media seemed baffled and hurt that the announcement was being greeted with derision by Muslims of all persuasions around the world. Wild rumors erupted that everyone from the Taliban to al Qaeda in the Arabian Peninsula had pledged loyalty to ISIS, only to deflate days or hours later.[31]

But their slogan, repeated by Adnani in the announcement, was "The Islamic State will remain," and the core supporters continued to work at selling the audacious idea of the caliphate, as the messaging machine ramped up. As recent months had shown, ISIS's media machine was increasingly organized and sophisticated, but the quality was often wildly inconsistent.

Just days after the announcement, the new "caliph" showed his face for the first time. In a carefully staged ISIS propaganda video, Abu Bakr al-Baghdadi was seen climbing to the pulpit of a mosque in Mosul, where he delivered a perfunctory speech to a subdued crowd.[32] Will McCants, a scholar with the Brookings Institution and expert on Islamist politics, commented that the speech was "jihadi catnip." [33]

While the words were powerful, the man was distant. The speech hit many classic jihadi tropes, but Baghdadi's delivery was flat and unexciting. Nevertheless, it continued to build on the now-prevalent theme that ISIS was powerful and already victorious.

Unlike its predecessors, ISIS did not seek a far-off dream of the caliphate. The caliphate was here and now. Echoing a phrase used by Adnani in the announcement, Baghdadi referred to the caliphate as the "abandoned obligation" of this era. It was another subtle but effective inversion of al Qaeda and other Islamist terrorist groups, whose messaging often spoke of jihad as the "forgotten duty."[34]

Strangely, ISIS's message was less nihilistic than the "less extreme" al Qaeda, whose scholars were known to argue that fighting was the only thing that mattered and could not end until the Day of Judgment, regardless of whether the jihadists were winning or losing. This was the argument of someone who expected to lose.[35]

Baghdadi and his minions were having none of it.

"Here the flag of the Islamic State, the flag of (monotheism), rises and flutters," he intoned. "Its shade covers land from Aleppo to Diyala. Beneath it, the walls of the (illegitimate rulers) have been demolished, their flags have fallen, and their borders have been destroyed. Their soldiers are either killed, imprisoned, or defeated. The Muslims are honored."

Despite the tepid response, ISIS continued to flood the Internet with more propaganda. Concurrent with the release of Baghdadi's speech, the Al Hayat Media Center published the first issue of *Dabiq*, a new English-language magazine (in an online format).[36] It was subsequently released in several other languages.

Dabiq was a small town in Syria, near the border of Turkey, which figured heavily in an Islamic end times prophecy that predicted that Muslims would defeat "Rome," which jihadis had long reimagined as a reference to the Western powers, in the area of Dabiq, before going on to conquer Constantinople, present-day Istanbul.[37] The prophecy was quoted at length in the opening pages of the magazine.[38]

The lead story, unsurprisingly, was the declaration of the caliphate, proclaimed in colorful banner headlines—"A new era has arrived of might and dignity for the Muslims," echoing Baghdadi's speech, which was excerpted at length.

The magazine was remarkable in several respects. It called for *hijra*, religious emigration inspired by the travels of the Prophet Muhammad, and not just for fighters. In an article in *Dabiq*, ISIS asked for "doctors, engineers, scholars" and "people with military, administrative and service expertise." Although jihadist groups were frequently bureaucratic, none had so publicly recruited middle managers before.

The fifty-page magazine also featured religious justifications for ISIS's ascension to caliphate status and reports on its military victories, including the now routine pictures of mangled enemy corpses. It borrowed a page from al Qaeda propaganda and quoted Western terrorism analysts to boost its credibility. Over the course of 2014, Al Hayat issued three more issues of the magazine.

As the summer gave way to fall, ISIS continued to flood the Internet with propaganda, and Western media outlets increasingly took note. "Slick" was the word of the hour, endlessly repeated in news stories and broadcasts. (A search for "slick," "video," and "ISIS" on Google yielded more than 5 million hits in November 2014.)

The tipping point arrived in late summer.

THE BEHEADINGS

On August 19, ISIS released a video titled "A Message to America." Clocking in at just under five minutes, it opened with a clip of President Obama announcing the administration's plans to carry out air strikes against ISIS.[39]

The scene cut to an image of James Foley, an American reporter who had been kidnapped in Syria in 2012. He had been transferred among various rebel groups, and ultimately ended up in the hands of ISIS. The United States had attempted to rescue him just a month earlier, unsuccessfully.[40]

Foley was kneeling in the desert sun, arms bound behind him, dressed in an orange jumpsuit meant to invoke the garb worn by jihadist prisoners of the United States in Guantanamo Bay and in Iraq during the American occupation. As with the drone imagery in *Clanging of the Swords Part 4,* it was yet another inversion by ISIS, usurping another powerful image associated with American domination. A masked ISIS fighter in black stood next to him.

A small, black microphone, of the sort used in Western news broadcasts, was clipped to the collar of his shirt.

Foley began to speak in a clear, steady voice.

"I call on my friends, family, and loved ones to rise up against my real killers, the U.S. government," he said. The video had been filmed using multiple cameras and it cut seamlessly from one angle to the next. "For what will happen to me is only a result of their complacency and criminality."

Foley painfully reproached his family, including his brother, a member of the U.S. military, referencing U.S. strikes against ISIS.

"I died that day, John; when your colleagues dropped that bomb on those people, they signed my death certificate."

Foley said he wished he had more time.

"I guess, all in all, I wish I wasn't an American."

The ISIS fighter then took over. He spoke in a British accent, accusing the United States of aggression against ISIS.

"You are no longer fighting an insurgency," he said. "We are an Islamic army."

The fighter bent to Foley and put a knife to his throat and began to saw. The video cut away before blood began to flow. When the picture resumed, the camera panned over Foley's dead body, his head severed and placed on the small of his back.

In the final scene, the fighter reappeared, gripping another hostage, an American journalist named Steven Sotloff, by the collar of his orange jumpsuit.

"The life of this American citizen, Obama, depends on your next decision," the fighter said as the video concluded, an excruciating cliffhanger that promised more agony to come. ISIS had learned from the *Salil as-Sawarim* series, the power of telling a spare, minimal story, framed by horrific violence.

The video exploded onto the Internet, as ISIS supporters took to social media to make sure their message was delivered not just to American policy makers, but to anyone whose attention they could reach (see Chapter 7).

In the weeks that followed, the short script would repeat itself over and over again, one hostage after another executed as the world watched in horror, again following the blueprint in *The Management of Savagery*, whose author specifically advised the taking of hostages to send a lesson about "paying the price" to anyone who would oppose the jihadis' campaigns. "The hostages should be liquidated in a terrifying manner, which will send fear into the hearts of the enemy and his supporters," the author wrote.[41]

By October, ISIS had beheaded three more Westerners, each installment concluding with a new hostage whose life was placed on the line. The target audience expanded past the United States with the execution of British aid workers Alan Henning and David Haines. Many ISIS Twitter users crashed hashtags for British television shows and directed harassing tweets and videos at British prime minister David Cameron's official Twitter account.[42]

The fifth video broke out of the format, dropping the "Message to America" title. The hostage was Abdul-Rahman Kassig, an American military veteran and a convert to Islam who had been working with aid organizations to assist suffering Syrians.

The fifteen-minute video included revoltingly graphic footage of a mass beheading of captured Syrian soldiers, a sharp contrast to the previous videos that had cut away at the start of the act of violence. The killings were carried out by a number of unmasked European foreign fighters, including from the United Kingdom, France, and Germany, ensuring massive news coverage in multiple countries. It ended with a message from the British executioner and an image of Kassig's severed head.[43]

Kassig's execution was not shown, and he did not deliver a statement. It's possible he refused to cooperate with the script, or that he was killed through some other happenstance (such as a rescue attempt or an air strike) before he could be executed. In another break from the previous installments, the video did not end with a new threat against a new hostage. The series had concluded, at least temporarily.

If these victims shared any common quality other than the English language and their white faces, it was their uncommon goodness. Each victim had been carrying out work that ultimately helped Syrians suffering in the civil war. The American journalists, James Foley and Steven Sotloff, were among the few who braved the terrible risks of reporting on the ground during the conflict. David Haines and Alan Henning were aid workers selflessly helping Syrians in dire need. Abdul-Rahman Kassig was a former U.S. soldier who had converted to Islam and trained as a medic so that he could minister to gravely injured Syrians. It seemed that no one was safe against the knives of ISIS, no matter how kind or how much they had done for Muslims, no matter if they were Muslims themselves.[44]

Another Western hostage, British journalist John Cantlie, surfaced in a separate series of video episodes titled "Lend Me Your Ears." Seated in a room in an orange jumpsuit, Cantlie recited scripted ISIS talking points at length.

The series took an unsettling turn in November, when Cantlie appeared in the role of a "reporter," in an ISIS video shot on location in Kobane, near the border between Syria and Turkey, where ISIS was battling Kurdish fighters for control of the city.[45]

The orange jumpsuit had been traded for a black button-down shirt, as Cantlie provided an account of the battles there, which was considerably more favorable to ISIS than the mainstream media's version of events, which Cantlie derided. The video was considerably more natural than "Lend Me Your Ears" episodes, leading to dark speculation that Cantlie was suffering from brainwashing or Stockholm syndrome—or worse, that he had simply gone over to ISIS.[46]

The propaganda tsunami continued unabated in other areas as well, as bloody weeks turned into bloody months. It was not unusual to see five or six distinct pieces of ISIS propaganda uploaded to the Internet in a single day. The quality and sheer volume of ISIS messaging dwarfed that of al Qaeda and its affiliates. Releases issued regularly from its regional hubs. Longer videos of varying quality were released, with titles such as "The Flames of War" and "The Resolve of the Defiant," in a growing number of languages.

New speeches from Adnani and Baghdadi emerged sporadically. Like Ayman al Zawahiri and Osama bin Laden before them, the top leaders of ISIS had operational security concerns that equated visibility with risk. Unlike al Qaeda, however, ISIS had compensated with a stream of content celebrating the lower ranks. Because ISIS operated in the open, compared to its secretive progenitor, it perceived little risk in allowing the rank and file to show their faces and tell their stories.

With so many fighters, it could pick and choose. Adnani was a talented speaker, Baghdadi much less so, but among their soldiers were many charismatic individuals. They might not be qualified to lead, but they could certainly sell.

A constant stream of communication resulted. ISIS was constantly seen to be active and vital, while al Qaeda lurked in silence. The latter's works, whatever they might be, were carried out in darkness, at a snail's pace. And some jihadis began to wonder openly on social media if those works even mattered.

Even the content of its infrequent releases paled in comparison to ISIS. An al Qaeda Central effort to create an English-language magazine, *Resurgence,* had taken months to produce a single issue, and when it arrived, it was 117 pages of dull. "*Resurgence* is a humble effort to revive the spirit of Jihad in the Muslim Ummah," an editor's note read. But the revival had decided months ago that it couldn't wait for al Qaeda.

As 2014 continued its bloody march, new realities took hold. The

United States was committed to a gradually expanding campaign against both ISIS and al Qaeda cells in Syria. And the first wave of shock and horror created by the bloody video beheadings of the summer had slowly hardened into something like resolve, alongside a terrible resignation, a recognition that the ISIS rampage would not shrivel under the first Western assault.

The Islamic State would—for now—remain, and it had placed its unedited and unfiltered message in front of exponentially more people than al Qaeda evér dared dream. Jihadist propaganda had had a history measured in decades, but it had long been obscure and limited to an audience of mostly true believers.

Suddenly, the stuff was everywhere, intruding on the phones, tablets, and computers of ordinary people who were just trying to go about their daily business online.

Although ISIS's skillful storytelling was an important factor in this process, it was not the entire story. As part of its quest to terrorize the world, ISIS had mastered an arena no terrorist group had conquered before—the burgeoning world of social media.

JIHAD GOES SOCIAL

How extremists use technology is no great mystery. Any high-tech tool that you use—from a desktop PC to a smartphone—is fair game for extremists, too.[1]

Unless a terrorist group is ideologically opposed to technology itself, it will generally use every available tool to do its work. Jihadists are no exception. Their morality may be centuries behind the times, but their technical skills expand to fit their available resources.

During the 1980s, jihadists produced propaganda films on videotape and printed sophisticated four-color magazines that were reasonable facsimiles of *Time* or *Newsweek*.[2] They didn't distribute them on the Internet. Instead, they went out via mail, or were handed out inside or outside a mosque. In dedicated centers around the world, including in the United States, those who were interested could go to find out more about the movement.

They discussed all this content, not over Facebook, but in person, after viewing a video together in a darkened room; not in YouTube comments, but after listening to an incendiary cleric speak before a roomful of people, everywhere from Cairo, Egypt, to Tucson, Arizona, and most points between.

And as media technology shifted, so did the extremists. Expensive magazines and newsletters with their associated postal costs, such as the *Al Hussam* (The Sword) newsletter published out of Boston, moved to email (like the *Islam Report,* out of Florida). These were pragmatic decisions. It cost about $1,000 a month to publish *Al Hussam* on paper. It cost virtually nothing to email *Islam Report*.[3]

Jihadis switched to digital video, around the same time early-adopting consumers did, and for similar reasons. It was cheaper and easier to distribute the same content in a downloadable file than on a videotape or DVD.

Social media wasn't much different. By 1990, white supremacists were using dial-up bulletin boards to communicate. As chat rooms became popular on services such as Yahoo! and AOL, radical recruiters signed up in droves, making friends and influencing people from a distance. As it became cheaper and easier to set up and maintain topic-centered message boards using software like vBulletin, jihadis and other extremists shifted again, with thousands of users taking to the new format.

After September 11, the message boards became the preferred social networking tool for jihadists. These message boards, more commonly referred to as online forums or just "the forums," are Web pages where a user can register, under a real or assumed name, to discuss topics of mutual interest.[4]

The forums are generally very structured environments, which suited jihadists in the post-9/11 era of justified paranoia about spies and security. Each forum features several major themes for discussion, under which users can start a "thread" on a specific topic of interest. For example, a major theme might be Syria, and a thread might be focused on the latest military action by a specific group.

After a thread starts, other users chime in to post their opinions. Users can reply to specific posts or simply type into the thread directly. Popular or controversial threads can grow to include hundreds of posts, but most peter out after a couple dozen.

The forums also have clear hierarchies. At the top is the person who owns the forum—the person or group that registered the forum's Internet domain name and has de facto control over the technical aspects of the site. The owner generally has the power to delete the entire message board, delete individual threads and content, accept new users, and ban or assign authority to existing users.

Beneath the owners are the administrators, also called moderators. Administrators have most of the powers of the owner, except for the ability to completely delete the forum, but they can be overruled by the owner. Administrators usually have their own hierarchy as well, with a small number in charge of the big picture, and a larger number of deputies to keep up with all the activity.

The general membership of the forum also has tiers of membership, which are indicated in users' profiles and also usually displayed next to their usernames when they post. Tiers can be based on different factors. Some forums allow users to score points based on popularity. Others allow advancement based on the number of posts by a user, or how long they have been on the board. Some accord special status to users who financially contribute to the forum's upkeep.

Most of the perks for advancing up the ladder are purely ornamental—social status and bragging rights, as well as adding a competitive element that motivates members to be active rather than passive.[5]

But extremist forums also have inner circles. Some topic areas are restricted to trusted members, who are involved in the offline work of terrorist groups, whether planning attacks or coordinating media releases. The forum's owners and administrators can designate users for special access, or they can restrict sets of ordinary users from routine access if they have concerns about security.[6]

At the highest levels, the forums have reportedly been used for direct communication among important offline jihadi leaders.

In 2013, a virtual "conference call" among jihadi leaders around the world took place within a closed section of an al Qaeda–linked forum to discuss an allegedly impending terrorist attack, although shifting language in media reports about the event left many questions about exactly what transpired.[7] It never became clear exactly what the plan was and how close it ever came to execution.

In a letter captured during the raid that killed Osama bin Laden, American al Qaeda spokesman Adam Gadahn complained bitterly about the content of the forums, suggesting the terror group's control of the forums was considerably less than perfect.[8]

The highly regimented forum system allows for a great deal of control, if not from al Qaeda itself then from its partisan moderators, but it can also stifle dissent and create resentment for those who feel excluded from the ranks of the elite. In addition to these internal social pressures (see Chapter 3), the forums were highly vulnerable to attack by hostile intelligence services, which could penetrate them for surveillance, or knock them offline entirely when it was convenient.[9]

In part because of these pressures, but mostly because terrorists follow the same technological trends that everyone else does, jihadist supporters began in recent years to filter out of the forums and start accounts on open social media platforms like YouTube, Facebook, and Twitter.

An early adopter on the open social media side was Yemeni-American jihadist cleric Anwar Awlaki. Born in New Mexico and raised largely in Yemen, Awlaki had returned to the United States to study engineering, but soon felt a call to Islamic ministry. His English was perfect, but more important, he was an eloquent, passionate, masterful storyteller.[10]

Through a combination of communication savvy and his careful cultivation of an ambiguous relationship to terrorism over the course of many years, Awlaki established himself on social media years before the broader jihadist community made the transition.[11]

He maintained a Facebook page and an active blog, where he communicated with readers in the comments section. Any given posting could prompt hundreds of responses.[12]

But YouTube was the social platform where Awlaki's videos achieved notoriety and elevated the issue of terrorist social media to the attention of the public and policy makers.

During his early career, Awlaki was a rising star in the world of mainstream American Muslims, keeping his dark side carefully hidden. While he successfully presented himself as a voice of moderate Islam, he secretly met with al Qaeda operatives and other radicals. Prior to September 11, he had been investigated for possible links to terrorism, and in the months preceding the attacks, he met with some of the hijackers in both San Diego and Falls Church, Virginia. His dark side was not confined to terrorism. San Diego police and later the FBI Washington Field Office investigated his patronage of prostitutes, including minors.[13]

But to the outside world, for a long time, he was simply an inspiring speaker. He had recorded dozens of lectures, some hours long, on a variety of religious topics. Few of his talks openly discussed radical Islamic concepts, but many contained elements that could be leveraged in that direction. Initially, his lectures were distributed on more than fifty CDs, but as more and more media moved online, they migrated to YouTube.[14]

Although YouTube has many social features, it is at heart a content delivery system. A wide variety of terrorist groups had been using YouTube to post and distribute propaganda. The conversation focused on reach—how easy it was to find and share terrorist videos and how many people were watching.

After years of pressure from politicians, particularly U.S. senator Joe Lieberman,[15] YouTube added an option for users to flag terrorist content.[16] If a review by the company found that a video "depicted gratuitous violence, advocated violence, or used hate speech" it would be removed. If not, YouTube would continue to defend "everyone's right to express unpopular points of view" and "allow our users to view all acceptable content and make up their own minds."[17]

But Awlaki's lectures didn't easily fit into the box. His material was wildly popular, and not just with terrorists. His spoken lectures routinely racked up hundreds of thousands of hits. By 2010, his content could be divided into three general content categories: early period, not especially radical; early to middle period, not unambiguously radical; and late period, very radical to openly terrorist.

As the cleric became more overtly associated with terrorism, the staggering amount of his content on YouTube presented a dilemma. Should the service remove lectures that were not obviously radical just because the lecturer had graduated to the most-wanted list? What if the lecture was overtly radical and anti-American, but did not openly advocate violence? What if they advocated generally for military jihad but not for specific acts of violence?

YouTube—with its roots as a fun-loving amateur video-sharing service—was ill-equipped to deal with this question. Its parent company, Google, ran a search engine that was arguably the single most powerful tool on the planet for driving Internet traffic, and the technology giant also owned a popular service to publish and host blogs, which was used by all manner of extremists. To take on the role of "values police" opened many cans of wriggling worms.[18]

Awlaki was not a static target. Increasingly, his name was associated with more than words. The cleric had exchanged emails with Fort Hood army psychiatrist Nidal Hasan, who killed thirteen people in a 2009 shooting spree on the base.[19] Later that year, al Qaeda in the Arabian Peninsula (AQAP) unsuccessfully attempted to bomb a Detroit-bound airliner on Christmas Day. Awlaki had not only inspired the would-be bomber but had met with him at a terrorist training camp.[20] In 2010, AQAP tried to detonate two cleverly disguised bombs, again unsuccessfully, on a UPS cargo plane. Awlaki's involvement was broadly telegraphed in the pages of the terrorist group's English-language magazine, *Inspire*.[21]

At last, YouTube gave in and announced it would more robustly remove Awlaki's content from its website, although his earlier nonviolent material was allowed to remain. It also announced it would ban accounts owned by government-designated foreign terrorist organizations, or used to support them.[22]

It was the dawn of a new age in which global corporations would imagine themselves as platforms for the ideal of free speech, only to be dragged kicking and screaming into a role brokering which values would be acceptable and which would not.

The problem was not unique to terrorism. For instance, YouTube had quickly devised algorithms to block pornography and then implemented even more stringent digital fingerprinting techniques to not only block child pornography but report those who posted it to the police, a practice soon adopted by other online providers.[23] And an army of lawyers convinced it to swiftly and aggressively address copyright violations.

Terrorism presented a particularly sticky dilemma. Terrorism was not only an inherently political activity, but it was one for which no consensus definition existed. Countries like Bahrain or Egypt might define terrorism very broadly, for instance, to include some legitimate political dissenters (as well as undisputed terrorists), and sometimes even experts on regional politics couldn't say for certain which was which. Angry activists, on the other hand, accused countries from Israel to the United States to Russia of perpetrating terrorism themselves through military actions and policies.

But regardless of the big-think debate, public outrage fueled scrutiny, and scrutiny led to changes, at least if a company was big enough and its terrorist users active enough to make headlines.

Literally every social media platform of meaningful size hosted some number of violent extremists. But most scrutiny was directed at the top. The easy availability of white supremacist "hatecore" music on the once-popular social media service Myspace, for instance, generated little public interest, in part because the platform was seen as fading into obsolescence and in part because newspaper reporters were far less interested in covering white nationalists than jihadists.[24]

San Francisco–based file-sharing service Archive.org was often the very first place where jihadi media releases appeared, but few outside of counterterrorism circles paid the clunky-looking website much heed, and even jihadis wasted no time transferring their videos from Archive to YouTube once they were published.[25]

Headline-friendly services such as Facebook and Twitter took the brunt of the criticism, in part because they were becoming extremely popular venues where terrorist recruiters and supporters could operate, and in part simply because they were popular. Everyone knew about Facebook and Twitter; fewer knew or cared about Tumblr, the blogging service that hosted its fair share of jihadi outlets.

Facebook and Google, while generally favoring free speech, were also publicly traded companies with concerns about liability and a desire to create safe spaces for users, especially the young, who were vulnerable to a range of online predators of which violent extremists and recruiters were only one part.

Twitter stood apart. A privately held company until late 2013, Twitter's founders and executives were perceived as libertarian-leaning advocates for free speech. Twitter more aggressively resisted broad government requests for information than most, and its rules for users contained few restrictions on speech.[26]

The company refused to discuss its criteria for suspensions, brushing aside queries with a boilerplate response.[27] With the exception of spam and direct personal threats against individuals, users could get away with a lot. And Twitter's "who to follow" recommendations for new users made it easy for would-be radicals to jump right in and start making connections with hardened terrorists, a process that was much more difficult in the 1980s and 1990s.[28]

The Taliban was one of the first jihadist-oriented organizations to embrace Twitter. In January 2011, its official media outlet created a Twitter account,[29] soon followed by its spokesman, Abdulqahar Balkhi.[30] Balkhi quickly became part of a sensation when NATO's International Security Assistance Force (ISAF) account began publicly sparring with him.[31] Toward the end of 2011, the Somali jihadist insurgent group al Shabab followed suit, and it soon racked up tens of thousands of followers.[32]

As highly visible insurgencies, rather than shadowy terrorist cabals, Shabab and the Taliban needed to manage public relations. They used their accounts to brag about military victories, harass their enemies, and rally supporters from their respective regions and around the world.

It wasn't all upside. In 2012, as described earlier, American al Shabab member Omar Hammami broke with the group over differences in methodology and accusations of corruption in Shabab's upper ranks. He took to social media to publicize the charges, airing Shabab's dirty laundry and launching an extended conversation with Western counterterrorism analysts, including a long series of public and private exchanges with coauthor J.M. Berger.[33]

While journalists and academics had, over the years, cultivated sources within terrorist groups, the advent of social media had opened the door to different types of interactions, exchanges that could involve daily or weekly conversation over the course of months. Social media also brought with it new risks, the danger of becoming too publicly or privately friendly with sources, at the risk of giving them a higher profile or being perceived as validating their views.

The more inherently secretive al Qaeda also established a presence on Twitter, along with some of its affiliates, but more covertly, resulting in a more limited reach. This lack of connectivity helped fuel the beginnings of dissent, as Hammami and other internal al Qaeda dissidents took to social media to air their grievances, only to be met by conspicuous silence (see Chapter 3).

After a slow beginning, Facebook took an aggressive stance against violent jihadists starting in 2009, actively monitoring, seeking out, and terminating pages and groups devoted to terrorist content, even when they were hidden from public view by privacy settings. It also terminated the accounts of key users who participated in such activities.[34]

Many of those suspended users simply sat down at their computers the very next day, created new accounts, and started all over again. So what was the point?

WHACK-A-MOLE

The phrase "whack-a-mole" had been used since the early 1990s to describe one of the major challenges of counterterrorism writ large.[35] A children's arcade game, Whac-A-Mole (sans *k*) features a table-sized playing field covered with holes. Toy moles pop out of the holes at random, first one at a time, then more and more, coming faster and faster.

The self-evident object of the game is to whack the moles with an included mallet as soon as they pop up. Inevitably, the moles begin to come faster than the player can whack them, and the player loses.

The dynamics of fighting terrorist groups are similar. Cracking down on a successful terrorist organization rarely led to the end of its associated movement. Take one cell out and new ones sprouted from the remains of the first. The CIA more elegantly described the problem in a secret 1985 internal analysis titled "The Predicament of the Terrorism Analyst," which compared the splintering of violent extremist groups under government pressure to the many-headed Hydra of legend—cut one head off and two more grow to take its place.[36]

While the Hydra metaphor continues to have its fans, "whack-a-mole" made for more colorful sound bites. With the dawn of the twenty-first century, it quickly became ubiquitous as a phrase to casually dismiss the value of efforts to counter or suppress terrorist and extremist use of the Internet.

The debate started with the Internet service providers that al Qaeda used to host the forums. While the forums were operationally important, they were specialized. A terrorist forum didn't come to you, you had to seek it out, sometimes armed with personal references. Some forums had lower barriers to entry, but a would-be al Qaeda member had to work his way through a series of such communities, earning trust and establishing credibility each time, which took time.

The social impact of the forums was relatively limited, while the counterterrorism benefits of allowing the forums to operate with only sporadic interference were clear. Although the forum administrators were usually based overseas, the United States offered the cheapest, easiest, and most reliable servers to host the content.

The fact that al Qaeda message boards were hosted by American companies incensed many people for reasons both political and patriotic, and some mounted public shaming campaigns in an effort to get those Internet service providers (ISPs) to take the forums down.[37] But if a forum was hosted on a server based in the United States, it was fairly simple for the government to get a subpoena and start collecting highly sensitive data. None of this was visible to Internet users in general, and so the debate remained relatively low-key.[38]

Both the ecosystem and the calculus changed dramatically with the rise of a new generation of social media platforms. The forums were gated communities; open social media services like Facebook and Twitter were town squares, where people wandered around meeting each other and seeking out those with similar interests.

Compelling evidence suggests social media taken as a whole tends to discourage extremism in the wider population,[39] but for those already vulnerable to radicalization, it creates dark pools of social connections that can be found by terrorist recruiters and influencers. On Twitter or Facebook, it was easy to seek out or stumble onto a radical or extremist account or community, and even easier for terrorist recruiters to seek prey within mainstream society.

"I see the cyber jihad as very, very important, because Al Qaeda, the organization, became mostly an ideology," wrote Abu Suleiman al Nasser, a prominent forum member who shifted to Twitter, in a 2011 email interview. "So we try through the media and Web sites to get more Muslims joining us and supporting the jihad, whether by the real jihad on the ground, or by media and writing, or by spreading the idea of jihad and self-defense, and so on." [40]

BIG BUSINESS

Virtually all extremist and terrorist groups have staked out ground on social media, from al Qaeda to Hamas, Hezbollah, the Tamil Tigers, the Irish Republican Army, and Babbar Khalsa (a Sikh militant group).[41] In a 2012 study commissioned by Google Ideas, coauthor J.M. Berger documented thousands of accounts related to white nationalist and anarchist movements on Twitter, and participation in those networks has soared in the intervening years.

As terrorists made the transition to social media, public pressure mounted. Twitter stoically sat out the debate, rarely commenting but making its libertarian views on speech well known. "One man's terrorist is another man's freedom fighter," an unnamed Twitter official told *Mother Jones* magazine.

"We take a lot of heat on both sides of the debate," said Twitter CEO Dick Costolo, in one of the company's extremely rare public statements on the matter.[42]

YouTube and Facebook, on the other hand, quickly learned the frustrations of whack-a-mole. Although the debate over terrorist suspensions frequently revolved around the intelligence question, terrorist content on social media was a business issue first and a cultural issue second. Intelligence concerns were, at best, a distant third.

The reason: Social media is run by for-profit companies, which are neither government services nor philanthropic endeavors (even if technology evangelists sometimes lost sight of the latter fact). The owners and operators of the platforms made the vast majority of decisions about which accounts would be suspended. Government intervention represented a tiny fraction of overall activity.

Each social media service had its own rules about abusive and hostile behavior that every user was obligated to follow or else risk being banned. The companies had no motivation to carve out exceptions for terrorist users who violated the rules, nor were they much inclined to treat their users as a resource for the intelligence community.

They did, however, have reason to worry about news headlines like "Despite Ban, YouTube Is Still a Hotbed of Terrorist Group Video Propaganda" and "Facebook Used by Al Qaeda to Recruit Terrorists and Swap Bomb Recipes, Says U.S. Homeland Security Report."[43]

After uneven beginnings, YouTube began to enforce its ban on terrorist incitement in a steady but less than robust manner. It responded quickly to user reports about terrorist videos, but it didn't deploy its full technological arsenal against them.

For example, Google could have written software to recognize the logos of terrorist groups and flag them for review. It did not. More significant, Google had developed the technological capability to prevent multiple uploads of a video that had already been flagged as a violation of its terms of service. The technology was invented to deal with copyright violations, but as of November 2014, it had not been deployed for use against terrorist videos.[44]

Facebook became proactive and began knocking down pages, groups, and users as a matter of routine, sometimes before ordinary users had a chance to complain about them. In an attempt to get around this, jihadis set up private, members-only Facebook groups to discuss bomb-making formulas and potential terrorist targets, but blatant plotting soon became a sure ticket to swift and repeated suspensions.[45]

As companies formulated policies for dealing with the influx of terrorist content, a cottage industry of open-source terrorism analysts blossomed almost overnight. Some analysts outside government preferred the one-stop shopping offered by the jihadist forums, which helped weed out noise and authenticate terrorist releases, but others found the insular environment difficult to crack, often requiring the creation of secret identities and undercover profiles to gain access to the juiciest and earliest content.

In contrast, social media seemed to offer ripe fruits for easy picking, especially on Twitter, where many jihadist organizations were now routinely distributing new releases describing their battles and claiming credit for attacks.[46]

Many among this new breed of social media analysts had a high opinion of the intelligence provided by low-hanging terrorist accounts. The new analysts broke down into several subcategories—academics, government contractors, government officials, journalists, and a burgeoning contingent of semiprofessional aficionados.

Some outside government confused terrorist press releases—by definition, the message the group wanted to promulgate—with verified information or operational intelligence. The easiest terrorist sources to find presented stage-managed messages, including outright lies. Many highly visible accounts belonged to stay-at-home jihadists far from the front lines.

Among global government officials and intelligence workers responsible for counterterrorism and countering violent extremism—the people fighting terrorism as opposed to those who study it—attitudes were different, especially as the months turned into years. Agencies were quick to recognize the power of so-called Big Data analytics in relation to the massive social networks that were forming in front of their eyes, but few had the capabilities to exploit the new pool of information on a large scale. In a majority of cases, social media was most useful to law enforcement and intelligence agencies not as a vast hunting ground but as a resource for discovering more information about suspects they had already identified.

In the United States, the government sometimes asked companies to suspend accounts. Some of the time, at least, the social media provider had some discretion in responding to such requests. Some European countries applied existing hate speech laws to social media platforms.[47] Other countries, in the Middle East and South Asia, took a more aggressive stance against speech they considered objectionable (terrorist or not).[48]

At times, government agencies asked to keep social media accounts active, when they were part of an ongoing investigation or when their intelligence value clearly outweighed their utility to terrorism. While shrouded in secrecy, these cases appeared to be rare and highly targeted.

Two fairly direct analogues cut to the heart of the intelligence argument for allowing terrorists to operate entirely unimpeded on social media.

The first is to substitute "terrorism" for virtually any other kind of crime or flagrant violation of the social contract. To pick an extreme example, allowing child pornographers to operate online without impediment would undoubtedly yield tremendous intelligence about child pornographers. Yet no one ever argues this is a reasonable trade-off.

In less emotionally loaded terms, the same could be said about the online operators of Nigerian oil scams, Ponzi schemes, and phishing attacks, or online purveyors of drugs, contraband, or prostitution. None of these problems are solved by online interdiction; the moles keep popping up. But no one ever argues that these social media accounts should be immune to suspension.

The second analogue is to real-life activity. Anyone who studies intelligence and law enforcement knows it is sometimes valuable to allow criminals and terrorists to remain at large for a period of time, under close surveillance, in order to gain information about their activities. But when the system is working properly, such surveillance culminates in concrete actions to prevent violence and disrupt the criminal network's function.

Of course, the very best intelligence on terrorism is produced by investigations that follow a successful terrorist attack, but no one would argue that the intelligence gained outweighs the cost. Online, the costs and gains are degrees of magnitude smaller and considerably more ambiguous. But it is wrong to assume they do not exist or matter, and that the equation is always, or even usually, weighted toward intelligence.

Although reasonable people can disagree about where to draw the lines, there is no reasonable argument for allowing terrorists complete freedom of action when alternatives are available.

WHACKED

The whack-a-mole metaphor was also flawed on its face, for two reasons: It assumes zero benefit to removing the moles temporarily, and it assumes the moles will never stop popping up.

Suspensions diminished the reach of a terrorist social media presence, degrading the group's ability to recruit and disseminate propaganda and forcing terrorist users to waste time reconstructing their networks. The suspensions didn't eliminate the problem, but they created obstacles for terrorists.

Killing civilians and destroying infrastructure are not typically a terrorist organization's end goals. Rather, they are a means to provoke a political reaction. Although people understandably forget sometimes, terrorism is ultimately intended to send a message to the body politic of the target, rather than being a pragmatic effort to destroy an enemy, although there are exceptions.

Therefore depriving terrorists of media platforms at key

moments—such as the release of a beheading video—disrupts their core mission.

Suspending the accounts that distribute such content requires mole whackers to keep whacking, but it also requires the moles to keep finding new holes from which to emerge, making it more difficult to land a message with the desired audience (see Chapter 7).

At the start of 2013, the debate reached a watershed. Al Qaeda's affiliate in Somalia, al Shabab, had grown fat and complacent on Twitter, where it maintained an official account (@HSMPress) that tweeted in English and had amassed 21,000 followers.

In addition to reporting its alleged military activities, in tweets that ranged from spin-laden to fantasy, the account frequently posted taunts and threats directed at Western and Somali governments.[49]

In January 2013, al Shabab tweeted a threat to execute a French prisoner it had captured. This was a rare example of a threat direct and specific enough to violate Twitter's extremely permissive rules, and the account was suspended after users reported the violation.[50]

The mole soon popped up under a new name.

"For what it's worth, shooting the messenger and suppressing the truth by silencing your opponents isn't quite the way to win the war of ideas," the account tweeted on its return, a deeply ironic statement coming from an insurgent group notorious for executing and imprisoning its internal dissenters.[51]

On the surface level, the suspension had cost nothing in intelligence value—for analysts who had the foresight to save copies of Shabab's original Twitter account. The old information was still accessible, albeit no longer conveniently online, and the new account continued the stream of press releases.

And in this case, the suspension improved the intelligence outlook. All Twitter accounts naturally accrue followers over time, not all of whom are especially interested in the account's content. The suspension wiped out a tremendous amount of analytical noise, and the low-hanging fruit of al Shabab's official tweets meant little compared to the value of the social network that had sprouted up around the HSMPress account.

Analysts who delved deeper could look at who followed the new account and deduce with some accuracy who was a member of al Shabab, by examining the relationships and interactions among the accounts, as well as their content. Similar capabilities were also being developed by analysts using Facebook and other social networks.

Such social network analysis required a critical mass of data, but the list of users following the original al Shabab Twitter account had grown large, and the data had become noisy. Some of the followers were curiosity-seekers, drawn in by headlines. Some were Somalis not associated with Shabab. Others had only a casual interest. Many were journalists and terrorism analysts.

After the new account surfaced, several hundred users rushed to follow within the first several hours. Analysts had previously been forced to sift through 21,000 accounts to pan for gold, but the new account had far fewer followers in its earliest hours, and the first ones to show up were among the most motivated. It was relatively simple to analyze the new accounts, removing the journalists and analysts. Most of what remained were hard-core al Shabab supporters and members on the ground. The suspension of the account had made it easier to glean real intelligence, not harder.[52]

Although it was nearly impossible to keep extremists from returning again and again to social platforms, it now became clear that suspensions were not an exercise in futility. A suspension cost the terrorists time. It deprived them of an easy archive of material. They had to reconstruct their social networks and reestablish trust, often exposing themselves to scrutiny in the process. Other users in their social networks were suspended and came back under new names and using different kinds of camouflage. It was not always obvious who your friends were.

There were also clear numeric costs. It might take a Facebook page weeks or months to build up a following of thousands of users, work that could be erased in an instant. An analysis of the pace at which al Shabab's second Twitter account accrued followers suggested it would take months, if not years, to regain all the followers it had lost.[53]

In September 2013, al Shabab commandos seized control of the Westgate Mall in Nairobi, Kenya, in an attack that lasted almost four days and left sixty-seven victims dead.[54] Its resuscitated Twitter account began live-tweeting details of the attack in progress. Although the account had previously tweeted terrorist attacks within Somalia, the media latched on to the activities of the high-profile account as the siege dragged on.[55]

After users complained about the account using Twitter's abuse-reporting forms, the mole was whacked, and a new account popped up.[56] Because it was breaking news, users flocked to the new account, which was whacked again in short order. Another popped up, and was whacked. The process continued for days. Each time, al Shabab was online for shorter and shorter periods.[57]

Finally, something remarkable happened. The mole stayed down.

It was unclear whether Twitter had permanently banned it using technical tools it normally reserved for spammers, or the terrorists had simply surrendered. What was clear was that it was over. Al Shabab had been denied the use of Twitter. The moles had definitively lost.

But that wouldn't stop the argument from reviving yet again in 2014, as ISIS burst onto the scene.

CHAPTER SEVEN

THE ELECTRONIC BRIGADES

"This is a war of ideologies as much as it is a physical war. And just as the physical war must be fought on the battlefield, so too must the ideological war be fought in the media."

—*Nasser Balochi, member of ISIS's social media team* [1]

The World Cup took Twitter by storm in 2014. More than 672 million tweets were posted referencing the global sporting event, peaking at more than 600,000 tweets per minute at the height of the excitement.[2]

But on June 14, Arabic-speaking fans who turned to Twitter for the latest scores discovered that their party had been crashed by ISIS. Mixed in with the highlight pictures and discussions of scores were shocking images of ISIS fighters executing hundreds of captured and unarmed Iraqi soldiers, and other atrocities.[3]

The next day, as ISIS consolidated its hold on Mosul (see Chapter 2), worried Iraqis took to Twitter amid rumors that the militants were closing in on the capital city. When they searched in Arabic for "Baghdad," they were greeted by ISIS banners containing the threat "we are coming" and images of a black flag flying over the Iraqi capital.[4]

ISIS had found a new way to put its message before the public—a Twitter app.

The app was the brainchild of J, a Palestinian living with his family in Gaza (we are withholding his name since he has not been publicly identified by investigators). J was a Web developer, graphic designer, and programmer who claimed to have been educated at Harvard and a "Los Angeles School of Arts" (which could not be confirmed through public records searches).[5]

J was associated with a large number of websites and social media accounts, under a variety of names and aliases, such as Azzam Muhajir and @DawlaNoor (a play on the Islamic State's name in Arabic). He had a day job as a commercial app developer. In his spare time, he split his days between issues related to Gaza and ISIS, but connections within his social network pointed to a heavy—and official—involvement in the latter.

J began experimenting with apps for Twitter and for smartphones that use Google's Android operating system. Some provided inspirational quotes from the Quran that could be read on a phone or pushed out to a user's Twitter account. Others appeared to be work for hire, such as a commercial app selling jewelry.[6]

In April 2014, J rolled out an app called the Dawn of Glad Tidings, devoted exclusively to ISIS content. It contained two components.

The first was an Android smartphone app that let users read headlines from a series of officially sanctioned ISIS news feeds. It was capable of collecting users' phone numbers and data about what networks the user connected to, which in turn could reveal where they were based and when they accessed the app.[7] The app also served advertising, which may have profited J or ISIS or someone else entirely—the ultimate destination of the revenue was unknown. In addition to reading stories on their phones, users could post them to Twitter, and J was working on adding Facebook functionality.[8]

The second component was a Twitter app, computer code that could take control of a consenting user's account to automatically send out tweets. An ISIS supporter could use their own account, which would function normally otherwise, or set up an empty account that tweeted nothing but content sent out by the person running the code.[9]

Prominent official ISIS members and supporters signed up for and formally endorsed the app as a trusted and official source of news.[10] The Dawn of Glad Tidings automatically sent out links to official ISIS news releases and media, and hashtags that the ISIS social media team wanted to promote.

A hashtag is a word or phrase preceded by the # sign, in order to make it a clickable Twitter search term. So, for instance, if an event in Syria is making news, users might tweet #Syria so that other users can easily find related tweets. Hashtags are also used by Twitter and outside services to identify "trending" topics—what's hot—in order to suggest content to other users. The more a hashtag is tweeted, the more often it shows up on "trending" lists, resulting in more tweets and more people reading tweets that contain the tag.

At its peak, the app was a formidable force, sending groups of hundreds of tweets at periodic intervals carefully timed to avoid raising red flags with Twitter's automatic antispam protocols.

A typical day might feature six or seven major broadcasts highlighting one to three official ISIS propaganda releases, such as video from an occupied area or photos of captured weapons. The app also promoted ISIS releases in advance, further evidence of its connection to the organization's official structure. Virtually every tweet included at least one hashtag and a link for new users to sign up for the app.[11]

The Dawn of Glad Tidings app was functional from April to June

2014. Although its existence had been reported before in counterterrorism circles,[12] the story broke widely in June after ISIS exploded
into the news with its capture of Mosul.[13]

Twitter and Google soon suspended the app. Google removed
all other apps by the author, while Twitter flagged the Web page
where users could sign up with a warning that the site could be dangerous to their privacy.[14]

The volume of tweets from a monitored group of more than
2,000 pro-ISIS accounts dropped almost 50 percent overnight when
the Twitter app was shut down, with hundreds of accounts falling
entirely silent.[15]

When questioned by the app's users, J promised it would soon
return, but heavy fighting broke out between the Israelis and Palestinians soon after the suspension, and he found himself distracted by
the explosions rocking his neighborhood (which he tweeted about).[16]

In September, another supporter took up the slack and began
creating accounts that tweeted systematically, controlled by simple
scraps of computer code known as bots, usually designed to perform
repetitive tasks. Bots have a variety of uses, many of them positive.
For instance, there are bots that monitor Wikipedia and tweet any
time a page is edited by a congressional staffer, in order to promote
accountability.[17] Other Twitter bots are primarily spam machines,
sometimes set up to look like real users. Hackers use such bots to
get unsuspecting Internet users to click on links that can infect their
computers with viruses or worse.[18]

The new ISIS bots fells into the spam category. Most of them
tweeted in English, but the content suggested that the developer
might be from Indonesia (a region where ISIS enjoyed wide support).
By December 2014, thousands of new bots were operational. To help
avoid detection, the new bots did not advertise their existence or try
to attract new users. They were created in medium-sized clusters
with similar names. For instance, some eighty bots were all named
some variation of "IS Ghost." Another cluster had Twitter handles

with variations on the phrase *pagdade* (a homonym for Baghdad that pointed to the developer's Indonesian origins).[19]

The bots mostly tweeted links to official ISIS releases, such as the propaganda video "Flames of War" or videos of the beheading of Western hostages, projecting the appearance of broad support for ISIS on Twitter in excess of reality.

But social media tactics and trickery were only part of ISIS's arsenal. The group has won legitimate support online, while benefiting from intense global interest in the Syrian civil war. The world had a ringside seat to the conflict, although the information flowing over social media was sometimes heavily edited.

THE TWEETED REVOLUTION

Starting in 2010, the Arab world was rocked by a series of popular protests known as the Arab Spring, beginning in Tunisia and Egypt, where citizens lobbied for an end to longtime dictatorships and the birth of participatory government.

Social media played an important role in publicizing the issues at play in those countries, and in organizing and publicizing the protests. Young activists used Twitter, Facebook, and YouTube to publicize corruption by the leaders of Egypt and Tunisia, according to a study by the Project on Information Technology and Political Islam. The content of social media reflected and amplified conversations from the streets, the study found, and it provided a fast way to mobilize tens of thousands of people for protests.[20]

For months, the popular uprisings flooded the squares of Cairo and Tunis with protesters demanding an end to decades of ironclad dictatorships, and social media was used to document every stage of the revolutions. When Egyptian president Hosni Mubarak sent the Egyptian Army to roust protesters out of Cairo's Tahrir Square, the event was chronicled on Twitter for a rapt worldwide audience, pushing the international community to condemn the crackdown.

And when army officers dramatically defected to join the protesters, the world watched and cheered.[21]

In these early, heady days, it seemed as if a revolutionary wave of positive change was washing over the region thanks to the emergence of this new technology.[22]

Fueled by the fall of iron-fisted regimes in those countries (Tunisian president Zine al-Abidine Ben Ali stepped down in January 2011, Mubarak in February), Syrian activists opposed to the brutal regime of President Bashar al Assad adopted some of the same tactics.[23]

But the Arab Spring froze into winter, and the longtime dictatorships in Tunisia and Egypt were replaced not by progress but by newly imperfect regimes.[24] In Syria, a vision of nonviolent regime change gave way to a violent crackdown and then civil war in early 2011. From the start, social media played a crucial role in disseminating and sometimes distorting information about the conflict.[25]

Syria had been a key way station for jihadists entering Iraq during the U.S. occupation, with the Assad regime turning a blind eye (in the most charitable interpretation) to frequent border crossings by militants associated with what was then al Qaeda in Iraq. There was little question that this activity was permitted by the Assad regime as a passive-aggressive hindrance to the U.S. occupation, and it may have provided more active support.[26] But the networks that had supported these efforts now turned against the regime, and foreign fighters began to flow back into Syria in greater and greater numbers.

Virtually everyone involved in the conflict began working social media to advance their agendas, almost from the start. Anti-regime activists continued to put out information about regime atrocities in very organized ways, while the regime turned to sophisticated disinformation tactics, using hackers to compromise the websites of opponents; professional trolls to unleash a steady stream of abusive tweets and posts, as well as disinformation; and "honeypots," friendly-seeming accounts offering access to valuable information or affection, but actually intended to seduce critics into giving up compromising personal information and computer passwords.[27] In addition to the organized disinformation, which proliferated wildly, rumors and genuine misunderstandings could be found in ample supply.[28]

The flood of new information created new opportunities and complex challenges for journalists, academics, and intelligence officials. On the one hand, open-source intelligence was being generated at a speed and volume unprecedented in the history of conflict. In Afghanistan during the 1980s, for instance, it might take weeks or months for videotape of a muhajideen battle to travel back to the United States for viewing, where it enjoyed only limited circulation.

In Syria, the turnaround could be hours or days, and the audience was immense. Many sources were inconsistent or unreliable. As competition grew into conflict among Syrian rebel factions, activists often produced "evidence" of each other's dirty deeds.[29] Supporters of some factions, particularly jihadists, would post images of conflict they had found on the Internet and claim they represented recent events. These images could go viral quickly, before anyone checked their veracity.[30]

The major fighting factions quickly established official media accounts on a number of platforms for disseminating "authenticated" propaganda and activity reports, and smaller fighting factions (of which there were many) soon followed suit. The major players included the secular Free Syrian Army, the Syrian al Qaeda affiliate Jabhat al Nusra, the independent jihadists of Ahrar al Sham (later folded into the Syrian Islamic Front), and what was then known as the Islamic State in Iraq and Syria.

ISIS set up its first official Twitter account as an official "media foundation" under the name *al I'tisaamm*, an Arabic reference to maintaining Islamic traditions without deviation. Its first Twitter handle was @e3tasimo, established in October 2013 to little fanfare and scant notice from the media, although it quickly gained more than 24,000 followers.[31]

The official account tweeted out videos and other propaganda at a steady but slow rate. Individual accounts for members of ISIS were more active and accrued more followers as a result. One of the most prominent accounts, using the handle @reyadiraq, claimed to be unaffiliated with ISIS, a deflection tactic that supporters would try over and over again, with little success.

In late February, after a steady diet of increasingly grisly documentation of ISIS activities, including live tweeting of the amputation of an accused thief's hand in Aleppo, Syria, with accompanying photos, Twitter suspended "Reyad," by which time he had accumulated more than 90,000 followers.[32]

The account returned in March under the name @dawlh_i_sh, a play on the Islamic State's acronym, but even after months, it never regained its full follower strength. It was suspended again at the end of the summer of 2014, with just under 28,000 followers, and did not return in a clearly identifiable form. Whacking the mole once again seemed to have some lasting effect.[33]

The @e3tasimo account was also suspended by Twitter in late 2013 or early 2014, for reasons that were not entirely clear. ISIS attempted to re-create the account several times in January, but five or six new accounts in a row were suspended almost immediately. The pattern suggested that a government request was behind the takedown, but (as detailed later in this chapter), Twitter was restricted from disclosing such requests under certain circumstances.[34]

After a pause, the account returned on February 20 as @wa3tasimu, and Twitter did not intervene. Through March, the new account accrued more than 18,000 followers. In contrast, its chief rival, Jabhat al Nusra, had more than 50,000.[35]

But beyond the follower counts, there were oddities in ISIS's social media profile, hidden patterns that betrayed a hidden purpose.

Day after day in the month following ISIS's return to Twitter, a strange effect became visible. Each group used its formal name as a hashtag to identify media releases and supporter content. Despite its huge follower deficit, ISIS's hashtag was consistently tweeted more often than al Nusra's, by about four to one.[36] A data-driven analysis of the followers of both accounts helped reveal the hidden dynamics.

Over the years, information scientists had observed a pattern of activity in online communities sometimes known as the 90-9-1 rule. Generally speaking, focused online communities tend to break down in predictable groups. As a rule of thumb, about 90 percent of users will be mostly passive, about 9 percent are active, and 1 percent are very active. In social media networks, this dynamic also applies to the distribution of influence, defined as the ability to prompt interaction and participation by other users.[37]

The followers of both al Nusra and ISIS roughly broke out into the same pattern, but the devil was in the details. ISIS users in the 9 percent group were measurably more active than their counterparts following al Nusra. ISIS had a few thousand active online supporters who were more enthusiastic—and more organized—than their counterparts in the 9 percent group of al Nusra supporters.[38]

This was no accident; it was strategy. ISIS had a name for these users—the *mujtahidun* (industrious).[39] The *mujtahidun* could be observed repeatedly using specific tactics to boost the organization's reach and exposure online.

For instance, media releases followed a predictable pattern. After being posted and authenticated by official ISIS members, a second-tier group of several dozen online activists would retweet the link with a hashtag, then retweet each other's tweets and write new tweets, all using the same hashtag. Other activists would upload the release to multiple platforms, so that it could be found even when Internet providers pulled the content down. After that, a third tier—the *ansar muwahideen* (general supporters)—would repeat the process on a larger scale.[40]

Similarly, online hashtag campaigns were designed by activists on jihadist forums, largely out of sight, then implemented on Twitter in the same systematic way, with key users repeatedly tweeting the same hashtag and each other, and the next tier retweeting the previous tier and each other. The technique would routinely result in hundreds of similar tweets with hashtags at coordinated times, sometimes referred by participants as a "Twitter storm." Using the most inclusive criteria, around 3,000 users were part of this social media battalion (as ISIS called it) at its height, although some of those accounts were automated bots.[41]

The coordination was designed, in part, to game the systems that identified trending topics on Twitter. By concentrating their tweets in a short period and repeatedly tweeting the same hashtag, the media battalion could cross the threshold that would trigger trending alerts that would be displayed by Twitter on its website, as well as by third-party services such as "Active Hashtags," an automated Twitter account with more than 160,000 followers that highlighted trending topics in the Arabic language.[42] A strong performance could also influence search results, as seen during the World Cup and the march on Mosul.

There was a cascading effect to these efforts. Each time a hashtag ranked as trending, it was exposed to more people, generating still more activity. When an ISIS hashtag appeared on the Active Hashtags account, for instance, the tweeted announcement was retweeted an average of 72 times—making the tag trend even higher just on the basis of its appearance in that tweet, without accounting for those who might click the link and take a legitimate interest in the content.[43]

The jihadis soon developed their own version of the account, @al3r_b, which purported to retweet the most important "Muslim" news, but whose content consistently favored ISIS and its prominent online supporters. Tweets picked up by @al3r_b would be retweeted as many as nine times more than other tweets by the same user. (Twitter suspended the account in September or October 2014).[44]

It was a classic case of "fake it till you make it" marketing—the boost in visibility and exposure created an appearance of momentum that gradually turned into real momentum and a growing base of support, especially within the online jihadist communities where ISIS was now directly competing with al Qaeda for legitimacy and resources (see Chapter 8).

Smart—if deceptive—social media strategies boosted ISIS across the board. In contrast, the al Qaeda–controlled al Nusra had simply replicated the old style of media distribution on the new platform of Twitter, with a focus on relatively simple propaganda videos and fund-raising channels, where it outperformed ISIS significantly.

Ultimately, al Nusra's organically grown social network was no match for ISIS's engineered network features, such as the *mujtahidun* and the Dawn of Glad Tidings app. When the app rolled out in April, it automated and enhanced the efforts of the *mujtahidun*, resulting in a huge surge in the group's visibility.

None of this online activity existed in a vacuum, and much of it was strategic. In March, for instance, one highly organized Twitter campaign featured a hashtag demanding that ISIS emir Abu Bakr al-Baghdadi "declare the caliphate." It was an unprecedented tactic by an extremist group, essentially providing an online focus group to test its messaging before making it official, allowing ISIS to fine-tune the actual announcement when it arrived months later.

The Dawn app also coordinated to offline activities, reaching new heights just as ISIS rolled into Mosul in June. At its peak, during the attack on Mosul, the app generated about 40,000 tweets in one day (including retweets of the app's content by other users). The termination of the app on June 17—just twelve days before the announcement of the caliphate—struck a blow against ISIS's messaging strategy at a critical moment.[45]

As a result, supporters had to work harder for lesser gains. At one point, ISIS activists resorted to posting lists of tweets that users could cut and paste in an effort to simulate the app's function, but these efforts could not offset the loss of automation.[46]

When ISIS announced its caliphate on June 29, it took the major media almost twenty-four hours to discover, authenticate, and report the story. The messaging apparatus was not unstoppable and it was not all-powerful. It was down, but far from out.

BEYOND THE OFFICIAL ACCOUNTS

Individual foreign fighters with all of the factions in Syria could be found on social media by the hundreds, at first, and soon by the thousands. While they were represented on a number of platforms (the Dutch fighter named Yilmaz, mentioned earlier, accrued a massive following on Instagram before being suspended), a significant proportion of activity gravitated toward Twitter.

Due to its simplified interface, Twitter was well suited to situations where users had limited Internet access—tweets could even be posted and read via SMS text, which could be sent over any functional cell phone network and did not require an Internet connection. Additionally, Twitter was still reluctant to suspend accounts for terrorist content, which allowed users to accrue more followers and spend less time rebuilding networks than on other platforms, such as Facebook.

In some ways, the fighters used social media like anyone else, to chat with friends and post the mundane details of their lives, often in their native languages. The tabloid media, particularly in the United Kingdom, had a field day breathlessly reporting on the fighters' selfies with kittens and cravings for Nutella, as well as "terrifying" threats posted by accounts of questionable significance (for instance, claims that ISIS had a "dirty bomb" or that its operatives would infect themselves with Ebola and enter the United States).[47]

Not everyone who tweeted in support of ISIS was actually linked to the group, and not everyone who looked like a foreign fighter was one in real life. But all of them quickly learned that the global media reliably pounced on whatever they said, and the more outrageous the better.

Google searches for "ISIS" soared in July 2014, after the announcement of impending U.S. air strikes in Iraq ignited a Twitter storm of threats from ISIS supporters. Using the English-language hashtag #AMessagefromIStoUS, at least hundreds of Twitter users directed a barrage of threats both vague and specific at Americans, promising retribution on U.S. soil if the United States attacked ISIS. Activity spiked again in September, after ISIS released videos of the beheading of American journalists.[48]

Not all of this activity was confined to ISIS. Fiery clerics took to the social "airwaves," exhorting supporters to action and praising their faction of choice. Dozens of prominent Persian Gulf fundraisers took to Twitter and Facebook, where they posted bank transfer information to "help the Syrians," which in many cases meant funding non-ISIS jihadist fighters, although they avoided explicitly naming the recipient of the funds. Their followers swelled into the hundreds of thousands, with clear signs of covert activity constantly bubbling beneath the surface (for instance, clusters of accounts that all had established links with each other but never tweeted, or accounts with tweets marked as private that had posted thousands of tweets but had no followers).[49]

Then there were the recruiters. Prior to 9/11, jihadist recruiters had done much of their work in "brick-and-mortar" settings, with former foreign fighters traveling from city to city to tell potential recruits about their experiences and urge them to join the conflict du jour.[50] In Syria, a new dynamic emerged. Fighters could do the work of recruitment without ever leaving the front lines, a phenomenon Shaarik Zafar of the U.S. National Counterterrorism Center dubbed "peer-to-peer recruiting."[51]

Potential fighters could follow actual fighters from their home countries on Twitter, talk to them, ask questions, and eventually receive guidance about how to join the fight. In addition to Twitter and Facebook, many fighters signed up for ask.fm, the question-and-answer website where they entertained queries that ranged from banal to practical.

Recruits might travel on their own initiative to Turkey near the Syrian border, then log on to Twitter and ask for someone to come and pick them up. Incredibly, it seemed to work on a regular basis.[52]

People specifically tasked with recruitment also stalked the vulnerable online,[53] although the old ways did not completely fade. Many groups maintained dedicated recruitment networks on the ground. For instance, ISIS had operatives recruiting in Minneapolis, once a major pipeline for al Shabab fighters.[54]

Individuals worked the community, promising money and marriage to young men (and women), some of whom belonged to gangs that had adopted street names based on famous jihadist figures.

All of them also followed each other on Twitter, where the recruiters could keep tabs on what was happening and communicate privately with those who seemed willing. One fighter who was closely connected to that online recruiting network was Douglas McAuthur [sic] McCain, a Minnesotan killed fighting for ISIS in late summer of 2014, who maintained multiple Twitter accounts that followed and were followed by members of the recruitment network back home.[55]

But some professional radicalizers and recruiters simply moved their whole portfolios online, where they could operate more privately, away from the target's friends and family, a practiced tactic.[56] The primary work of the recruiter was building relationships, after all, and social media was made for that. Dozens of men and women on Facebook who listed their profession as *dawah* (an Arabic word for evangelical preaching) could be found working their way through

Muslim social circles, seeking the vulnerable and providing them with connections that would lead them to Syria.

Evidence of these networks could be found in the case of Nicholas Teausant, a California native indicted in March 2014 for attempting to join ISIS. Although his social network connected him to legitimate radical communities supporting ISIS in the United Kingdom, Teausant was diverted by the intervention of an FBI informant and arrested. Just days later, two men from North Carolina with connections to the same social network online were arrested for planning to travel to Syria.[57]

ISIS social media operatives liked Facebook, with its rich media capabilities and multiple network options (such as fan pages and moderated groups that resembled the old forums), and they took a sophisticated approach to establishing its presence.

Fan pages for Abu Bakr al-Baghdadi quickly accrued more followers than those of better-known extremists such as Anwar Alwaki. ISIS laid other plans signaling the announcement of its caliphate, creating accounts on Twitter (@islamicstatee) and Facebook two weeks in advance.[58]

One English-language page, Bilad al Shaam (a reference to Syria under the historical Umayyad caliphate), was created, suspended, and rebuilt dozens of times. Each time it returned, with a number denoting how many iterations it had gone through and sometimes a jab at the "Facebook thugs." Each time it met with a quick end. Another popular campaign using the slogan "We Are All ISIS" launched on Facebook before expanding to other social media. At least forty-eight iterations of the Facebook page were created after repeated terminations, according to Jeff Weyers, an analyst closely tracking extremist use of Facebook.[59] Facebook kept whacking the more visible moles, and terminating the accounts of bomb-making instructors and active terrorist plotters, but it was more difficult to intercept the recruiters, who often presented themselves simply as devout Muslims, avoiding obvious indicators of their affiliation and doing most of their work through private interactions.

But Facebook's vigilant policing had still paid benefits. While they had by no means exterminated the infestation, jihadis began to express frustration with the platform. As on Twitter, a core group of *mujtahidun* helped populate pages with content and likes, but the frequent suspensions limited ISIS's ability to reach an outside audience where fresh recruits could be lured.

Throughout 2014, more and more ISIS supporters moved their main activity to Twitter, where they could reliably expect to operate free from interference. While it was easier for jihadis to operate on Twitter, the social media strategies of ISIS were fueling public and private pressure on the libertarian social media platform. A showdown loomed.

TWITTER VS. ISIS

As ISIS rose in prominence, Twitter once again came under scrutiny for its practices. Unlike Facebook and YouTube, which allowed users to flag terrorist content for review, Twitter initially offered few reporting options.

"Users are allowed to post content, including potentially inflammatory content, provided they do not violate the Twitter Rules and Terms of Service," its guidelines read.[60]

At the time, Twitter's abuse reporting form was lengthy and restrictive, asking for substantial information on the user filling out the report and recommending that people just block accounts they didn't like. Blocking is a procedure that prevents other users from "mentioning" the blocker and thus showing up on their Twitter timeline, but it did not, at the time, prevent the blocked user from other activity, including reading the blocker's tweets by going directly to their Twitter profile page.[61]

There were other ways to get around blocking as well, raising issues that were more problematic in areas other than counterterrorism. For instance, blocking was virtually no impediment to stalkers or sexual harassers.

The platform's policy on violence was similarly narrow. Only "direct, specific threats of violence" were explicitly banned in the "Twitter rules." That generally meant naming an individual and threatening specific bodily harm against him or her. When al Shabab was first suspended for threatening to execute a hostage, it had crossed that line.

During the Westgate Mall siege (Chapter 6), Twitter took a broader view of its existing policies, with threats against Kenya during an ongoing terrorist attack against Kenyans apparently being specific enough to merit a response. Or perhaps the prospect of headlines such as "Are mass murderers using Twitter as a tool?" made the difference.[62]

After Westgate, the operating environment on Twitter slowly began to change. Several hundred al Shabab members maintained accounts on Twitter. Slowly and steadily, many of those accounts began to disappear. Three of the most important accounts in the Shabab network with the largest follower counts were among the first to go. They came back, smaller, but were soon suspended again.[63]

In addition, a number of tiny accounts began to blink out of existence, one or two at a time, often including new followers of the few prominent Shabab accounts that remained active. These were not noisemakers engaged in highly visible social media campaigns; some had only dozens of followers.[64]

This pattern suggested the suspensions were the result of government requests, although it was unclear which government. Some government requests came packaged in a form that prohibited Twitter from disclosing whether the requests had taken place and whether it had complied with them.

"The data in these reports is as accurate as possible, but may not be 100% comprehensive," Twitter's "transparency" page noted laconically. Its blog post on transparency was considerably blunter, noting that, within the United States alone, it was prohibited from reporting on suspension requests that were received in the form of official "national security letters" (NSLs) and certain types of Foreign Intelligence Surveillance Act (FISA) warrants.

Twitter's complaints about transparency had significant merit, but its efforts to win the right to even disclose generalities about such requests were rebuffed by the government.[65] In October 2014, Twitter filed suit against the government seeking the right to disclose more information. As of this writing, the lawsuit was still in progress.[66]

Even hobbled by these disclosure restrictions, Twitter made it clear the pace had changed. During the first six months of 2013, Twitter reported receiving 60 requests from governments and other entities[67] around the world. During the second half of 2013, that number skyrocketed to 377, an increase of more than sixfold, which did not include the exempted requests noted above. The number of documented requests increased by another 14 percent in the first six months of 2014.[68]

Initially, Twitter's suspensions of ISIS accounts were similarly ambiguous. When the official ISIS account was first suspended in February 2014, there was no obvious provocation to which it could be attributed.

The second time the official account was suspended, around the end of May, the reason was clearer. The suspension closely corresponded to the release of *Salil al-Sawarim 4*, Arabic for *The Clanging of the Swords Part 4,* the latest installment in a series of increasingly sophisticated and violent propaganda videos (see Chapter 5). The video showed hundreds of executions in graphic detail.

The tactics ISIS used for distribution online were designed to inflate the appearance of its popularity. The Dawn of Glad Tidings app blasted out thousands of tweets promoting the video,[69] which quickly racked up large numbers of views on YouTube (likely also fueled by repeated clicks from the *mujtahidun* and bots that could automatically access the video over and over again without involving a human viewer, although this could not be conclusively proven).[70]

Regardless, the fake-it-till-you-make-it principle applied and ultimately resulted in the video being widely viewed and discussed, with some Western analysts calling it the "most successful" jihadi video in history. That is almost certainly true, but ISIS's manipulations played a critical and generally underestimated role in inflating its importance. There is no way to know how many people actually viewed the video.[71]

The second official ISIS account was suspended almost immediately after the release of *The Clanging of the Swords 4,* with a speed that again suggested a government hand. But many other Twitter accounts remained online, including top influencers with more followers than the official account, such as ISIS media distributor Asawirti (Interpreter) Media, popular Chechen foreign fighter Abu Walid al Qahtani, and a notorious English-speaking tweeter using the name Shami (Syria) Witness (a user based in India who was arrested in December 2014).[72] Each had tens of thousands of followers, and the calculated ISIS distribution strategy was in full effect.

The timing of the release, and its focus on the mass execution of Iraqi soldiers taken prisoner, was significant. ISIS was employing social media as a tool for military and psychological offense. It foreshadowed actions with deliberation and strategic intent. On the surface, *The Clanging of the Swords* appeared to be just another ISIS video production, albeit a very successful one, but it came into play on the ground just a couple of weeks later.

Starting in early June, ISIS forces stormed through northern Iraq. When they reached Mosul on June 9, Iraqi troops defending the city turned and ran, some stripping their uniforms off as they fled. Some Western analysts and many ISIS supporters credited the video for inspiring the fear that led to this stunning retreat.[73]

Within a week, several official regional ISIS Twitter accounts had been shut down, including one of the only sources providing information on the attack against Mosul.[74] While Twitter refused (or was prohibited) from discussing the reasons for the shutdown, government requests were again suspected. On June 17, Twitter suspended the Dawn app, likely for violating its terms of service and related to news coverage rather than government requests.[75]

These setbacks came less than two weeks before ISIS's next big move, the declaration it was changing its name to simply the Islamic State and claiming the mantle of the caliphate.

While the losses weakened ISIS's distribution of content on the day of the announcement, June 29, and in the days to follow, the announcement was big news in the jihadi world, and ISIS supporters were fired up. Their burst of hyperactivity helped offset the disadvantages and distribute a string of important media releases, including an unprecedented video of emir Abu Bakr al-Baghdadi giving a sermon in conquered Mosul (along with translations in multiple languages) and the first issue of an ambitious English-language magazine named *Dabiq,* after a town in Iraq featuring prominently in Islamic prophecies (see Chapter 5). After Baghdadi showed his face for the first time, thousands of Twitter and Facebook fans began to use his image in their profiles.

But the announcement still underperformed relative to social media campaigns earlier in the year. ISIS supporters were extremely disappointed in the reaction of Muslims in general and their fellow jihadis in particular.

One *mujtahidun* complained that no one had showed up for a Twitter storm he announced. "Where are the others? Let's terrorize the *kuffar* on #Twitter. Is it too much difficult? Kuffar is doing their best to fight us. What about us?" Others complained petulantly that if people didn't want to swear allegiance to ISIS, they could at least refrain from mocking the would-be caliphate.[76]

The regional accounts—ISIS had one for each of its major geographical holdings—trickled back after a few weeks, with sporadic resuspensions and respawns. As the summer stretched on, whack-a-mole continued at a simmer, as did ISIS's aspirations for global support (see Chapter 8). Some ISIS accounts went down; most remained.

Then, on August 13, things began to ramp up. The official ISIS regional accounts were again suspended, but this time they were knocked down as soon as they came back, sometimes within minutes of returning. This continued for some hours, through dozens of iterations, until the message finally became clear. While some of the big influencers remained online, official ISIS accounts would no longer be tolerated on Twitter.

Some dozens of smaller accounts went down as well, including several prominent foreign fighters tweeting in English. Almost a week later, Twitter suspended the biggest and most influential ISIS accounts, including Abu Walid al Qahtani, Asawirti Media, and "al Khansa'a," a female former al Qaeda supporter who had become a powerful voice supporting ISIS and helping to organize its online recruitment of women (see Chapter 4). Al Qahtani went dark for a time, while the latter two kept popping out of holes to get whacked again periodically.[77]

Other jihadi accounts, including some associated with al Qaeda and Jabhat al Nusra, were also suspended, but sporadically. The main focus was ISIS.

A comedy of errors ensued. While the number of suspensions had climbed to more than one hundred, they still represented only a fraction of the active ISIS supporters on Twitter. Nevertheless, panic began to spread among ISIS tweeters.

Some, like Shami Witness, made their accounts private on the theory that it would help insulate them against suspension. Others changed their screen names and user handles (because surely that would fool Twitter). Many changed their profile pictures from ISIS's characteristic black flag emblem to pictures of flowers and kittens. One changed his screen name to "Syrian Food" and profile pictures to a restaurant. Many users subsequently abandoned these tactics, as their effectiveness was questionable and they made it much more difficult to do the work of social media activism.

Stalemated with Twitter, ISIS began trying to reconstitute its official accounts on alternative social networks. It moved to an obscure Twitter alternative called Quitter, where it was quickly suspended. It went to a pro-privacy social network called Friendica, which killed the accounts quickly and posted a message to anyone who came looking: "Islamic State not welcome on friendica.eu." [78]

ISIS then moved to Diaspora, an open-source social network specifically designed to let individuals and groups host the service using their own infrastructure, which in principle should have insulated the accounts from suspension on a purely technical basis. But the social network's designers and users found a way to take them down yet again.

Finally, in what must have been desperation, ISIS moved its accounts to VKontakte, a popular Russian equivalent to Facebook. This was, in many ways, an amazing turn of events. VK, as it is popularly known, had some months earlier lost a struggle for independence against pro-Putin forces in the Russian marketplace. There was good reason to suspect that sharing user data with VK was functionally indistinguishable from simply handing it over to the FSB, the Russian intelligence service. Hundreds and hundreds of ISIS followers did just that, until even VK grew weary of them and suspended the official accounts.[79]

ISIS had not been idle during all this tumult. On August 19, it released the now-infamous video of the beheading of American journalist James Foley.[80] ISIS Twitter accounts hit the ground running, distributing the video using the English hashtag #AMessagetoAmerica and directing tweets to the accounts of random Americans in an effort to spread them among the American public. ISIS supporters also paid spammers based in the Persian Gulf region to send out tweets that included the hashtag.

Twitter, apparently on its own initiative, began to take down accounts that were spreading the video and graphic images included in it. The sweep of this crackdown was so broad it took down the accounts of some journalists and analysts who had tweeted the content (they were restored later).[81] Still, only about fifty ISIS accounts were suspended in the first twenty-four hours after the video released. It was more than Twitter had suspended in one sweep before, but still a tiny fraction of all ISIS-supporting accounts.[82]

The move may have been empowered by a new policy Twitter announced just hours after the video was released.[83] The policy had been in the works for days prior and was prompted by the suicide of actor Robin Williams on August 11. Internet trolls had tweeted Photoshopped images that they claimed showed the actor's corpse.[84]

"In order to respect the wishes of loved ones, Twitter will remove imagery of deceased individuals in certain circumstances," Twitter wrote, adding that it "considers public interest factors such as the newsworthiness of the content and may not be able to honor every request." [85]

Although news reports attributed the policy to the Williams incident, there were hints that Twitter might have known the Foley video was in the works. The crackdown on ISIS had started prior to the video's release. In the thirty days preceding the Foley video, Twitter had suspended at least eighty ISIS accounts, including all of its official outlets. A number of ISIS accounts had foreshadowed the release by tweeting images from the 2004 beheading of Nicholas Berg.

In keeping with its typical silence, Twitter refused to comment on why specific accounts were suspended, but in the wake of the Foley video, it referred curious reporters to the family request policy.[86] Some ISIS supporters took the hint and stopped tweeting the video and images. Others persisted and were suspended, often multiple times.

In the weeks following the Foley video, Twitter continued suspending the accounts of ISIS supporters, usually dozens at a time, with periods of inactivity between.

News coverage and Twitter intervention seemed to track uncomfortably with the race of the victims depicted. When ISIS publicized the executions of Iraqis and Syrians, news coverage and organizational responses were minimal, but the beheadings of white Western journalists—which continued throughout the fall—led to more Western media coverage and more suspensions by Twitter.[87]

ISIS supporters joked about being Twitter *shahids* (martyrs), and when they created new accounts, they began listing the number of times they had been previously suspended in their profiles. As both the suspensions and the beheadings continued, it became more difficult for ISIS to push its message to the widest possible audience. It turned more heavily to manipulative tactics such as bots and purchased tweets.

The number of suspensions began to climb. More than 400 accounts went down in one seven-hour period in late September. Between September 1 and November 1, at least 1,400 ISIS-supporting accounts were suspended—a very conservative estimate.[88] As frustration mounted, many ISIS users took to threatening to kill Twitter executives, sometimes by name. The social media platform had been weaponized against itself.[89]

In December 2014, Twitter announced it would overhaul the process of reporting abusive or violent behavior, making it easier to report accounts or specific tweets that violated Twitter rules and preventing users from viewing the profile pages of someone who had blocked them. Some online advocates continued to insist more changes were needed, and Twitter said its policies would continue to evolve.[90]

In the meantime, the suspensions were starting to have an effect on ISIS. The number of retweets that an average ISIS supporter could expect to receive dropped significantly. From August, when the major crackdown began, through the end of October, the average number of retweets for every tweet by a monitored ISIS supporter (excluding bots) plunged 42 percent, from 5.02 to 3.49. The percentage of tweets by ISIS supporters that received no retweets at all climbed from 57 percent to 62 percent.[91] Other metrics (such as the number of followers and number of tweets per day) also appeared to drop dramatically.

While these analyses pointed to an impact from the suspensions, it is important to note the difficulty of creating a reasonable comparison set. One especially large challenge stems from the fact that there is no definitive estimate of how many ISIS supporters are active on Twitter.

In late 2014, we attempted to answer the question of how many ISIS supporters were active on Twitter, in a research project commissioned by Google Ideas and coauthored by J.M. Berger and Jonathon Morgan. As of press time, we estimated that at least 45,000 pro-ISIS accounts were online between September and November 2014, along with thousands more pro-ISIS bot and spam accounts.

This represents only an initial finding. The research project will be completed between the writing of this book and its publication, and the complete results will be published by the Brookings Institution and will also be available at Intelwire.com by the time this book is published. Once a baseline set of ISIS supporters has been identified, it will be possible to conduct better research on the effect of suspensions.

ISIS STRIKES BACK

Although analysts continued to debate the merits of whack-a-mole, ISIS supporters delivered their verdict loud and clear.

"As the accounts of the caliphate's supporters become scattered, their effectiveness rises and falls, and the control the supporters have decreases," wrote Shayba al Hamd, an ISIS social media activist, on September 12, calling the campaign "devastating" and a "dirty war."[92]

"The Crusaders tremble at the media power of *Dawla* [the Islamic State], which has taken up permanent residence in the depths of Twitter," he wrote.

ISIS strategy documents diagnosed the problem as emanating from the top down, with the official accounts targeted first and the industrious *mujtahidun* second. The *ansar* supporters were less vulnerable, and a fourth tier, "the silent supporters," was therefore required to step up and become more active. If Twitter closed a tier, he wrote, the next tier would simply rise up to take its place (not addressing the fact that the supply of tiers appeared finite).

"Why must you return to Twitter?" al Hamd asked rhetorically. He compared it to war: If the frontline fighters desert, what hope has the army?

Other ISIS users specifically pointed to the amount of time and energy that they were now wasting on rebuilding the same networks over and over again, and the fact that even the hard-core *mujtahidun* were growing weary of promoting newly resurrected accounts for days on end.[93]

Some devised elaborate countermeasures, based on ISIS social media experts' beliefs about how Twitter decided whom to suspend and whom to permit. In addition to a brief bout of camouflaging accounts, some periodically changed their online names. Others blocked anyone following them who "looked suspicious" (these might be anonymous accounts or journalists and terrorism researchers who failed to take steps to hide their own presence online).

Twitter largely remained silent about its ISIS problem, even when the group's supporters began threatening to kill and behead their employees, sometimes by name and photo.[94] But it quietly made a change to its terms of service, which allowed it to request a valid phone number to verify the identity of any user. A spree of suspensions followed almost immediately. ISIS noted this, and guided users to services that provided false phone numbers that could be used to verify accounts.

Some also made more elaborate plans. ISIS had, for some time, been recruiting and training hackers, some with links to broad international cybernetworks that later repudiated them.[95] These activists, including many on the social media team, were part of the "Islamic State Electronic Army" and were active on both Twitter and Facebook.

The army had a "brigade" devoted to media operations, which included many key members of its social media team (known as *i'lamiy nasheet,* or "the energetic journalists") and a "technical brigade" that worked on hacking and security operations.[96] The two sometimes overlapped and collaborated, for instance to design the bots and apps that were so important to ISIS's social media success.

One unsigned technical brigade strategy document suggested supporters should hijack older accounts with significant numbers of followers that had been abandoned by their Western users, providing detailed instructions; examples of success were not abundantly detected in the wild (that is, actually being implemented on Twitter). The document then provided overly complex instructions on how to gain followers.[97]

All of these strategies were predicated on the incorrect assumption that such tactics would also provide protection. Most of the proposed countermeasures were stabs in the dark. At best they might slow down the process of suspension. At worst they contributed to a growing sea of confusions.

At first, the technical brigade recommended that someone whose account had been suspended should return with the same name and a number added to the end (a tactic also used on Facebook), to make it easier for supporters to find each other. It was also recommended that users create multiple backup accounts and let their followers know where to find them if they were shut down.

It soon became clear that this was just making it easier for Twitter to suspend them, so they reversed course and told users to come up with entirely different names. The subtleties of this process were lost on some suspended users, who opened each new account by proudly announcing how many times they had been suspended before. Some ran into the dozens.

At the time of this writing, the ultimate outcome of the battle for Twitter supremacy was still a work in progress, but one thing was clear—ISIS was far from ready to concede the online battleground, and it had chosen Twitter as the field on which it would make its stand.

It seems strange that Twitter could lose control of a system it owned and operated in its entirety, even as that tool was being used to threaten the company's own employees and executives. Yet while the suspensions had hurt the organization's efforts and taken away some of the tools that had made ISIS notorious, the electronic brigades were adapting.

New generations of bots emerged weekly, some of them carefully calibrated to avoid Twitter's countermeasures. Out of 85 "ghost" bots detected on September 15, 2014, only 25 had been suspended by early November, despite tweeting links to some of ISIS's most graphic material.[98]

The ghosts were a calculated affair, with very specific profiles, small follower counts, and an intentionally limited reach; they could easily be missed by anyone scouring the Internet for ISIS, and they were lesser targets for suspension. But their tweets would still help trend hashtags and distribute content. Other similar clusters of bots were set up in the same manner, using different technical specifications. ISIS had learned from its experience with the Dawn app. The new generation of bots were smaller, less visible targets with no single point of failure.

While the effects of these visible ISIS countermeasures set off new rounds of whack-a-mole complaints, ISIS had been forced to spend more energy on smaller returns as a result of pressures and setbacks its own members described as "devastating." [99, 100]

If the pressure continued, the network would continue to suffer. If the pressure eased, the network would recover, at least in part, its members shuffling back online to rebuild and regroup, and start the whole process again.

They are weeds.

No gardener expects weeds will simply give up after being uprooted once. Gardening is a process; it requires care and maintenance. A constant gardener does not let weeds overrun the plot.

CHAPTER EIGHT

THE AQ-ISIS WAR

ISIS was born from the crucible of America's "war on terrorism," and al Qaeda looms over these pages like a shadow. The road from al Qaeda in Iraq to ISIS has, at every step, revealed a clear pattern of deliberate differentiation.

Jihadist groups have a long history of splintering and separation. In some cases, this process involves competition among factions that sprang from the same source. In Algeria during the 1990s, this dynamic reached disastrous proportions, from the terrorists' point of view. The ensuing *fitna* (infighting) resulted in the emasculation of every group involved, and became a widely cited case study in failure for jihadists everywhere.[1]

The separation of ISIS from al Qaeda, while born out of strife and irreconcilable differences, did not have to result in war. Certainly, al Qaeda did not want such an outcome, and it has repeatedly pleaded with ISIS to submit to an arbitrated reconciliation. ISIS not only rebuffed these overtures, it upped the ante, and with the declaration of the caliphate, it demanded that al Qaeda submit to its authority.

As a result, the global jihadist movement has split into two major factions. Al Qaeda and its declared affiliates continued to operate under the nominal leadership of Ayman al Zawahiri. ISIS and a growing number of global affiliates have staked their loyalties to Abu Bakr al-Baghdadi.

The two groups are now locked in a battle for supremacy and for the loyalties of unaffiliated groups and the members of existing organizations.[2]

It is easy to misunderstand the stakes in this battle. ISIS has adopted the rhetoric of the absolute—al Qaeda must submit and become part of its caliphate—and the two compete, to some extent, for loyalty, funds, and recruits.

But most important, this conflict is about vision. The "winner" of the war between al Qaeda and ISIS will wield tremendous influence over the tactics and goals of the next generation of jihadists. Understanding the contours of the battle will help reveal the shape of things to come. In that respect, the question of who "wins" is incredibly important, not just to the region but to the world. The West has too often found itself fighting the last war, when the next war is taking shape before its eyes. Faced with the expansionist, populist rise of ISIS, we cannot afford to keep making that mistake.

But before we can forecast what lies ahead, we must first understand what is happening now. Who is winning the battle for leadership of the global jihadist movement?

THE BATTLE FOR *BAYAH*

Terrorist groups are often amorphously linked to one another, with cooperation and coordination taking place across a spectrum of activities.

For instance, in 1998, Osama bin Laden and the leaders of other jihadist terrorist groups announced the formation of the World Islamic Front, an alliance to fight the United States, but each signatory to the statement had a different relationship to al Qaeda.

Al Qaeda folded the Egyptian Islamic Jihad organization, led by Ayman al Zawahiri, into itself. Islamic Jihad did not become al Qaeda in Egypt, however; it was simply subsumed into al Qaeda.[3] On the other hand, the Islamic Group (the Egyptian jihadist group responsible for the World Trade Center bombing with support from AQ) remained somewhat independent and eventually drifted away from al Qaeda in most meaningful respects, even taking part in the political process that emerged after the Arab Spring.[4]

After September 11, the power dynamics began to shift. Although the amorphous links continued, the leaders of some groups now pledged their loyalty to the emir of al Qaeda, Osama bin Laden, and subsequently to his successor, Ayman al Zawahiri.

The oath of loyalty, known as *bayah*, is the principal mechanism of control in the al Qaeda network, adding a religious obligation to relationships that historically would rise and fall when the prevailing winds changed. *Bayah* is extended from leader to leader, not group to group, so when the players change, it must be renewed. A pledge offered must be accepted by the leader of al Qaeda before it is valid.

On paper, at least, al Qaeda itself is subordinate to Mullah Omar, leader of the Taliban, reportedly through a loyalty oath from bin Laden to Omar during the late 1990s, which was affirmed last year by al Qaeda's current leader, Ayman al-Zawahiri, and again in the summer of 2014, in a print publication attributed to al Qaeda.[5]

But the pledge to Mullah Omar was largely theater. It is difficult to point to any examples of al Qaeda following Omar's commands or directions, and relatively easy to find examples of its disobedience. Jihadi accounts of the relationship between al Qaeda and the Taliban describe a fractious mess entered into under protest.[6]

Under the emir of al Qaeda are the organization's official affiliates. As noted previously, the list includes al Qaeda in the Arabian

Peninsula (AQAP, mostly in Yemen), al Qaeda in the Islamic Maghreb (AQIM, mostly in North Africa), al-Shabab (mainly in Somalia), the al Nusra Front (in Syria), and al Qaeda in the Indian Subcontinent, announced in 2014.

In February of 2014, Zawahiri disavowed ISIS, which was at the time considered an al Qaeda affiliate, although there is some dispute about whether its current emir, Baghdadi, ever swore the loyalty oath.[7]

The media latched on to the idea that ISIS was disavowed because it was "too extreme for al Qaeda." While it's true that al Qaeda saw ISIS as too extreme, it's more accurate to say Zawahiri fired ISIS for its public defiance of his wishes and commands.[8]

ISIS shed no tears over the separation. It was already functionally independent from al Qaeda in most respects, and the dismissal played into its long-term plan—the presumptive declaration of a new Islamic caliphate. Cut loose from its parent, ISIS moved forward with the declaration within short months. When the time came, it proclaimed that all previous loyalties were voided by the new development and demanded that jihadi groups around the world swear loyalty to Baghdadi.[9]

When the world's Muslim militants failed to drop to their knees, the online supporters of ISIS were baffled and disappointed.[10] The realist leadership of the group probably knew that the announcement would not produce immediate breakthroughs, but it may have been disappointed at the volume of the first wave of rejection.[11]

Given how tightly ISIS has synchronized its media strategy, it was telling that the group could not arrange even a single high-profile pledge within the first week after the announcement. The controversial declaration was no fait accompli. And even as of this writing, none of the official al Qaeda affiliates had yet broken with the core.

But over time, the so-called caliphate began to draw concrete support.

Its first new constituents were small-timers, and most had thrown their support behind ISIS earlier in 2014, after the rift with al Qaeda became overt. Many of these new pledges were from malcontents within the AQ affiliates. Some individual fighters and small groups simply deserted the affiliates and joined ISIS in Iraq and Syria.[12]

Others declared the formation of breakaway groups. It was difficult to gauge the size of these splinters; most involved a handful of people who signed their names and purported to represent larger groups of followers.

A prominent cleric with al Qaeda in the Arabian Peninsula, Mamoun Hatem, openly declared his support for ISIS. A number of other AQAP figures on social media have also endorsed it, and it is believed that a significant number of AQAP fighters lean in that direction. In November 2014, AQAP issued a blistering statement condemning ISIS and its declaration of the caliphate, which included a tacit admission that the Yemeni affiliate was fracturing over divided loyalties.[13]

An important early splinter emerged within the North Africa affiliate, al Qaeda in the Islamic Maghreb, from a group identifying itself as the "central division."[14] Over time, the appeal of ISIS has broadened in the areas where AQIM operates.

Very few old-school establishment al Qaeda supporters and clerics have come down in favor of ISIS, with the notable exception of Abu Bakar Bashir, an Indonesian cleric and the spiritual leader of the former Jemaah Islamiyah, a now-defunct organization with long-standing ties to the original al Qaeda.[15]

Bashir pledged allegiance to Baghdadi from a prison cell. But his decision split the successor group to Jemaah Islamiyah, with Bashir's sons denouncing the defection and breaking away with some number of supporters.[16]

AQAP's Hatem might have been more pragmatically important, but Bashir brought prestige, and he reflected a very large base of enthusiastic Indonesian and Malaysian ISIS supporters, many of whom were very active on social media.

In the Philippines, the leader of the Abu Sayyaf Group, founded with money from al Qaeda decades earlier, pledged to the leader of ISIS in September. The group had devolved into a criminal gang over the decades, and many observers suggested the pledge was simply an opportunistic bid to increase the ransoms they demanded for kidnapped Europeans. But even before the caliphate declaration, ISIS had enjoyed significant support from young people in the island nation.[17]

A small group of known al Qaeda figures in Afghanistan issued a statement supporting ISIS, and the venerable Afghan militant group Hezb-e-Islami signaled that it was considering the Islamic State's claim to the caliphate.[18] In neighboring Pakistan, unruly Tehrik-e-Taliban (TTP) factions had begun splintering over a number of different fracture lines. One of several points of contention was the rise of ISIS. TTP had to fire its spokesman after he publicly pledged allegiance to Baghdadi. Other commanders soon joined him, and there were signs of interest from other Pakistani radicals.[19]

Boko Haram, a hard-line jihadi group in Nigeria, declared an "Islamic caliphate" in its own territory after ISIS's announcement, but the rambling statement[20] by its notoriously incoherent leader, Abubakar Shekau, was decidedly unclear as to whether he was placing the territory under the umbrella of ISIS, and subsequent announcements only confused the issue.[21]

In Africa, members of Ansar al-Shariah in Tunisia (AST) displayed significant sympathies for ISIS, and analysis of its social media networks pointed to operational links. But the leadership remained steadfastly silent. ISIS counts large numbers of Tunisians among its foreign fighters, more than any other single nationality, and authorities claim to have arrested thousands more[22] who were trying to join the fight.[23]

English-speaking radical communities have been particularly

critical to ISIS's support base. Two of the most important English-language Muslim radical organizations have aligned with ISIS, including Authentic Tauheed, led by Jamaican national Abdullah Faisal, and the network formerly known as al Muhajiroun, led by British cleric Anjem Choudary.

Faisal is best known in the West as the spiritual leader of the defunct Revolution Muslim, an online collective of al Qaeda supporters, most of whom are now in prison.[24] Although he rarely makes headlines, Faisal has been a loud, active voice in radicalization for decades, with a consistent presence online via audio lectures and the Paltalk forum. Years ago, he once condemned American al Qaeda cleric Anwar Awlaki for not being radical enough (albeit this was before Awlaki came out of the terrorist closet).

After the announcement of the caliphate in June, Faisal weighed in strongly in favor of the ISIS's caliphate, buttressing it with his "scholarship" and a series of rousing lectures. He later formally pledged his loyalty.[25]

Choudary led the radical group al Muhajiroun, which was banned in Britain, and a series of successor organizations that were, to a greater or lesser extent, the same group under a different name. Despite this, he remains at large as of this writing and functions as ISIS's primary cheerleader in the Western media.[26]

The al Muhajiroun network, by any other name, has been one of the most important funnels for hundreds or more British foreign fighters in Syria and Iraq, with many of them now fighting under the ISIS banner and maintaining a robust presence on social media.

Other important English-speaking clerics are widely followed by ISIS supporters, including Musa Cerantonio, an Australian, and Ahmad Musa Jibril, an American. Cerantonio is openly affiliated with ISIS,[27] whereas the Michigan-based Jibril is broadly popular with English-speaking fighters, despite the fact he does not openly call for violence and has not endorsed ISIS.[28]

COALITION OF THE *WILAYAT*

Support built slowly but steadily in the weeks and months after the declaration of ISIS's so-called caliphate, but these public expressions were not the endgame. On November 13, 2014, an important new plank in ISIS's plan for expansion became clear, another innovation in the jihadist milieu.

The media had been captivated for days by unfounded rumors that Abu Bakr al-Baghdadi had been killed in an air strike. When ISIS released a new audio recording of its would-be caliph, observers flocked to analyze it for clues about when it was recorded, trying to discern if Baghdadi still lived. Most passed by the real news in the speech.

"Glad tidings, O Muslims, for we give you good news by announcing the expansion of the Islamic State to new lands, to the lands of [Saudi Arabia] and Yemen, to Egypt, Libya and Algeria," Baghdadi said. "We announce the acceptance of *bayah* of those who gave us *bayah* in those lands, the nullification of the groups therein, the announcement of new *wilayat* (provinces) for the Islamic State, and the appointment of [leaders] for them."

ISIS's organizational structure in Iraq and Syria was based on the *wilayat*, essentially provincial subdivisions each with its own governor. With the acceptance of *bayah* and the naming of governors outside of Iraq and Syria, Baghdadi was signaling that these new pledges were more than just business as usual.

The pledges had been announced on November 10, but their importance wasn't clear until the speech placed them in context. Although many had offered their *bayah* to Baghdadi, this marked the first time he had definitively accepted any in public.[29]

Highlighting the substantiality of these new *wilayat*, a number of additional groups that had pledged to ISIS were omitted from the announcement, including prominent organizations in Southeast Asia. Their *bayah* had been accepted, but they had not consolidated their leadership and infrastructure enough to be granted formal standing.[30]

The Egyptian terrorist group Ansar Bayt al Maqdis announced its alignment with ISIS, concurrent with the designation of the group as a *wilayat* in Egypt. The merger had been rumored for weeks.[31] ABM was an active jihadist group that had emerged after the Arab Spring. Most of its attacks were carried out in the Sinai Peninsula, but the group also had ties to Gaza.

Within days, the rechristened Sinai Wilayat of the Islamic State had issued a new video under the ISIS flag, a significant upgrade to the group's previous offerings, and displaying ISIS's distinctive mix of high production values and graphic violence.[32]

More important, the video was distributed by known members of the ISIS media team, the same channels that had released Baghdadi's announcement.[33] This was marked contrast to al Qaeda, which had never visibly coordinated with its affiliates so closely.

This was not business as usual.

Each new *wilayat* penned a statement, distributed by ISIS, outlining its reasons for pledging. In Libya, three *wilayat* were specified—in the regions of Barqah, Fazzan, and Tripoli. Barqah included the town of Derna, which had supplied many foreign fighters to al Qaeda in Iraq during the U.S. occupation. More recently, large numbers of Derna residents had made very public pledges to ISIS.[34]

In Algeria, the pledge emanated from Jund-al-Khalifa, an AQIM brigade based in the Tizi Ouzou region that had splintered from the group in March 2014 and formally offered its allegiance to ISIS in September, when it had offered a concrete token of its loyalty, beheading a French hostage on video, just days after ISIS spokesman al Adnani had issued a blanket call for such actions in response to U.S. air strikes in Iraq.[35]

The other *wilayat* were less clearly defined, with the pledges from Saudi Arabia and Yemen signed simply as from the *muhjahideen* of each country. Neither specified where the *wilayat* were located, nor did they indicate that they represented existing groups. But evidence of ISIS's presence in the Arabian Peninsula soon emerged from an unlikely source—al Qaeda.

In a statement issued November 21, 2014, AQAP's top religious official, Harith bin Ghazi al Nazari, issued a statement sharply condemning ISIS for its declaration of the caliphate and its announced expansion into Yemen. In the statement, al Nazari accused ISIS of "dividing the mujahideen" around the world and in the Arabian Peninsula. He also called on Baghdadi to recant its claim on Yemen and other regions, a step that would not be necessary unless Baghdadi's call to join ISIS had been heeded by a significant number within the al Qaeda affiliate's ranks.[36]

In some ways, the announcement was the debut of the first ISIS affiliates, but more accurately, it appeared to be an expansion of the proto-state itself beyond contiguous borders. Where al Qaeda's affiliate system had emerged in fits and starts over time, with little evidence of a clear agenda, ISIS was making a definitive statement about both expansion and control. Al Qaeda was not well structured to support and control the affiliate system, and as a result, the affiliates had nearly undone it.

ISIS would not make the same mistakes. It was creating an "archipelago of provinces," in the words of jihadism scholar Aaron Zelin, who was early to assess the implications of ISIS's plan. The *wilayat* abroad would share connective tissue of control and governance, but would exist in noncontiguous spaces.[37]

Precisely how ISIS intended to control these remote territories was unclear at the time of this writing, but subsequent statements indicated that the *wilayat* designations were only extended to those groups that had demonstrated they had implemented the infrastructure of control.

Given the large number of smaller groups that had pledged ISIS without being designated *wilayat*, the selective designations strongly suggested that a formal architecture existed for the new concept of governance.[38]

It was dawn of the era of distributed warfare, in which affiliated insurgent armies could arise in geographically distant regions but still answer to a single authority.

The full ramifications of the new paradigm were still nebulous at the time of this writing, but the unlikely coalition that had arisen to fight ISIS was able to function largely on the basis of an extremely limited engagement.

The new *wilayat* held territory and conducted operations on the sovereign soil of coalition members. Direct military confrontation between the West and ISIS in nations like Saudi Arabia and Egypt would be virtually impossible, and efforts to conduct such a war would further destabilize the region.

THE OLD GUARD

As of the writing of this book, ISIS had still not managed to score an outright win over al Qaeda in its core network—the official affiliates and the most prominent jihadi scholars—despite its considerable gains and the weakness signaled by AQAP in its November 21 statement.

All the top leaders of al Qaeda's affiliates had sworn *bayah* to Zawahiri, and for as long as he lived, they were religiously obligated to maintain that loyalty.

Some ISIS supporters advanced arguments about when such an oath could be rendered void, but this was a slippery slope. If the leaders disrespected their oaths to Zawahiri, their own followers might feel free to disrespect them.

There was legitimate cause for concern about opening that door. The increasingly spectacular fragmentation of the Pakistani Taliban in 2014, along several different lines of dispute over tactics and leadership, demonstrated both the fragility of many established jihadist organizations and the opportunities they afforded ISIS.

"Our groups were in crisis; now [ISIS] has provided them with a powerful framework that is transforming their narrative," Muhammad Amir Rana, director of the Pak Institute for Peace Studies, told the *New York Times* in November.[39]

Within the al Qaeda affiliates, as well, ISIS had sown deep divisions, or highlighted dissent that already existed.

In Syria, where literal shots had been fired and animus toward ISIS was arguably greatest, the al Nusra Front struggled with a loss of enthusiasm from the broader global jihadist support network and a string of defections from the lower ranks. (The calculus of defection was complicated, as moderate rebels deserted or defected to and from both al Nusra and ISIS, and concrete numbers were impossible to determine.)[40]

Al Nusra was plagued by a steady stream of rumors and disinformation about a possible merger with ISIS, which were greeted with a credulousness that spoke volumes about al Nusra's weakness. While most of the rumors were sourced to Syrian rebel factions with well-known axes to grind, there were some contacts between the groups, which appeared to end unceremoniously when ISIS demanded al Nusra simply submit and swear loyalty.[41]

Further afield, each of the affiliates issued statements after the declaration of the caliphate that split the difference, affirming allegiance to Zawahiri, with stronger or weaker language, while noting and sometimes praising the successes of the Islamic State, reflecting fears that too strong a stance opposing ISIS would split their own organizations down the middle. Pleas for reconciliation among the factions appeared with clockwork regularity from al Qaeda's partisans and were just as regularly ignored by ISIS.[42]

The razor's edge walked in these statements was sharpened when it became clear that the United States was preparing to take military action against ISIS, making it even more difficult to criticize the would-be caliphate.[43] The pressure built throughout 2014, finally cresting in the November 2014 AQAP statement, the first time an affiliate unequivocally condemned ISIS's actions.

Despite pressures from every side, Zawahiri received apparently unsolicited declarations of loyalty from Mokhtar Belmokhtar, leader of a terrorist faction separated from AQIM in Africa, and the Caucasus Emirate, a Chechen insurgent group. These were unqualified wins for the al Qaeda leader but did not represent great strength, especially in regards to the Caucasus Emirate, many of whose former members had joined ISIS as fighters in Syria. (The group began visibly splintering over ISIS in December 2014 as this book went to press.)[44] Zawahiri had not acknowledged either group as a formal part of the al Qaeda network by the end of December.

In fact, Zawahiri had remained almost entirely silent on the subject of ISIS and its presumptive caliphate as weeks stretched into months, to the considerable frustration of his supporters.[45] His public silence did little to offset the growing perception that the core al Qaeda had been weakened and thrown off balance by ISIS's dramatic military advances and its audacious demand for the allegiance of the world's jihadists.

In September, al Qaeda finally released the first new message from Zawahiri since the caliphate announcement. It was strangely tangential, announcing the formation of a new al Qaeda branch in the Indian subcontinent. Although the region, which included Pakistan and India, was flush with preexisting jihadist organizations, none were named as participants in the new venture.

Some observers rushed to portray the move as an attempt to counter the perception that ISIS had rendered al Qaeda irrelevant.[46] While the rise of ISIS may have been a factor in the timing and the framing of the announcement, Zawahiri claimed the affiliate had been in development for two years. And as analyst Arif Rafiq noted, there was perhaps a more likely explanation: The new branch assured a continuing presence for al Qaeda in the region if Zawahiri was killed and al Qaeda Central relocated to another part of the world.[47]

Nevertheless, ISIS was the elephant in the room. Throughout the fifty-five minutes of Zawahiri's typically dry and long-winded rhetoric, he made no explicit reference to ISIS or the challenge it presented, furthering the impression that Zawahiri was out of touch or simply too weak to deal with the crisis. The response from al Qaeda supporters online was muted, and Zawahiri fell silent once more.[48]

Zawahiri's absenteeism was driven in part by operational security concerns. Despite the rise of ISIS, he remained the world's most wanted terrorist. Zawahiri may believe that ISIS will self-destruct due to its own excesses and that his best play is to minimize any infighting or splintering of al Qaeda until that happens. And he is not necessarily wrong about that.

But the weakness of this position leaves room for ISIS to exploit one of the fundamental risks that terrorist organizations face—decapitation.

As previously noted, *bayah* extends from leader to leader, not organization to organization. When one of the leaders of an affiliate is killed, the new leader of the affiliate is required to make a new oath to Zawahiri and have that oath acknowledged in order to stay in the network.

In the event of a death at the leadership level, an al Qaeda affiliate could choose to drop its official affiliation with al Qaeda and realign with ISIS, or even opt for independence from both. And in the event of the death of Zawahiri himself, all bets are off. All of the al Qaeda affiliates would have the option to switch allegiances.

It is by no means certain this would happen. Al Qaeda survived its first major test in the post-ISIS era in September 2014, when a U.S. air strike killed Ahmed Godane, the leader of the Somali al Qaeda affiliate al Shabab.

Al Shabab moved swiftly to replace Godane, and its new leader immediately pledged continued allegiance to Zawahiri. However, the insurgent group remained under heavy pressure in Somalia, and the long history of infighting among Somali jihadist groups left the question only temporarily settled.[49]

It is decidedly unclear whether other affiliates would stay in line in the event that their own leaders or Zawahiri is killed. The current U.S. strategy against terrorism, which is heavily focused on decapitation, could eventually prove to be ISIS's greatest asset. If a drone strike kills the leader of AQAP or AQIM, the uncertainties of succession could result in powerful new allies for ISIS.

TERROR RECRUITS AND LONE WOLVES

In the eyes of many Westerners, the competition between al Qaeda and ISIS is a battle for survival and relevance. During the Arab Spring and after the death of Osama bin Laden, pundits as well as some serious students of terrorism were happy enough to write al Qaeda's obituary, if prematurely.

As ISIS commanded a greater and greater share of the headlines, many observers decamped into opposing factions, arguing either that ISIS had made al Qaeda mostly or completely irrelevant, or on the other side, that ISIS was an unsustainable flash in the pan, and al Qaeda remained the chief global terrorist threat.

In the heat of this debate, many glossed over the fundamental reality of terrorism. Asymmetrical warfare is defined by asymmetry. Any terrorist ideology that can attract five recruits and the contents of their checking accounts can make headlines for months. A terrorist group with twenty willing recruits and half a million dollars can make headlines for years. Although ISIS was dominating the headlines and attracting more recruits, al Qaeda was still quite capable of carrying out terrorist attacks.

Extremist and terrorist groups do fade, but it can take an extraordinarily long time for them to fade completely away. Consider the Ku Klux Klan, which was supplanted in the 1980s and 1990s by the more violent and extreme racist neo-Nazi movement. The Klan did not cease to exist, nor did it cease to carry out plots and violence. But the center of gravity for the white supremacist movement shifted away from the KKK, and it has not returned.

At this stage, either al Qaeda or ISIS could entirely collapse or be subsumed into the other as a result of the conflict, but neither of those outcomes is necessarily likely. The risk of total collapse is likely greater for ISIS, which is younger and less risk-averse than al Qaeda, but at this stage, there is a good chance both will continue in some form.

The battle is not simply between the organizations but between the visions they represent for the future of the jihadist movement.

Al Qaeda represents the intellectual side of the jihadist movement. While its ideology runs counter to hundreds of years of Islamic scholarship, it is nevertheless carefully constructed and has been articulated over the years in considerable detail.

Al Qaeda's vision for the restoration of the Islamic caliphate is framed squarely in the long term. Its most frequently cited theme is a classic extremist trope—the defense of one's own identity group against aggression. Its most charismatic leaders are dead. Those who remain are prone to deliver long hectoring speeches while sitting barely animate in a chair.

The net result of all these elements is most visible in recruiting. Despite its distorted worldview and its willingness to kill civilians, al Qaeda's recruitment message is ultimately intended to appear "reasonable" and to resonate with a wide audience of thinking people.

Al Qaeda and other old-school jihadists often exploited tragic and evocative situations to attract fighters. In Bosnia, for instance, mainstream Western media paid close attention to the unfolding genocide, with a steady drumbeat asking, "Why aren't we doing more?" Al Qaeda asked the same question. The decision to go to Bosnia and try to help did not seem especially radical in a mainstream context. But when volunteers arrived in the country, they were exposed to and allied with jihadists with a much more extreme agenda. For many, violent radicalization was not the reason for fighting in Bosnia, it was the outcome.

In Syria, the same dynamic unfolded, at first. Analyst Aaron Weisburd noted in November 2013 that the desire to participate in the Syrian conflict was not especially "extreme" for either Shi'a or Sunni foreign recruits. The statement was striking coming from someone known for his hard-nosed and unyielding pursuit of violent extremists online. Radicalization, he wrote, would depend on where the fighter landed, and with whom he surrounded himself.[50]

Al Qaeda's broad foreign fighter model—the 2013 model—was to attract people to relatable causes, then radicalize them later. This approach is more likely to result in foreign fighters who are relatively discriminating and possess some manner of moral compass; people who are more likely to set limits on their actions.

These are the foreign fighters studied by Hegghammer (Chapter 4)—those who were certainly more likely than the average person to engage in terrorism, but still not all that likely to engage in terrorism.

Al Qaeda's focus on that wider and more legitimate audience also worked against its efforts to attract individual jihadists, the so-called lone wolves. Over the last decade, al Qaeda–inspired lone wolves have frequently focused on military or government targets, although not without exception.

Many nonnetworked terrorists who were inspired by al Qaeda openly discussed their discomfort targeting civilians, even though al Qaeda was famous for the tactic and frequently encouraged it. These self-radicalized recruits experienced cognitive dissonance and made a choice they believed was morally defensible, even if it meant the target would be more difficult to strike.[51]

The same trend can be seen at the organizational level. While al Qaeda in the Arabian Peninsula has been the most active affiliate in pursuing traditional civilian-focused terrorism against Western targets, it has devoted remarkably few resources to this goal, spending only a handful of men and a tiny fraction of its war chest. It has to date focused the vast majority of its resources on fighting the Yemeni government and Shi'a movements.[52]

Of course, al Qaeda has seen more than its share of bottom feeders over the years. Terrorist groups naturally attract a certain number of thugs and violence junkies. But there is now a more natural home for members of that demographic—the Islamic State.

ISIS too has an articulated ideology with texts and an underlying high-level analysis.[53] Its so-called caliph holds a doctorate in Islamic studies—considerably more religious education than Osama bin Laden. When it is expedient, ISIS indulges in religious argument, for example, to justify its capture and sale of sexual slaves.

But its messaging betrays a different kind of sophistication. Where al Qaeda framed its pitch to potential recruits in more relatable terms as "doing the right thing," ISIS seeks to stimulate more than to convince. Its propaganda and recruiting materials are overwhelmingly visceral, from scenes of graphic violence to pastoral visions of a utopian society that seems to thrive, somehow, in the midst of a war zone.

Its calls to religious authority turned heavily toward the apocalyptic. For instance, an article in *Dabiq* that justified the enslavement of Iraq's Yazidi minority by ISIS cites a prophecy saying that slavery will return before the end times begin. Such themes are surely not unique in the modern jihadist movement, but they are now being deployed loudly and effectively (see Chapters 5 and 7).

As discussed in previous chapters, ISIS also distinguishes itself with a projection of strength and an appeal to populism—the gates are open for anyone who wants to join. All of these elements have coalesced into a unique offering in the world of extremism.

Identity-based extremism is frequently concerned with themes of purification, and the message of ISIS was extremism itself, purified. No more rationalizations about self-defense; instead, talk of revenge. No more subtle and embedded assumptions of weakness. Instead, aggression, shocking violence and strength. No more talk of a generational war to restore the caliphate. It was here, now.

After the Arab Spring, the Muslim Brotherhood had taken power in Egypt and was almost immediately confronted by political failure. Mark Lynch, director of the Institute for Middle East Studies at George Washington University, wrote that the Brotherhood "was profoundly shaped at every level—organization, ideology, identity, strategy—by its clear understanding that taking power was not an option. Removing that constraint proved more radically destabilizing than might have been rationally expected."[54]

Al Qaeda's organization, ideology, identity, strategy, and messaging were also predicated on the expectation that it would not take power. It stood for an idealized future that its leaders did not expect to see realized in their lifetimes.

ISIS rejected this fundamentally defeatist model and saw an opportunity to implement the future now. The result was profoundly destabilizing to its progenitor. What message could al Qaeda craft to compete with ISIS's continual declarations of victory? Zawahiri's months of silence spoke volumes.

ISIS's model had a potent attraction, and foreign fighters flocked in record numbers to join the movement. But its gravity also drew debris into its orbit. In the West, individual jihadists—the lone wolves—began to act out. But its messaging also resonated with people at risk of committing violence, whether or not those people were truly engaged with its goals and ideology.

Some resembled spree killers more than terrorists, such as Alton Nolen, a Muslim convert in Oklahoma who beheaded one coworker and stabbed a second at the food store from which he had recently been fired. Nolen's social media accounts pointed to a confusing mix of sexual repression and radical Islam. The attack came soon after a spree of ISIS beheading videos; the connection to Nolen's attack was unclear but fueled intense speculation both in the media and among jihadis.[55]

In November 2014, a man walked into a California mall and asked to have a hat embroidered with "We Love ISIS." Store employees alerted police, who found assault rifles and thousands of rounds of ammunition at his home. He was detained on a psychiatric hold after telling police he was a veteran suffering from post-traumatic stress disorder.[56]

Others showed signs of being more deeply engaged with the ideas of ISIS, such as Zale Thompson, who attacked New York City police with an ax after spending months reading jihadist content online.[57] In the province of Quebec, Martin Couture-Roleau drove his car into two soldiers before being killed by police. His social media accounts showed close associations with French-speaking ISIS supporters.[58]

While the spike in violence by individual actors was cause for concern, ISIS's predilection for violence had also irrevocably changed the nature of the Syrian civil war, shifting the calculus of risk from foreign fighters.

In November 2013, the impulse to travel to Syria and get involved in the conflict was not necessarily extreme.

By November 2014, the landscape had changed radically. Jihadist groups were fighting each other and the moderate Syrian rebels. After being targeted by U.S. air strikes, Jabhat al Nusra went on the offensive against U.S.-backed rebel factions, driving them out of key strongholds.[59]

In the portions of Iraq and Syria where ISIS reigned, a charnel-house atmosphere mixed bizarrely with antiseptic images of nation building, weighted almost equally. Who would be attracted to this disturbing contradiction?

ISIS's media push has moved the radicalization window far afield, eschewing the al Qaeda model of attracting fighters first and radicalizing them later. With its heady media mix of graphic violence and utopian idylls, ISIS sought recruits and supporters who are further down the path toward ideological radicalization or more inclined by personal disposition toward violence.

Once these pre-radicalized fighters and their families arrive in Iraq and Syria, they are exposed to an environment seething with traumatic stress, sexual violence, slavery, genocide, and death and dismemberment as public spectacles.

Among returning foreign fighters of previous generations, perhaps one in nine would eventually take up terrorism on returning to their homelands.[60] The fighters of ISIS are a new and untested breed. If they and their families someday attempt to return to their home countries, they will be unimaginably different from their predecessors.

ISIS didn't invent ultraviolent jihad. There have been many examples in the past, but they have led to consequences. In the horrific 1997 Luxor massacre in Egypt, sixty-two tourists (including women and children) were literally cut to pieces by dissident members of the Egyptian Islamic Group. The backlash led the group to moderate its overall approach.

The Abu Sayyaf Group has long beheaded hostages, sometimes on video, but its brutality and indiscriminate targeting have increasingly led to the perception that it is a criminal enterprise with expedient jihadist trappings.

But ISIS has crafted a novel formula for mixing brutal violence with the illusion of stability and dignity, and it has moved the bar for recruits.

Its combination of successful ground strategy, aggressive messaging, and an appeal to strength over weakness has proven uniquely powerful and energized at least tens of thousands of ardent supporters.

The challenge that lies ahead for the group is whether it can sustain all three elements over time and whether its extraordinary capacity for violence will eventually alienate even its core supporters.

And if it survives the first two challenges, it will be faced with a third—whether its deliberate cultivation of ultraviolence as a core element of its society will lead it ever further into darkness, into a pit of horror that cannot be escaped.

CHAPTER NINE

ISIS'S PSYCHOLOGICAL WARFARE

Terrorism is psychological warfare. Its most immediate goals are to bolster the morale of its supporters and demoralize and frighten its victims and their sympathizers. For the audience, the radius of fear dwarfs that of injury and death. Terrorists also aim to make us overreact in fear. While they don't always get what they want, terrorists often succeed at these two vital goals: spreading fear and provoking reactive policies.

Terror can make us strike back at the wrong enemy, for the wrong reasons, or both (as was the case for the 2003 invasion of Iraq). We want to wage war, not just on terrorism, but also on terror, to banish the feeling of being unjustly attacked or unable to protect the blameless. We want to wage war on evil. Sometimes the effect of our reaction is precisely that which we aimed to thwart—more terrorists and more attacks, spread more broadly around the world. While some politicians wanted to see Iraq during the allied invasion as a roach motel, we see it more like a hornet's nest—with allied bombs and bullets spreading the hornets ever further, throughout the region and beyond.

People often ask, how afraid should we be? Our answer is that it depends on who you are, where you live, and your role in society. If you are a national leader, ISIS should scare you a lot. This applies, firstly, to the leaders of Iraq and Syria as well as to the leaders of nearby countries. ISIS is already spreading ethnic and/or sectarian conflict into the Arabian Gulf as well as in Algeria, Lebanon, Jordan, Libya, Turkey, and beyond. Unrest in Yemen will likely make it vulnerable to exploitation by ISIS, especially since the organization already enjoys wide support inside the ranks of the local al Qaeda branch.

As we have seen, an estimated 17,000 foreign fighters have traveled to Syria and Iraq to join jihadi groups.[1] Jihadist organizations in the Gulf, North Africa, the Caucasus, and Southeast Asia which once looked to al Qaeda for leadership have officially declared their allegiance to ISIS. Individual supporters of ISIS are spread around the world, including the United States, Canada, Europe, Australia, India, Afghanistan, and Pakistan.[2]

ISIS established new *wilayat* (provinces) in Saudi Arabia, Yemen, Egypt's Sinai Peninsula, Libya, and Algeria, noting that "while the eyes of the world were all blinded and spellbound by the sorcerous media 'covering' the battle for [Kobane], the eyes of the Islamic State were scanning East and West, preparing for the expansion that—by Allah's permission—would put an end to the Jewish State, [the Saudi monarchy], and the rest of the apostate [tyrants], the allies of the cross."[3] And ISIS and its sympathizers will continue to strike out at the West.

There are three broad categories of likely perpetrators outside of Syria and Iraq (not only in the West, but around the world): recruits who return from the battlefields to bring their holy war back home; homegrown or self-recruited actors, inspired by ISIS and its ideology, perhaps over social networks, or commissioned by its money; and an ISIS-led attack, perpetrated by hardened terrorists emanating from its strongholds. So far, we have seen successful examples of the first two and aspirational examples of the third. Among them: A French national returned from Syria and killed four people at a Jewish museum in Brussels.[4] A young teen claimed to have been paid by ISIS to commit an attack in Vienna.[5] A lone actor in Ottawa, Canada, left a video recording of his ideological and political grievances before an attack on Parliament Hill, which left one soldier dead.[6]

Western returnees have been horrified by what they saw in the Islamic State and appear to have little interest in attacking their home countries, at least for now.[7] (The infighting among jihadi groups sparked by ISIS also alienated and ultimately drove out some fighters on all sides.) But even if only a tiny percentage take up violence in their native lands, it will have a large effect on how people perceive their safety.

People willingly engage in dangerous activities, imagining, often wrongly, that they are in control of their fate. But they expect their government to protect them from organized violence. Thus governments may feel compelled to act in response even to low-level attacks. While there is no evidence in open sources that ISIS could mount an attack of the scale and complexity seen on September 11, it currently commands many times more money and men than al Qaeda did in 2000, and a large-scale attack cannot be ruled out. ISIS has demonstrated clearly that it has both the inclination and the practical capacity for bold, aggressive action. But spectacular terrorist attacks are rare. They require coordination and communication among operatives, rendering them vulnerable to penetration and interception by law-enforcement personnel. As such, the risk is difficult to predict.

More reliably predictable are small-scale attacks in the West (such as those discussed in Chapter 4), which have noticeably increased in tempo since ISIS began to advocate for them. This is likely to continue and may very well get worse. We may see random beheadings, or shoot-outs at shopping malls, or subway attacks. The prospect can be frightening, especially for law enforcement, intelligence agencies and political leaders, all of whom share a mission to protect citizens from violence.

But the likelihood that any given individual will be caught in such an attack is vanishingly small. You are significantly more likely to die in a car accident, especially if you fail to wear a seat belt, than to be attacked by ISIS. Wear your seat belt.

IT HAS LONG been observed that the things that frighten us most are often quite different from those most likely to harm us. Consider the risks you're exposed to on an ordinary day. When you got up this morning, you exposed yourself to risks at nearly every stage of your progression from your bed to the office. Even lying in bed exposed you to hazards. One in four hundred people are injured doing nothing but lying in bed or sitting in a chair. The odds of dying by falling off a bed or other furniture are one in 4,283.[8] Most people are far more frightened by a terrorist attack than by a swimming pool or the drive to work, even though the latter are far more likely to kill us.

Perception of risk is highly correlated with levels of news coverage.[9] Inevitably and often inadvertently, the media tends to facilitate terrorists' theatrical performances. Terrorists know this. As noted previously, Ayman al-Zawahiri, the current leader of al Qaeda, once wrote that more than half the battle against the West and for "the hearts and minds of our Umma" is "taking place in the battlefield of the media."[10]

In their technical assessments, experts focus on probabilities and outcomes, but the *perception* of risk depends on other variables. There is little correlation between objective risk and perception of danger.[11] People tend to exaggerate the likelihood of "available" events that are easy to imagine or recall, when a visual or aural image seems taped to the brain.[12] Terrorism is often "available" in the sense that risk analysts' use the term, in large part because of media coverage.

Images matter. Most of us can't get the images of September 11 out of our heads: the crash of the planes into the steel and glass tower, followed by the sight of tiny figures leaping, as if in a dream of flight, to murderously concrete ground. And now ISIS is taking the imagery one step further by using social media to broadcast images of deliberately brutal beheadings into our homes and minds.

Surveys conducted by Daniel Kahneman and Amos Tversky showed that people evaluate choices with respect to the status quo. These findings have been repeatedly replicated: We overvalue losses relative to gains; we will pay more to avoid the loss of something we already have than we would to acquire it.

We also overestimate the likelihood of rare events, and underestimate the likelihood of more common ones.[13] We are at risk of overreacting to relatively minor incidents because they represent a loss relative to the status quo and because of our tendency not to distinguish adequately between ten deaths and ten thousand.[14]

Risk analysis involves attempting to generate statistical, rather than emotional judgments. What is missing from risk analysts' assessment is that terrorists' determination to harm us, their malice and forethought, coupled with our lack of agency, strongly influence our perception of risk. The chair that breaks beneath us has no agency and harbors no malice, therefore we assess the importance of that risk differently.

Kahneman, who won a Nobel Prize for two extraordinarily elegant and influential papers he wrote with Tversky in the 1970s, revisited his earlier work in 2011, applying it directly to our topic, this time describing his emotional reaction and his struggle to maintain a "rational" approach.

He writes:

I visited Israel several times during a period in which suicide bombings in buses were relatively common—though of course quite rare in absolute terms. There were altogether 23 bombings between December 2001 and September 2004, which had caused 236 fatalities. The number of daily bus riders in Israel was approximately 1.3 million at that time. For any traveler, the risks were tiny, but that was not how the public felt about it. People avoided buses as much as they could, and many travelers spent their time on the bus anxiously scanning their neighbors for packages or bulky clothes that might hide a bomb.

I did not have much occasion to travel by bus, as I was driving a rented car, but I was chagrined to discover that my behavior was also affected. I found that I did not like to stop next to a bus at a red light, and I drove away more quickly than usual when the light changed. I was ashamed of myself, because of course I knew better. I knew that the risk was truly negligible, and that any effect at all on my actions would assign an inordinately high "decision weight" to a minuscule probability. In fact, I was more likely to be injured in a driving accident than by stopping near a bus. But my avoidance of buses was not motivated by a rational concern for survival. What drove me was the experience of the moment: being next to a bus made me think of bombs, and these thoughts were unpleasant. I avoided buses because I wanted to think of something else.

My experience illustrates how terrorism works and why it is so effective: it induces an availability cascade. An extremely vivid image of death and damage, constantly reinforced by media attention and frequent conversation, becomes highly accessible, especially if it's associated with a specific situation such as the sight of a bus. The emotional arousal is associative, automatic, and uncontrolled, and it produces an impulse for protective action. We may "know" that the probability is low, but this knowledge does not eliminate the self-generated discomfort and the wish to avoid it.[15]

DREAD OF EVIL

Another factor, not yet studied by risk analysts such as Kahneman, is the impact of *evil* on our perception of dangers. Theologians, psychologists, and moral and political philosophers, among others, have various perspectives on what constitutes evil, its causes, and how to fight it. Philosophers traditionally identify three kinds of evil:

- **Moral evil:** Suffering caused by the deliberate imposition of pain on sentient beings.
- **Natural evil:** Suffering caused by natural processes such as disease or natural disaster.
- **Metaphysical evil:** Suffering caused by imperfections in the cosmos or by chance, such as a murderer going unpunished as a result of random imperfections in the court system.

The use of the word evil to describe such disparate phenomena is a remnant of pre-Enlightenment thinking, which viewed suffering (natural and metaphysical evil) as punishment for sin (moral evil). Drowning is more likely to be the result of "natural evil," than "moral evil," while terrorism is an example of the latter. It is moral evil that most frightens us.

Before September 11, philosopher Susan Nieman wrote, we had grown used to complex villains, whose evil was less immediately apparent than bin Laden's. We were in the habit of thinking about evil in Hannah Arendt's terms—ordinary people contributing, like cogs in a wheel, to evil outcomes.[16]

And now we are faced with an enemy that seems psychopathic in its theatrical acts of violence, but extraordinarily clever in knowing what will most horrify and disgust us. Horror, William Miller tells us, is "fear-imbued" disgust for which "no distancing or evasive strategies exist that are not themselves utterly contaminating." [17] The horror we feel at the image of beheadings is hard to escape.

We have grown unused to visible displays of cruelty. In his monumental study of the decline of violence, *The Better Angels of Our Nature*, Steven Pinker demonstrates that institutionalized cruelty began to decline in the West by the end of the eighteenth century. Beginning in the late seventeenth century, after the Thirty Years War and the Treaty of Westphalia, Europeans gradually stopped killing people on the basis of their holding the "wrong" supernatural or religious belief. [18] In the eighteenth century, the Humanitarian Revolution led to a growing respect for human lives. Pinker attributes this revolution to the growth of writing and literacy rates. When a person reads, she learns to empathize with individuals beyond her family or tribe or nation. It is a "technology for perspective-taking," Pinker argues. [19]

Empathy is the antidote to human cruelty. In *The Science of Evil*, Simon Baron-Cohen defines empathy as consisting of two stages. The first involves the ability to identify what someone else is thinking or feeling; the second involves responding to their thoughts and feelings with an appropriate emotion. [20]

Some people are born with less empathy than others. Absence of empathy can be a trait (as in biologically based psychopathy) or a state. [21] Empathy can be temporarily and sometimes necessarily shut off, as when a surgeon needs to cut into flesh to save a life.

But empathy can also become attenuated, such as when a person is too often severely frightened, too often victimized, or too often involved in perpetrating violence. Frequent exposure to savagery is one way to reduce a person's capacity to feel. When a person is trained, or trains himself, to feel less empathy and its absence becomes a trait, he becomes capable of dehumanizing others, putting him at risk of acts of extreme cruelty. In our view, ISIS is using frequent exposure to violence as a technology to erode empathy among its followers.

But empathy alone is not enough to explain the decline in violence, Pinker argues. The Enlightenment added another variable: the recognition that there is a universal human nature, and that like everything else, this too can be studied.[22] Reason allows us to move beyond our personal experiences, and to frame our ideas and experiences in universal terms. This leads us to recognize the ways our actions might harm others. The interchangeability of perspectives is the principle behind the Golden Rule and its equivalents, which have been discovered and rediscovered in so many moral traditions.[23] ISIS rejects this universal moral principle, in a way that repulses and disgusts not only "children of the Enlightenment" but most observers,[24] including jihadi ideologues.[25]

That said, Scott Atran and Jeremy Ginges urge us to remember that ISIS is appealing to sacred values, not reason. Although "logically and empirically inscrutable," such beliefs can strongly influence behaviors, they argue. They find that "seemingly contrary evidence seldom undermines religious belief, especially among groups welded by costly commitment in the face of outside threats [see discussion of millenarian movements in Chapter 10]. Belief in gods and miracles also intensifies when people are primed with awareness of death or when facing danger, as in wartime." They also find that "cross-national analyses show that a country's devotion to a world religion correlates positively with existential insecurity."[26]

But appealing to sacred values could (and often does, at least in modern times) lead to peace, not terrorism and war. Sacred texts are filled with contradictions. Terrorists across religions find justification in religious texts to do what they want to do, in ISIS's case, rape, pillage, and plunder.[27] While an appeal to sacred values may make conflicts more intractable, why is ISIS drawn to the parts of the text that would seem to justify slavery, rape, and murder?

During the early 1930s, Albert Einstein and Sigmund Freud exchanged letters that were later published (although they were subsequently suppressed by Hitler). Einstein asked:

> How is it possible for this small clique to bend the will of the majority, who stand to lose and suffer by a state of war, to the service of their ambitions?

He further wrote, in partial answer to his own questions,

> Because man has within him a lust for hatred and destruction. In normal times this passion exists in a latent state, it emerges only in unusual circumstances; but it is a comparatively easy task to call it into play and raise it to the power of a collective psychosis.

Freud responded:

> When a nation is summoned to engage in war, a whole gamut of human motives may respond to this appeal—high and low motives, some openly avowed, others slurred over. The lust for aggression and destruction is certainly included; the innumerable cruelties of history and man's daily life confirm its prevalence and strength. . . . [T]he ideal motive has often served as a camouflage for the dust of destruction; sometimes, as with the cruelties of the Inquisition, it seems that, while the ideal motives occupied the foreground of consciousness, they drew their strength from the destructive instinct submerged in the unconscious. Both interpretations are feasible.

What "unusual circumstances" are most likely to bring forward this "lust for aggression and destruction"? Possible answers include political disenfranchisement (Chapter 2) and collective trauma (discussed at the end of this chapter).[28]

As we have noted, ISIS's psychological warfare is directed at its potential victims. But it is also directed at those it aims to control. It is deliberately attempting to blunt its followers' empathy by forcing them to participate in or observe acts of brutality. Over time, this can lead to secondary psychopathy, or a desire to harm others, and contagion of violence. Beheadings are one such tool for blunting empathy.

BEHEADING

In a detailed assessment of capital punishment, Rudolph J. Rummel estimates that nineteen million people were executed for trivial offenses between the time of Jesus and the twentieth century.[29] Offenses that were once punished by execution included stealing bread and criticizing royal gardens.[30] Public executions were common and often took on a celebratory atmosphere until their prominence diminished in the mid-nineteenth century with a growing awareness of their inhumane nature.[31] Today, many countries consider capital punishment of any kind as a violation of human rights, although it is still practiced in the United States, as well as some non-Western countries.[32]

Until fairly recently, beheading was a common form of execution throughout the world, because it was once viewed as more humane than other forms of execution. But decapitation is not easy. To ensure that the victim quickly loses consciousness and does not feel multiple swipes at his neck, a skilled headsman is required. Beheading devices, precursors to the guillotine, were used for criminals of noble birth.[33] The guillotine, considered more humane but also more efficient than decapitation by hand, was used on an industrial scale to execute thousands of people during the French Revolution's Reign of Terror, and more than 16,000 people in Nazi Germany. The very word *terrorism* comes from Reign of Terror, and thus beheading is intimately associated with terrorism.[34] The guillotine continued to be used in France until capital punishment was banned in that country in 1977,[35] and in Germany until 1966.[36] China and Japan also employed beheading—as a dishonorable death—until the twentieth century.[37]

Saudi Arabia is the only country in the world that still practices public beheadings.[38] Beheadings are performed on Fridays, outside of mosques in major cities. The punishment derives from the Wahhabi interpretation of the Islamic religious laws of Shariah.[39] The crimes of rape, murder, apostasy, blasphemy, armed robbery, drug trafficking, witchcraft and sorcery, and repeated drug use are punishable by beheading.[40]

Muhammad Saad al Beshi, one of Saudi Arabia's lead executioners, explained that it takes a great deal of skill to sever a head with a single stroke of the sword, to minimize pain. It is not something that can be done with a knife or a dagger, he said, and requires training.[41] To use unskilled headsmen is sadistic.

ISIS's style of execution—hacking away at the victim's neck—is not designed to minimize pain, but rather to maximize it. In an interview with captured ISIS fighters, Israeli journalist Itai Anghel said one ISIS executioner intentionally used a dull knife because he wanted the beheading to last longer and cause more pain.[42]

CHILD SOLDIERS

ISIS actively recruits children[43] to send them to training camps and then to use them in combat, including suicide missions. ISIS has used children as human shields, suicide bombers, snipers, and blood donors.[44] The U.N. Secretary General's Special Representative for Children and Armed Conflict reports that "ISIL has tasked boys as young as 13 to carry weapons, guard strategic locations or arrest civilians."[45] Human Rights Watch (HRW) found that hundreds of "non-civilian" male children had died in the fighting.[46]

ISIS strictly controls the education of children in the territory it controls. According to a teacher from Raqqa, ISIS considers philosophy, science, history, art, and sports to be incompatible with Islam.[47]

"Those under fifteen go to Shariah camp to learn about their creed and religion," an ISIS press officer in Raqqa told *Vice News*. "Those over sixteen, they can attend the military camp. . . . Those over sixteen and were previously enrolled in the camps can participate in military operations."[48] But in ISIS propaganda videos (discussed in more detail in Chapter 5), even younger children are shown being trained in the use of firearms.

This is a hallmark of a "total organization," which sociologist Erving Goffman defined as one that "has more or less monopoly control of its members' everyday life."[49] Pol Pot experimented with creating a utopia in Kampuchea (the name used for Cambodia when the Khmer Rouge controlled it) in the 1970s, using methods not that different from those employed by ISIS. The idea was to create an entirely new society, uncontaminated by the values the Khmer Rouge aimed to stamp out. Children were seen as the least corrupted by bourgeois values and would be educated "according to the precepts of the revolution," which did not include traditional subjects.[50] The children were both victims and perpetrators of terror.

According to the research of Mia Bloom and John Horgan, ISIS follows a trend of training ever-younger operatives. By doing so they hope to ensure a new generation of fighters. Leadership decapitation is significantly less likely to be effective against organizations that prepare children to step into their fathers' shoes.[51]

Residents of Raqqa reported to *Syria Deeply* that children are taught how to behead another human being, and are given blond dolls on which to practice.[52] One child told HRW interviewers, "When ISIS came to my town . . . I liked what they are wearing, they were like one herd. They had a lot of weapons. So I spoke to them, and decided to go to their training camp in Kafr Hamra in Aleppo."[53] He attended the camp when he was sixteen years old, but the leader told him he preferred younger trainees. Pol Pot, too, preferred younger trainees.[54] Like other "total organizations" (discussed in Chapter 10), ISIS aims to create a new form of man. Young children are easier to mold into ISIS's vision of this new man. As psychiatrist Otto Kernberg explains, "Individuals born into a totalitarian system and educated by it from early childhood have very little choice to escape from total identification with that system. . . . Totalitarian educational systems permit a systematic indoctrination of children and youth into the dominant ideology," especially when they are young.[55]

Another child, Amr, told the HRW interviewers that he had participated in a "sleeper cell" for ISIS at age fifteen, to collect information on the Syrian government's operation in Idlib. When he started working for ISIS full time, he was given a Kalashnikov rifle, a military uniform, and a bulletproof vest. He and the others in his unit, including other children, were encouraged to volunteer as suicide-bombers, and several hundreds of fighters did so. Amr said that he didn't want to be a suicide-bomber, so he delayed signing up, hoping his name would come up last. He told HRW that he felt social pressure to "volunteer" to die.[56]

Some of the children come with their parents from abroad, to grow up in what their parents see as a pure Islamic state. They learn to say that they are citizens of the Islamic State rather than from their country of origin.[57] The poorer neighborhoods of Ankara, Turkey, are reportedly a source of child recruits. One such neighborhood, Hacibayram, has become a recruitment hub for ISIS.[58]

HRW discovered that child soldiers are paid the equivalent of $100 per month, around half as much as adult fighters.[59] In Raqqa, ISIS pays parents and bribes children to attend the camps.[60] But the recruits are not always volunteers. Children of ethnic minorities, particularly the Kurds and Yazidis, have been kidnapped and forced to join ISIS. According to Syrian Observatory for Human Rights, in one case, more than six hundred Kurdish students were kidnapped on their way home from taking exams in Aleppo. Their captors gave the boys an Islamic "education," encouraging the children to join the jihad, showing them videos of beheadings and suicide attacks.[61]

A doctor told the HRW interviewers that he had treated a wounded boy between the ages of ten and twelve. The boy's job was to whip prisoners.[62] Army Lieutenant General H. R. McMaster is Deputy Commanding General for the Future of U.S. Army Training and Doctrine Command. His job is to assess threats of the future for the U.S. Army. He describes ISIS as "engaging in child abuse on an industrial scale. They brutalize and systematically dehumanize the young populations. This is going to be a multigenerational problem."[63]

Using children under the age of eighteen as soldiers is a war crime.[64]

LONG-TERM EFFECTS OF VIOLENCE

What can we expect the long-term effect on these children to be?

At any one time, an estimated 300,000 children around the world are used as soldiers.[65] A "child soldier" is defined as a person under eighteen who is associated with an armed group or armed force. The definition of child soldier includes not only those who participate in combat, but also cooks, porters, spies, and sex slaves.[66]

Researchers have been studying the reintegration of child soldiers for a number of years now, principally in Sierra Leone and Uganda. Individuals exposed to a single traumatic event may develop PTSD. Those exposed to repeated or prolonged trauma, as is the case for child soldiers, are at risk of developing complex PTSD,[67] or developmental trauma disorder,[68] wounds that are more difficult to treat.

A team led by Fiona Klasen that studied three hundred former Ugandan child soldiers found that the most common experiences were exposure to shootings, beatings, starvation, and witnessing of killing. More than half the children had killed someone. Three-quarters of the children also had at least one experience of domestic or community violence.[69] Approximately one-third of them were diagnosed with PTSD. Two-thirds were suffering behavioral and emotional problems, mostly anxiety and depression, not violence.[70]

Another team, led by Theresa Betancourt, evaluated child soldiers from Sierra Leone. There, too, approximately one-third showed PTSD symptoms.[71] A follow-up study showed improvement in PTSD symptoms four years later, with half as many reporting PTSD symptoms. Psychological adjustment was greatly improved when children received family and community support; while post-conflict stigma increased symptoms.[72] Longitudinal data on aggressive behavior in former child soldiers is not yet available.[73]

Psychologists who study the impact of trauma and violence refer to "moral injury" as a risk factor for further violence, post-traumatic stress disorder (PTSD), and major depression.[74] The term "moral injury" refers to pain or damage to the conscience caused by an individual's witnessing, failing to prevent, or perpetrating acts that violate deeply held ethical norms.[75]

But what kinds of transgressions cause moral injury? There is a large amount of literature demonstrating that ethical norms are often culturally or situationally specific. However, some acts are considered wrong by nearly all cultures and religions. One of these is murder. Another is the act of deliberately targeting civilians in war, which is banned by all major religions.[76] Thus, those who perpetrate acts of terrorism, as we have defined it, are susceptible to moral injury, and to acquired callousness, which is sometimes called secondary psychopathy. Thus, inducing followers to commit atrocities is part of the technology for reducing empathy.

It is more difficult to treat the aftermath of war for those who experience moral injury. PTSD, in turn, is a risk factor for further violence, especially among men.[77] Perhaps surprisingly, among military personnel, combat exposure and life threat are not the most significant risk factors for PTSD. When military personnel know that they have hit their target and killed someone—as is the case for close combat (such as ISIS's beheadings), they are at greater risk to develop posttraumatic stress disorder.[78]

We usually think of moral injury and PTSD as a problem for legitimate military personnel, not terrorists, and one might ask why it should matter to anyone other than the terrorists themselves that their actions put them at risk of PTSD. The reason we should care, in our view, is that widespread commission of atrocities could lead to a form of societal PTSD—both for victims of atrocities and for perpetrators. One of the results of continuously witnessing morally injurious actions, or of perpetrating them, is the blunting of feeling, and loss of empathy. Ironically, some child soldiers may avoid adverse mental health outcomes by developing an appetite for aggression; those who learn to take pleasure from killing appear to be less susceptible to PTSD symptoms, according to work in Northern Uganda and Colombia by Roland Weierstall and colleagues.[79]

Is ISIS deliberately trying to create a society with an appetite for violent aggression? It is impossible to know ISIS's conscious intentions in this regard, but either way, the end result of its rule in Syria and Iraq will no doubt be a deeply traumatized generation and a host of new challenges from within.

SLAVERY

Slavery was abolished in most countries by the end of the nineteenth century, although it is still practiced illegally in some countries.[80]

In a report issued in early October, the Office of the UN High Commissioner for Human Rights and the UN Assistance Mission for Iraq (UNAMI) reported that hundreds of women and girls were abducted from Yazidi and Christian villages in August 2014. By the end of August, UN officials reported that some 2,500 civilians from these villages had been abducted and held in a prison. Teenage children, both males and females, were sexually assaulted, according to villagers who managed to speak with the UN officials. Groups of children were taken away. Women and children who refused to convert were sold as sex slaves or given to fighters. Married women who agreed to convert were told that Islamic law did not recognize their previous marriages. They were thus given to ISIS fighters to marry, as were the single women who agreed to convert.[81]

The Yazidis are a mostly Kurdish-speaking population whose syncretic religion pulls from both Islam and Christianity. ISIS views the Yazidis as devil worshippers.[82] The Yazidis and other religious-minority groups are not "people of the book," and are therefore required to convert or die, according to ISIS' interpretation of Shariah law.

Matthew Barber, a scholar of Yazidi history at the University of Chicago, estimates that as many as 7,000 women were taken captive in August 2014.[83] According to ISIS, the practice of forcing the Yazidis and other religious minorities into sexual slavery is a way to prevent the sin of premarital sex or adultery, as well as a sign that the Final Battle will soon occur. In the fourth issue of *Dabiq*, an article titled "The Revival of Slavery Before the Hour" explains that polytheist and pagan women can and should be enslaved. Indeed, their enslavement is one of the "signs of the hour as well as one of the causes of al Malhalah al Kubra," the Final Battle that will take place in Dabiq.[84] Further, they wrote, "a number of contemporary scholars have mentioned that the desertion of slavery had led to an increase in *fāhishah* (sexual sins such as adultery or fornication), because the *sharīa* alternative to marriage is not available, so a man who cannot afford marriage to a free woman finds himself surrounded by temptation towards sin. . . . May Allah bless this Islamic State with the revival of further aspects of the religion occurring at its hands."[85]

Below are some of ISIS's answers about its theological justifications for sexual slaves and how to keep them:

"There is no dispute among the scholars that it is permissible to capture unbelieving women [who are characterized by] original unbelief [*kufr asli*], such as the *kitabiyat* [women from among the

People of the Book, i.e., Jews and Christians] and polytheists. However, [the scholars] are disputed over [the issue of] capturing apostate women. The consensus leans towards forbidding it, though some people of knowledge think it permissible. We [ISIS] lean towards accepting the consensus. . . ."[86]

"It is permissible to have sexual intercourse with the female captive. Allah the almighty said: '[Successful are the believers] who guard their chastity, except from their wives or (the captives and slaves) that their right hands possess, for then they are free from blame [Koran 23:5–6].' . . ."[87]

"If she is a virgin, he [her master] can have intercourse with her immediately after taking possession of her. However, if she isn't, her uterus must be purified [first]. . . ."[88]

"It is permissible to buy, sell, or give as a gift female captives and slaves, for they are merely property, which can be disposed of as long as that doesn't cause [the Muslim ummah] any harm or damage."[89]

"It is permissible to have intercourse with the female slave who hasn't reached puberty if she is fit for intercourse; however if she is not fit for intercourse, then it is enough to enjoy her without intercourse."[90]

ACCORDING TO ESTEEMED political psychologist Vamik Volkan, collective historical trauma can predispose a society toward violence, identity politics (in the form of hatred of an out-group), and the rise of paranoid leadership and ideologies. The memories of this collective trauma become part of a shared myth, and what Volkan calls a "chosen trauma."[91] Volkan also sees a role for societal humiliation and cultural group psychology in the Middle East as contributors to paths of mass radicalization.[92]

Within Iraq and Syria, ISIS has a rich vein of collective historical traumas on which to draw in consolidating its position and certainly the outcomes Volkan describes (violence, paranoia, and identity politics) correspond closely to the reality of ISIS today. Such traumas can lead to the selection of values, sacred or otherwise, that justify "purification" of the world. Once such paranoid leaders arise, they can neutralize "individual moral constraints against personal perpetration of suffering, torturing and murder," psychiatrist Otto Kernberg explains. [93]

In addition to whatever benefits ISIS can extract from the traumas suffered by Iraqis and Syrians (some of which were instigated by ISIS and its predecessors), it is also inflicting an ongoing collective trauma of nearly apocalyptic proportions on those same populations. The longer that ISIS rules its domain, the deeper and more catastrophic those traumas will become.

While ISIS may not articulate its reasons in this manner, we believe it is deliberately engaged in a process of blunting empathy, attracting individuals already inclined toward violence, frightening victims into compliance, and projecting this activity out to the wider world. The long-term effects of its calculated brutality are likely to be severe, with higher rates of various forms of PTSD, increased rates of secondary psychopathy, and, sadly, still more violence.

CHAPTER TEN

THE COMING FINAL BATTLE?

Many Muslims anticipate that the end of days is here, or will be here soon. In a 2012 Pew poll, in most of the countries surveyed in the Middle East, North Africa, and South Asia, half or more Muslims believe that they will personally witness the appearance of the Mahdi. In Islamic eschatology, the messianic figure known as the Mahdi (the Guided One) will appear before the Day of Judgment. This expectation is most common in Afghanistan (83 percent), followed by Iraq (72), Tunisia (67), and Malaysia (62).[1]

Historically, narratives of the apocalypse have occupied a relatively marginal role in Sunni Islam, as distinct from Shi'ism. For Sunnis, the Mahdi is not yet here. For most Shi'ites, the Mahdi has already been born, but is now hidden, and when he reveals himself, justice will prevail.[2] The 1979 Iranian Revolution is considered by some Shi'ites to be an early sign of the Mahdi's appearance. For both Sunnis and Shi'ites, the Mahdi's role is, in part, to end the disunity of the Muslim community and to prepare for the second coming of Jesus Christ, who is understood to be a prophet in Islam.

Jean-Pierre Filiu, an expert on Islamic eschatology, observes that popular pamphlets and tracts "colored with superstition" have always circulated, but "until recently [their] impact on political and theological thinking was practically nil" among Sunnis.[3] A conscious effort to connect these narratives to current events can be traced, however, to at least the early 1980s, when Abdullah Azzam, an architect of modern jihad, argued that Muslims should join the jihad in Afghanistan, which he considered to be a sign that the end times were imminent.[4]

For years, al Qaeda invoked apocalyptic predictions in both its internal and external messaging, by using the name Khorasan, a region that includes part of Iran, Central Asia, and Afghanistan, and from which, it is prophesied, the Mahdi will emerge alongside an army bearing black flags. Internal al Qaeda documents and communiqués from Osama bin Laden often listed his location as Khorasan, and more recently, an al Qaeda cell in Syria adopted the name.[5] These claims were, however, mostly symbolic.

ISIS has begun to evoke the apocalyptic tradition much more explicitly, through actions as well as words. Thus ISIS has captured Dabiq, a town understood in some versions of the narrative to be a possible location for the final apocalyptic battle, and declared its intent to conquer Constantinople (modern-day Istanbul), in keeping with prophecy.[6]

For ISIS, and AQI before it, an important feature of the narrative is the expectation of sectarian war. Will McCants, a historian of early Islam, explains: "The early Islamic apocalyptic prophecies are intrinsically sectarian because they arose from similar sectarian conflicts in early Islam waged in Iraq and the Levant. As such, they resonate powerfully in today's sectarian civil wars."[7]

Hassan Abbas, an expert on jihadi movements, observes, "ISIS is trying deliberately to instigate a war between Sunnis and Shi'a, in the belief that a sectarian war would be a sign that the final times have arrived. In the eschatological literature, there is reference to crisis in Syria and massacre of Kurds—this is why Kobane is important. ISIS is exploiting these apocalyptic expectations to the fullest," he said. It is also why it was so important for ISIS to establish a caliphate, he explains. That too is a sign in their worldview.[8]

WHILE MUSLIM APOCALYPTIC thought is diverse and complex, most narratives contain some elements that would be easily recognized by Christians and Jews: at an undetermined time in the future the world will end, a messianic figure will return to the earth, and God will pass judgment on all people, justly relegating some to heaven and some to hell.

Considerable diversity exists, however, in writings about what will *precede* this final judgment. David Cook is a leading authority on Muslim eschatology. Because the Qur'an "is not an apocalyptic book," he explains, writers have been forced to turn to supplementary materials—including the words attributed to Muhammad, the Bible, global conspiracy theories about Judaism, stories of UFO abductions, and theories about the Bermuda Triangle—when discussing "the confused period" that comes before these final events.[9]

Cook explains that the events in this period are typically described as Lesser Signs of the Hour and Greater Signs of the Hour. The Lesser Signs are "moral, cultural, political, religious, and natural events designed to warn humanity that the end is near and to bring people into a state of repentance."[10] These signs tend to be so general that it is possible to find indicators of them in any modern society (for example, crime, natural disaster, etc.).

The Greater Signs, by contrast, offer a more detailed account of the final days, and while there is considerable variation among these stories, a few elements are consistent: Constantinople will be conquered by Muslims; the Antichrist will appear and travel to Jerusalem; a messianic figure (in some instances Jesus, and in some instances the Mahdi) will come to earth, kill the Antichrist, and convert the masses to Islam. The world's non-Muslim territories will be conquered.[11]

Many contemporary writers concerned with the apocalypse resent the suggestion that they are somehow affiliated with or participating in terrorist violence, Cook observes. But it would be naïve to deny the increasing role that this literature has played in contemporary jihad. Since September 11, he says, these writers have come to focus increasingly on Iraq—thus relegating Afghanistan and Israel to positions of lesser importance—and have implied that the American invasion was a sign of the coming apocalypse.[12]

This isn't to suggest that Israel has become insignificant in these narratives; much of this writing is virulently anti-Semitic and assumes a worldwide Jewish conspiracy against Muslims. In the new formulation, however, America is understood to be "the more or less willing instrument of Israel."[13]

ISIS is using apocalyptic expectation as a key part of its appeal. "If you think all these mujahideen came from across the world to fight Assad, you're mistaken. They are all here as promised by the Prophet. This is the war he promised—it is the Grand Battle," a Sunni Muslim told Reuters.

Another purported sign is the movement into Syria of the pro-Assad Hezbollah militia, whose flag is yellow. "As Imam Sadeq has stated, when the (forces) with yellow flags fight anti-Shi'ites in Damascus and Iranian forces join them, this is a prelude and a sign of the coming of his holiness," Rohollah Hosseinian, an Iranian cleric and member of Parliament, explained.[14]

The New York Times interviewed dozens of Tunisian youth, who are disproportionately represented among foreign fighters with ISIS, and found that messianic expectation was part of the appeal. "There are lots of signs that the end will be soon, according to the

Quran," a twenty-four year-old said.[15] Almost none of the interviewees believed that ISIS was involved in mass killings or beheadings. "All of this is manufactured in the West," a twenty-eight-year-old taxi driver said.[16] All of the youth viewed the existing Arab governments as autocratic and corrupt. They complained that there were no pure scholars of Islam whose views were untainted by politics or allegiance to some form of earthly power; but at the same time noted that the absence of uncorrupted Islamic scholars could be yet another sign of the coming apocalypse. Another sign for these youth was ISIS's declaration of the caliphate.[17]

ABU MUSAB AL SURI, one of the most important strategists of jihad, whom we have discussed throughout this book, incorporated apocalyptic narratives in his writings. His famous book, *A Call to a Global Islamic Resistance,* is not only the template for "individual jihad," but contains many pages of apocalyptic predictions. Filiu observes that the book, advertised as "Your Path to Jihad," was meant to attract a very wide readership of ordinary Muslims, not just committed Salafis.

"As against al-Qaida's adventurism and centralized elitism, which in [al Suri's] view renders it vulnerable at its very core, Abu Musab al-Suri proposes a distributed network model of decentralized resistance that reflects and responds to the aspirations of ordinary Muslims."[18] To that end, according to Filiu, al Suri included a discourse on the apocalypse, which, as he shows, has become increasingly popular, especially after 9/11 and the allied invasion of Iraq.[19]

"There is nothing in the least theoretical about this exercise in apocalyptic exegesis," Filiu observes in regard to al Suri's apocalyptic writings. "It is meant as a guide for action: 'I have no doubt that we have entered into the age of tribulations. The reality of this moment enlightens us to the significance of such events. . . . We will be alive then, when Allah's order comes. And we shall obey what Allah has commanded.' " [20]

Zarqawi set about fulfilling al Suri's prophecies, even going so far as to publish communiqués detailing the fulfillment of specific predictions.[21] He used apocalyptic imagery more than any other contemporary jihadist, Cook explains, much more so than bin Laden or Zawahiri.[22] Baghdadi, the successor to Zarqawi, is taking the fulfillment of apocalyptic portents even more seriously than his predecessor.

In the summer of 2014, ISIS fought to capture Dabiq, a Syrian town close to the Turkish border, and released the first issue of its English-language magazine, called *Dabiq,* in July. Its editors explained that they anticipate that Dabiq will play a historical role in the period leading to the Final Day, but first it was necessary to purify the town and to raise the black flags of the caliphate there.[23] Now that allied forces have entered the battle, the jihadists anticipate that the final battle in Dabiq is drawing near, McCants explains, and both Shi'a and Sunni groups hope to achieve the privilege of destroying the infidels.[24]

In ISIS's November 2014 video announcing the death of Abdul-Rahman (Peter) Kassig, a twenty-six-year-old former U.S. Army ranger, a British executioner claimed that Kassig had been killed at Dabiq. He also said, "Here we are burying the first American crusader in Dabiq, eagerly waiting for the remainder of your armies to arrive." [25]

Why is ISIS's obsession with the end of the world so important for us to understand? For one thing, violent apocalyptic groups tend to see themselves as participating in a cosmic war between good and evil, in which ordinary moral rules do not apply.[26] Most terrorist groups worry about offending their human audience with acts of violence that are too extreme. This was true even for bin Laden and al Qaeda Central, who withdrew their support for the Algerian terrorist group GIA and admonished AQI for their violence against Muslims, as we have seen.

But violent apocalyptic groups are not inhibited by the possibility of offending their political constituents because they see themselves as participating in the ultimate battle. Apocalyptic groups are the most likely terrorist groups to engage in acts of barbarism, and to attempt to use rudimentary weapons of mass destruction. Their actions are also significantly harder to predict than the actions of politically motivated groups. The logic of ISIS is heavily influenced by its understanding of prophecy. The military strategic value of Dabiq has little to do with ISIS's desire for a confrontation there.

While most new religious movements that emphasize apocalyptic prophecy are not violent, the deliberate inculcation of apocalyptic fears often precedes violence. Two types of violence can occur: violence perpetrated by members against the membership, such as mass suicide; and violence against the outside world.

The American apocalyptic group Heaven's Gate is an example of a suicidal cult.[27] In 1997, 39 members committed mass suicide in an effort to join a group of aliens on their spacecraft, which cult members believed was following the tail of the Hale-Bopp comet. In 1993, more than 80 followers of David Koresh, the leader of the Branch Davidian cult, died in a fire they set themselves after a fifty-one-day standoff with federal agents.[28] Koresh had predicted, based on his reading of the book of Revelation, that his followers would achieve salvation as a result of violence at his compound.[29] The breakaway Catholic organization known as the Movement for the Restoration of the Ten Commandments of God anticipated the end of the world in the year 2000. Soon after adherents arrived at church on the anticipated end of the world, the church burned down. Ugandan authorities suspected mass suicide, but when they found signs that some adherents had been poisoned or strangled, they concluded that the cause of death was murder.[30]

It is not easy to determine which apocalyptic groups will turn violent, or which violent groups will turn even more so. Michael Barkun, a leading scholar on violent apocalyptic groups, explains:

> Predictions of violence on the basis of beliefs alone are noto-riously unreliable. Inflammatory rhetoric can come from oth-erwise peaceable individuals. It does appear, however, that apocalypticists are more likely to engage in violence if they believe themselves to be trapped or under attack. Both condi-tions are as much a product of their own perception as of outside forces." [31]

The group responsible for the 1979 Meccan Rebellion, a small sect led by Juhayman al 'Utaybi, is an example of a Muslim apoca-lyptic cult. Its leader, Juhayman, was a member of the Bedouin tribe that had participated in the Ikhwan Revolt in the 1920s, the aim of which was to return Saudi Arabia to its pure, Wahhabist roots. In No-vember 1979, Juhayman's followers laid siege to the Grand Mosque compound in Mecca, a sacred site in Islam, which they held for two full weeks. Hundreds of people died during the siege. Most of the perpetrators were summarily executed or imprisoned, and the Saudi government kept the details regarding the perpetrators' motivations secret.

Some twenty-five years later, Thomas Hegghammer, a Norwe-gian scholar of Islam, was able to piece together what occurred. The cult was inspired by the teachings of Nasir al-Din al Albani, a quiet-ist Salafi who advocated a return to the pure Islam of the Quran and the Hadith. In his view, most of the Saudi Salafis, who considered themselves to be followers of the "pious predecessors," were actually influenced by later interpretations rather than the original texts. Al Albani eschewed politics and violence, and the cult began with the same quietist tendencies.

Two years before the siege, the leader of the cult had escaped into the desert, having received a tip that the police were closing in on his group. While in the desert, he had a dream that his companion, Muhammad al Qahtani, was the Mahdi. Some of the members left the cult in response to the leader's messianic obsessions. But the rest of the group was determined to consecrate Qahtani as the Mahdi in Mecca, in the belief that this would precipitate the end of the world and the series of related events described in Muslim apocalyptic writings. Three hundred rebels attacked the Grand Mosque, taking thousands of worshippers hostage. Most of the civilians trapped inside were allowed to leave, but an unknown number were retained as hostages.[32] Then they awaited the arrival of the hostile army from the north, as promised by the eschatological tradition. The timing of the attack was propitious—the end of the *hijri* century, "the last pilgrimage of the 14th century according to the Islamic calendar."[33] ISIS reportedly circulates Juhayman's dissident writings.[34]

But the Meccan Rebellion is instructive in another way, which seems to have gone unnoticed by scholars. On the third day of the siege, al Qahtani, the supposed Mahdi, was killed. Juhayman solved this problem by ordering his followers not to acknowledge the death of the purported Mahdi. Years afterward, Hegghammer explains, some followers continued to believe that the Mahdi was still alive.[35] In other words, despite the failure of their leader's prophecy, at least some of Juhayman's followers refused to believe the truth of what had happened to the supposed Mahdi, and vowed to continue with their fight. This may prove instructive as it's conceivable that we could see ISIS follow this model if and when their own prophecies fail.

In a study that is widely seen as among the most important contributions to social psychology, a team of observers joined a prophetic, apocalyptic cult to determine what would happen to the group if the predicted events failed to materialize. Marian Keech (a pseudonym for Dorothy Martin), the leader of the cult, predicted the destruction of much of the United States in a great flood, scheduled for December 21, 1955. She told her followers that they would be rescued from the floodwaters by a team of outer-space men in flying saucers with whom she was able to communicate, she said, through telepathy. When the apocalyptic flood did not materialize, instead of walking away from the cult and its leader, most members continued as loyal followers, and commenced efforts to recruit new followers.

Out of this observation, the researchers, Leon Festinger, Henry Riecken, and Stanley Schachter, developed the theory of cognitive dissonance, which states that when individuals are confronted with empirical evidence that would seem to prove their beliefs wrong, instead of rejecting their beliefs, they will often hew to them more strongly still, rationalizing away the disconfirming evidence. All of us have experiences with cognitive dissonance in our ordinary lives: When we hear or see something we don't want to believe because it threatens our view of ourselves or our world, rather than changing our views, we may be tempted to persuade ourselves that there has been a mistake—the disconfirming evidence is wrong, we need new glasses, we misheard. When this happens in cults, members may try to recruit others to join them in their views.[36] Since then, a number of similar cults have been studied, many but not all of which followed this pattern. The vast majority survived the failed prophecy, but some employed other stratagems to cope with cognitive dissonance, such as "spiritualizing" the prophecy by claiming that life did not end, but changed significantly, on the day the world as we know it was predicted to end.[37, 38]

AMONG PROTESTANT APOCALYPTIC cults, there is an important distinction between pre-tribulation and post-tribulation fundamentalists. Pre-tribulation believers expect that Jesus will save them from experiencing the apocalypse through a divine rapture, the simultaneous ascension to heaven of all good Christians.[39] Post-tribulation believers expect to be present during the apocalypse. Christian militants who subscribe to post-tribulation beliefs consider it their duty to attack the forces of the Antichrist, who will become leader of the world during the end times.

William McCants explains that there is no analogous post-tribulation eschatology in Islam. "The Islamic Day of Judgment is preceded by a series of 'signs,' some of which occurred in Muhammad's own life time. The signs are mentioned in words attributed to Muhammad and usually have the formula, 'The Hour won't come until . . .' As you get closer to the Day, the signs become more intense. ISIS can't hasten the Day with violence but it can claim to fulfill some of the major signs heralding its approach, which might be tantamount to the same thing." [40]

Many new religious movements employ a set of practices for enhancing commitment. These include sharing property and/or signing it over to the group upon admission; limiting interactions with the outside world; employing special terms for the outside world; ignoring outside news sources; speaking a special jargon; unusual sexual practices such as requiring free love, polygamy, or celibacy; communal ownership of property; uncompensated labor and communal work efforts; daily meetings; mortification procedures such as confession, mutual surveillance, and denunciation; institutionalization of awe for the group and its leaders through the attribution of magical powers; the legitimization of group demands through appeals to ultimate values (such as religion); and the use of special forms of address.[41] Most terrorist groups employ at least some of these mechanisms. Violent cults develop a story about imminent danger to an "in-group," foster group identity, dehumanize the group's purported enemies, and encourage the creation of a "killer self" capable of murdering large numbers of innocent people. As we have seen, ISIS members engage in a number of these practices. Many Western recruits burn their passports as a rite of passage. ISIS flaunts its sexual enslavement of "polytheists" as a sign of its strict conformance with Shariah, and of the coming end times. The strict dress code is enforced in part by public shaming of women who don't comply.

Like other apocalyptic groups in history, ISIS's stated goal is to purify the world and create a new era, in which a more perfect version of Islam is accepted worldwide. This is a typical millenarian project, which always involves transforming the world into something more pure, either politically (as with the communists' "New Man") or religiously. Dr. Robert J. Lifton is a psychiatrist who has studied "totalistic"[42] groups since the 1950s, and he continues to write about them. "Increasingly widespread among ordinary people is the feeling of things going so wrong that only extreme measures can restore virtues and righteousness to society."[43] None of us is entirely free of such inner struggles; there is much that is confusing about contemporary life, in which many people are no longer tethered to traditional societies. But apocalyptic groups act on these feelings, "destroying a world in order to save it," in Lifton's words.[44] Lifton was referring to another violent millenarian cult, Aum Shinrikyo, which in the 1990s had attempted to acquire nuclear weapons and had succeeded in poisoning some five thousand people on the Tokyo subway, twelve of whom died.[45] But his words apply as well to ISIS. "Having studied some of the most destructive events of this era, I found much of what Aum did familiar, echoing the totalistic belief systems and end-of the-world aspirations I had encountered in other versions of the fundamentalist self. I came to see these, in turn, as uneasy reactions to the openness and potential confusions of the 'protean' self that history has bequeathed us."[46] ISIS is similarly apocalyptic in its views, and similarly unpredictable.

As we have seen, ISIS emerged out of an especially barbaric strain of al Qaeda, which was initiated by Abu Musab al Zarqawi rather than Osama bin Laden. One of the reasons for both Zarqawi's and ISIS's anti-Shi'ite savagery is their apparent belief in end-times prophecies. It is impossible to know whether Baghdadi and other ISIS leaders truly believe that the end times are near, or are using these prophecies instrumentally and cynically to attract a broader array of recruits. Either way, appealing to apocalyptic expectation is an important part of ISIS's modus operandi. And goading the West into a final battle in Syria is a critical component of the scenario.

CHAPTER ELEVEN

THE STATE OF TERROR

ISIS traces its lineage back to the founding of al Qaeda in 1988, but the heirs to Abu Musab al Zarqawi have wrought a creation that feels both old and new. It is a millenarian group whose goal is to "return Islam to an imaginary ideal of original purity,"[1] while creating a worldwide caliphate. Like all fundamentalist movements, it is an inherently *modern* movement. While they see themselves as turning back time to practice a truer, purer version of their religion, ISIS is reinterpreting its religion in an "innovative and radical way," to use Karen Armstrong's description of fundamentalism,[2] and exploiting every opening it can find. ISIS aims to cleanse the world of all who disagree with its ideology.

But ideology is not all of its appeal. "Some are flocking to ISIS not because of its ideology, but also because it represents to them a rallying force against establishments that have failed them, or against the west," Marwan Muasher explains.[3]

There have been many millenarian groups like ISIS throughout history, although ISIS trumps most for wealth and violence in the world today. While its military has had successes in Iraq and Syria, it is quite small compared to the world's real powers. No nation in the world has recognized it as a state.

ISIS flaunts its cruelty, and that literally shameless practice is perhaps its most important innovation. Its public display of barbarism lends a sense of urgency to the challenge it presents and allows it to consume a disproportionate amount of the world's attention.

President Obama has laid out a mission for an international coalition to "degrade and ultimately destroy" ISIS. "We can't erase every trace of evil from the world," Obama said, emphasizing that the effort would "not involve American combat troops fighting on foreign soil."[4]

The coalition's policy, for now, is limited to air strikes paired with a train-and-equip mission for Iraqi forces and the increasingly ephemeral "moderate Syrian rebels." In our view, the mission described by the president cannot be accomplished with the limitations he has set out. Less than a week after President Obama spoke, General Martin Dempsey, chairman of the Joint Chiefs of Staff, hinted that he might feel the need to recommend ground forces.[5]

Even ground forces would likely not be enough to completely destroy ISIS. Absent a military invasion that would somehow—improbably, magically—transform both Iraq and Syria into truly viable, pluralistic states in which Sunnis and Shi'a both feel secure, ISIS would likely remain, at least as a terrorist group, for many years to come.

Beyond the necessity to oversee *political* change in both Iraq and Syria, a tall order indeed, the international impact of ISIS must also be considered, as it inspires oaths of loyalty and acts of violence in nearly every corner of the globe. As with its military might, ISIS's potential to wreak terrorism has been limited until now, although the alignment of regional terror groups such as Jund al Khalifah in Algeria and Ansar Bayt al Maqdis in Egypt raise serious concerns going forward.

The broader problem is that jihadism has become a millenarian movement[6] with mass appeal, in some ways similar to the revolutionary movements of the 1960s and '70s, although its goals and the values it represents are far different.

Today's radicals are expressing their dissatisfaction with the status quo by making war, not love. They are seduced by Thanatos rather than Eros. They "love death as much as you [in the West] love life," in Osama bin Laden's famous and often-paraphrased words. In this dark new world, children are seen to reenact beheadings with their toys, seduced by a familiar drama of the good guys killing the bad guys in order to save the world. Twitter users adopt the black flag by the tens of thousands. And people who barely know anything about Islam or Iraq are inspired to emulate ISIS's brutal beheadings.

ISIS has established itself as a new paradigm, one that is more brutal, more sectarian, and more apocalyptic in its thinking than the groups that preceded it. ISIS is the crack cocaine of violent extremism, all of the elements that make it so alluring and addictive purified into a crystallized form.

ISIS's goals are impossible, ludicrous, but that does not mean it can be easily destroyed. Our policies must look to the possible, which means containing and hopefully eliminating its military threat and choking off its export of ideas.

Circumstances will almost certainly have changed in between the writing of these words and their publication.

But certainly the history of ISIS and al Qaeda before it show that overwhelming military force is not a solution to hybrid organizations that straddle the line between terrorism and insurgency. Our hammer strikes on al Qaeda spread its splinters around the world. Whatever approach we take in Iraq and Syria must be focused on containment and constriction, rather than simply smashing ISIS into ever more virulent bits.

We can speak more authoritatively about efforts to counter ISIS as an extremist group and ideology. Here we have specific suggestions that are likely to remain relevant despite whatever happens on the military front.

ISIS's military successes are formidable. But the international community has dealt with far worse. ISIS does not represent an existential threat to any Western country. Perhaps the most important way to counter ISIS's efforts to terrify us is to govern our reactions, making sure our policies and political responses are proportionate to the threat ISIS represents.

We asked Steven Pinker, who has written extensively on violence in society, to compare the atrocities of ISIS to those of the past. He wrote in an email:

> In terms of the sheer number of victims, they are nowhere near the Nazis (six million Jews alone, to say nothing of the exterminated gypsies, homosexuals, Poles and other Slavs, plus the tens of millions of deaths caused by their invasions and bombings). Mao and Stalin have also been credited with tens of millions of deaths. In the 20th century alone, we also have Pol Pot, Imperial Japan, the Turks in Armenia, the Pakistanis in Bangladesh, and the Indonesians during the Year of Living Dangerously.[7]

None of this minimizes the impact of ISIS. They kill their enemies and minorities who offend them with deliberate and brazen cruelty. They sell women and children into slavery and subject them to abominable sexual abuse. They kill anyone who opposes them and anyone who refuses to accept their bizarre system of belief, which has been rejected as morally wrong by jihadist clerics we once considered the worst of the worst.

Neither its leaders nor its bloodthirsty adherents see the slightest problem in publicizing and celebrating their atrocities. Some of this is calculated, at least at the leadership level, to frighten potential victims and to attract new psychopathic recruits. But this violence is now pervasively ingrained in the society ISIS is trying to build, with disturbing ramifications for the innocent children growing up in its charnel-house "caliphate."

Our horror and revulsion are appropriate responses to this regime of atrocities, and we can and should do what is in our power to help ISIS's victims, but we should measure our actions to avoid spreading its ideology and influence.

ISIS evokes disproportionate dread. As we have shown, the "availability" of ISIS's crimes, together with its evil, makes us prone to exaggerate the risk, and prone to react rather than strategize.

Political leaders and policy makers are particularly susceptible to ad hoc policy making with little regard to competing interests, in large measure because ISIS is so good at manipulating our perceptions.[8] Decision makers are pressured by a bias toward action, the understandable desire to respond swiftly and visibly to threats. Our political system and security bureaucracies incentivize theatrical action over caution and consideration of unintended consequences and the long term.[9] "Action is consolatory," Joseph Conrad tells us in *Nostromo*. "It is the enemy of thought and the friend of flattering illusions."

Any effort to make the world a better place can have the perverse effect of creating new risks—just as an aspirin can aggravate a stomach ulcer.[10]

We need not look as far back as the 2003 invasion of Iraq for a lesson in perverse effects. The 2011 intervention in Libya provides a more recent example. There were profoundly compelling humanitarian reasons to support the popular rebellion against Moammar Gadhafi. But it is nearly impossible to argue that either Iraqis or Libyans are better off than they were before our interventions. These military actions, which seemed imperative at the time, introduced a new risk, and an explosion of jihadism has engulfed both countries. In both places, ISIS has staked its claim to territories and mounted fighting forces.

The only thing worse than a brutal dictator is no state at all.

The rise of ISIS is, to some extent, the unintended consequence of Western intervention in Iraq. Coalition forces removed a brutal dictator from power, but they also broke the Iraqi state. The West lacked the patience, the will, and the wisdom to build a new, inclusive one. What remained were ruins.

If there is a final nail in the coffin of a full-scale military intervention to defeat ISIS, it is the incongruity of targeting the jihadists while Bashar al Assad remains in power. Assad's regime has tortured thousands of political prisoners to death. He has bombed hospitals and schools. An average of 5,000 Syrian refugees are fleeing every day, totaling more than 3 million registered refugees, most of them in neighboring countries. Jordan is overwhelmed by the refugee burden, and it is clearly incumbent on other nations to shoulder more of the burden. An additional 6.5 million people are displaced inside Syria.[11]

Arguably, the Western-led intervention against ISIS has already aided Assad. With the rebels fully engaged in infighting, Assad's forces have hit the same targets bombed by the coalition.[12] U.S. strikes against Jabhat al Nusra and Ahrar al Sham have resulted in more infighting among rebel factions and further marginalization of the secular groups.[13] As Charles Lister of the Brookings Institution wrote in December 2014 after interviewing dozens of rebel faction leaders:

> For the Syrian opposition, the Assad regime and ISIS are two sides of the same coin, but with Assad being "the head of the snake" and ISIS merely "the tail." The U.S.-led coalition's failure to target the regime is therefore perceived as tantamount to a hostile act against the revolution. Moreover, while surprising to outsiders, the al-Qaeda affiliate Jabhat al-Nusra is still to this day perceived by many as an invaluable actor in the fight against Damascus and as such, the strikes on its positions are seen by many as evidence of U.S. interests being contrary to the

revolution. Although this perception may be subtly changing, with one Syrian Salafist commander admitting that "Nusra is going down the wrong path," the strike on a headquarters of Syrian group Ahrar al-Sham late on November 5—confirmed to me by multiple Syrian and international sources—consolidated this impression that U.S. interests have diverged from those of Syria's revolution.[14]

Even if Western voters could be convinced to support a full-scale invasion to remove Assad, what would happen in the ensuing vacuum? The cautionary tales of Iraq and Libya loom large. In the words of Lieutenant General Daniel P. Bolger (ret.), who served as a senior commander in Iraq:

> The surge in Iraq did not "win" anything. It bought time. It allowed us to kill some more bad guys and feel better about ourselves. But in the end, shackled to a corrupt, sectarian government in Baghdad and hobbled by our fellow Americans' unwillingness to commit to a fight lasting decades, the surge just forestalled today's stalemate. Like a handful of aspirin gobbled by a fevered patient, the surge cooled the symptoms. But the underlying disease didn't go away. The remnants of Al Qaeda in Iraq and the Sunni insurgents we battled for more than eight years simply re-emerged this year as the Islamic State, also known as ISIS. . . .
>
> We did not understand the enemy, a guerrilla network embedded in a quarrelsome, suspicious civilian population. We didn't understand our own forces, which are built for rapid, decisive conventional operations, not lingering, ill-defined counterinsurgencies. We're made for Desert Storm, not Vietnam. As a general, I got it wrong. . . .
>
> Today we are hearing some, including those in uniform, argue for a robust ground offensive against the Islamic State

in Iraq. Air attacks aren't enough, we're told. Our Kurdish and Iraqi Army allies are weak and incompetent. Only another surge can win the fight against this dire threat. Really? If insanity is defined as doing the same thing over and over and expecting different results, I think we're there.[15]

General Bolger argues that we would have needed to occupy Iraq for three decades to create a viable state, echoing similar arguments made at the time by both Jim Webb and then Secretary of State Powell.[16] The problem is that if we're not prepared for a thirty-year occupation, we cannot create a viable state in Syria, and even that level of commitment comes with no guarantee of success. And if there is anything we ought to have learned from our mistakes in both Iraq and Libya, a failed state is the worst of all possible outcomes.

On August 14, 2014, Haider al Abadi took over from Nouri al Maliki as prime minister of Iraq. He faces a daunting task in stemming the chaos and healing a society profoundly riven by ethnic and religious strife, a fire that rekindled under Maliki and has been stoked continually since by ISIS.

We wish him well, but we do not—and should not—necessarily expect that the post–World War II boundaries of the Middle East will remain intact. The devolution of powers to the regions, with a limited central government, may be, as Leslie Gelb has long argued, the only policy glue that will prevent the outright breakup of Iraq.[17] Gelb has proposed that Sunni, Kurdish, and Shi'ite regions each be responsible for their own domestic laws and internal security. To some extent, this is a fait accompli for the Kurds.

"The Middle East is clearly in one of those pivotal moments," said General David Petraeus in July. "We're in a period of history where the organizing principles, the lines on the map drawn by British and French diplomats early last century, are being erased." [18]

How can we stop this carnage, without inadvertently assisting ISIS, Assad, or both? If a military operation only serves to create more insurgents than it takes out, it is not a useful operation. If we cannot practically impose a political and military solution on the region, we can at least learn from our past mistakes.

Instead of smashing ISIS in the same way we approached al Qaeda, Clint Watts of the Foreign Policy Research Institute proposes, we should consider "letting them rot," in some ways the modern equivalent of a medieval siege.[19] The rot may already be setting in. Reports in December indicated that ISIS's capitals in Iraq and Syria, Mosul and Raqqa respectively, are suffering under dramatically deteriorated living conditions.[20]

Rather than trying to displace ISIS with an external force, we should consider efforts to cut off its ability to move fighters, propaganda, and money in and out of the regions it controls, weakening its ability to use brute force and extreme violence to keep the local population in check. It would also force ISIS to fail based on its own actions instead of being displaced by outsiders, which would do much over the long run to discredit future efforts at jihadist nation building. Such a strategy would have to be probed for its own pitfalls and weighed against the moral conundrums it presents, especially as it pertains to the human costs that ISIS could impose on the population in the areas it controls. Targeted military action may be able to inhibit ISIS's ability to carry out genocide with impunity, but it will not entirely remove that ability. Our military approach will unavoidably need to evolve along with the situation on the ground.

THE EXTREMIST MIND

Fundamentalists see religious texts as inerrant guides to life. But even for those who see scripture as the literal word of God, the people who read it and interpret it are human and fallible, a concept fundamentalists are often unable to conceptualize as it applies to themselves, although they happily apply it to others.

This is not particular to ISIS or to jihadists; it applies to many violent fundamentalists across a range of ideologies, whom we have spoken with and studied. Readers bring their prejudices and pain to religious texts.

Salafism, like all fundamentalisms, is a response to the pain of modernity. Karen Armstrong, a former nun, has studied fundamentalism across different religions. She observes:

> Fundamentalist movements in all faiths . . . reveal a deep disappointment and disenchantment with the modern experiment, which has not fulfilled all that it promised. They also express real fear. Every single fundamentalist movement that I have studied is convinced that the secular establishment is determined to wipe religion out.[21]

What seems to be most appealing about violent fundamentalist groups—whatever combination of reasons an individual may cite for joining—is the simplification of life and thought. Good and evil are brought out in stark relief. Life is transformed through action. Martyrdom—the supreme act of heroism and worship—provides the ultimate relief from life's dilemmas, especially for individuals who feel deeply alienated and confused, humiliated, or desperate.

Although ISIS, like many fundamentalist groups, claims to be practicing the religion in its purest, most original form, this represents a longing, not a reality.

Peter Suedfeld, a psychologist and researcher, has studied the role of complexity in conflict, including how it plays into extremist narratives. His work and that of others supports our own observation that violent extremist messaging and narratives are less complex than similar messages from nonviolent movements, stripping narratives down to their bare essentials with little qualification or elaboration. (His research compared al Qaeda and AQAP messaging to that of nonviolent Islamists.)[22]

Integrative complexity, defined by Suedfeld as being able to examine problems from different perspectives and make cognitive connections drawing on those different perspectives, is not the same thing as intelligence. Extremists are sometimes exceptionally intelligent. Rather, it applies to flexibility of thought and the ability to see things from someone else's point of view. Studies have found that integrative complexity and empathy are closely correlated, with empathy being the emotional equivalent of the cognitive process.[23] Research by Jose Liht and Sara Savage of the University of Cambridge suggests that it is possible to promote integrative complexity among people vulnerable to extremist radicalization.[24]

This suggests two possible avenues for countering the appeal of ISIS and groups like it. First, we can attempt to continually reinforce messages that flesh out the nuance and complexity of the situations and conditions that extremists use to recruit, undermining the incorrect thesis that the problems faced by communities vulnerable to radicalization are easily reduced to absolutes.

In practice, this means refusing to characterize our conflict with ISIS in stark, ideological terms, an uphill battle in the current media and political climate, which tends to incentivize simple explanations. It is further complicated when ISIS theatricalizes dreaded risks such as beheadings to evoke a stripped-down primal response. In many ways, *The Management of Savagery* outlines a specific psychological campaign designed to provoke enemies into the same simplistic thinking that dominates jihadist thought—al Naji refers to the process as "polarization," and that is why those who argue that ISIS's public displays of brutality will backfire are wrong (up to a point). The object of ISIS's extreme displays of violence is to polarize viewers into sharply divided camps of good and evil, not to rally the general public around its actions.

The second prescription follows from the first. Our policies must not lend credence and support to ISIS's simplistic and apocalyptic worldview. When ISIS began beheading Westerners on video in September 2014, it did so with the intention of prodding the United

States into an ever-deeper engagement in Iraq, consistent with the blueprint in *The Management of Savagery*. ISIS made its intentions even clearer with the November video announcing the execution of hostage Abdul-Rahman (Peter) Kassig.

"We bury the first crusader in Dabiq, eagerly awaiting the remainder of your armies to arrive," said "Jihadi John," the anonymous executioner, in the conclusion of that video.[25] It was a transparent ploy to goad the West into a military confrontation in Dabiq, in fulfillment of a key apocalyptic prophecy to which ISIS has alluded again and again. If we take the bait, we arm ISIS with evidence that the end of the world—the ultimate moment of simplification—is indeed at hand. Aggressive military action by Shi'a militias, whether Iraqi or Iranian, also contributes to the apocalyptic narrative and plays into ISIS's desire for a simple, Manichean divide between good and evil, actualizing its narrative of an all-consuming battle between true believers and apostates.

One arena where we can fight the battle for nuance, however, is on the messaging front, the beating heart of ISIS's campaign to reduce the world's complexities to fit its black-and-white narrative. ISIS has devoted unprecedented resources to its messaging, and the West has thus far failed to craft a cohesive and comprehensive response.

MESSAGE DISTRIBUTION

For the first decade of its life, al Qaeda was publicity-shy. For the first five years of its existence, barely a handful of people in the U.S. government even knew its name.

ISIS, in contrast, is a publicity whore. While it is extremely important to keep its propaganda and social media activities in the proper perspective—no one was ever killed by a tweet—it's clear that ISIS considers messaging one of the most important fronts in its war with the world, and it's also the primary method by which ISIS extends its influence outside its physical domain. Western efforts to counter ISIS must account for both the content and distribution of its message.

As the discussion of social media in Chapters 6 and 7 suggests, there is a robust debate about how to handle terrorist use of social media in general. The problem lies at the center of an uneasy intersection of constituencies—corporations, governments, citizens, and extralegal organizations.

All media is social, but mass social media is a relatively new development in society. Throughout the twentieth century, there was a sharp distinction in the use of communications technology—platforms for broadcasting to large audiences were mostly monopolized by governments and corporations, while peer-to-peer communications infrastructure such as the postal service or telephone lines came with relatively clear expectations about privacy. Platforms that fell between these poles—such as anti-Semitic ham radio broadcasts[26]—had only a limited reach.

Today, social media platforms straddle the line between broadcasting like a television station and communicating peer-to-peer as if by phone. In most countries, neither the laws nor the expectations of the people have fully assimilated the difference.

Users of social media often expect that the same privacy and freedom they enjoy in their living rooms will extend to conversations they broadcast publicly over social media. Governments, generally, deal with social media using laws designed for telephone carriers, which usually exempt corporations from responsibility for how customers use their platforms—as opposed to a television station or even a newspaper, both of which face certain legal liabilities for content they broadcast.

The complexities and future challenges of this intersection go well beyond extremism, but they are particularly acute in that arena, in large part thanks to the aggressive ways in which ISIS has exploited gray areas and cutting-edge techniques for distribution.

The most obvious way that this plays out in the ISIS context is suppression, namely the suspension of social media accounts that distribute extremist content. Debates about how to deal with extremists on social media suffer from a chronic framing problem. Advocates of free speech see it as a censorship issue, as do some social media companies.

But most Western definitions of free speech do not include the right to unrestricted use of broadcasting platforms. There was little controversy in 2006 when the U.S. government designated Hezbollah's Beirut-based Al-Manar television station as a terrorist entity.[27] If al Qaeda Central set up a newspaper office in Manhattan, few would step forward to argue it should be allowed to run its presses.

But when ISIS broadcasts unsolicited beheading videos to thousands from Syria using the infrastructure of a company based in San Francisco, some free speech advocates object to any effort to suppress that activity—whether led by government or by social media companies themselves.[28]

As noted in Chapters 5 and 7, the same objections are rarely voiced when it comes to other crimes, such as posting child pornography on YouTube or hiring contract killers on Craigslist. While it is certainly true that ISIS is engaging in a form of political speech, its content also exceeds the bounds of the contract every user agrees to when he or she signs up for the service. Each social media platform sets terms of service for its users. When a company denies a user access to its platform for violating those terms, it's not exactly censorship. Or is it? Everyone participating in new technologies is engaged in a process of exploring these questions and defining the debate.

With concessions to the complexity of all of these considerations, it seems to us uncontroversial that ISIS's social media activity should—at a minimum—be subject to the same restrictions as any other antisocial user, especially when it commits violations that would put a nonterrorist user in danger of suspension, such as deploying spambots or threatening violence. While we believe additional study is necessary to fully evaluate the impact of such suppression techniques, the early data is very encouraging and ISIS supporters online certainly believe that suspensions degrade their ability to accomplish their terroristic goals.

That said, it is not so easy to implement a policy of suppression. Social media platforms are run by multinational corporations, not by any individual government, and they must navigate a bewildering morass of laws and regional customs in determining both their legal responsibilities and their ethical stands.

The problem of devising a consistent response is also complicated by a lack of transparency from both governments and companies, with the United States and Twitter as highly visible offenders. It is clear from Twitter's transparency reports that some accounts are suspended (or allowed to remain online) due to secret government requests. But Twitter's steadfast refusal to discuss details of its suspension polices—a tactic likely indicating its desire to make suspension decisions on an ad hoc basis—is also an obstacle to transparency and to open airing of the issues involved.

Despite these complications, ISIS has chosen to fight much of its battle with the West on social media. Through a combination of public infrastructure and private companies, the West effectively owns this battlefield, and our failure to control ISIS's messaging is a direct result of our failure to understand and act on that fact. Never before has there been a war where one side controlled the operating environment. Our power over the Internet is the equivalent of being able to control the weather in a ground war—it is not a complete solution, but it should offer an overwhelming advantage if used correctly.

There is a legitimate intelligence interest in allowing extremists to use social media up to a point, and equally legitimate concerns about allowing them to openly radicalize new followers without interference. It is not difficult to see that some balance between these competing interests is desirable. The best outcome for policy makers is an environment that hinders extremists' efforts without forcing them to abandon social media entirely. The current environment on Twitter is arguably approaching that ideal, which allows Internet service providers to accommodate some of their also-legitimate concerns about censorship and free speech.

The hindrance model discourages casual engagement with extremism on social media by increasing the cost of participation and reducing the reach of radicalizers. This yields benefits both in the realm of countering violent extremism, by shrinking the pool of available recruits, and in intelligence work, by removing some of the noise that is created by people who are only lightly engaged with ISIS's ideology.

We recommend that a conference be dedicated to airing these issues publicly, with participants from both the public and private sector, with an eye toward establishing some consistent, reasonable practices and clearly defining areas that require more study or the resolution of more complicated questions.

HOW TO DEAL WITH ISIS'S MESSAGE CONTENT

Governments around the world have invested considerable funds under the heading of countering violent extremism (CVE), which can be loosely defined as the use of tools other than killing and incarceration to combat terrorist and extremist groups.

These initiatives take a wide variety of forms—too wide, as most practitioners would agree. After September 11, vast pools of money became available for CVE, which resulted in many people repurposing their pet projects under that heading.

On top of that, well-intentioned efforts at community building have been generously funded as CVE despite a near-total lack of evidence that they actually prevent violent extremism in any meaningful way—town halls and soccer leagues, as the joke in the practitioner community goes. Similar dynamics apply on the grand stage of world politics, where nation-building exercises such as foreign aid, jobs programs, education initiatives, and democratic reforms are taken on faith as ways to inoculate countries and regions against violent extremism. The fact that Germany and the United Kingdom each appear to have provided more foreign fighters to ISIS than Somalia should call some of those assumptions into question.

While there is arguably little downside in trying to do good works for communities and nations, there is a risk that promoting such projects as CVE will result in a future consensus that CVE as a general idea does not and cannot work, or worse, that it is simply a budgetary boondoggle for funding pet projects.

There are many challenges in demonstrating that "positive" CVE initiatives work, but we can see very clearly the tools that ISIS uses to radicalize potential recruits and recruit those who are already radicalized. Rather than spending our resources on uncertain and potentially wasteful wagers on nation building, the more obvious course is to thoroughly catalog what ISIS is doing to achieve its goals and disrupt both its distribution, as discussed above, and the integrity of its messaging content.

The State Department's Center for Strategic Counterterrorism Communications has worked to do this on Twitter by mocking and discrediting ISIS messaging and challenging ISIS supporters directly, both in Arabic and English. The initiative has received decidedly mixed reviews from many analysts.[29] We believe it is a step in the right direction, albeit one that can be refined and improved.

The ISIS propaganda machine is a calculated affair. It has five major goals, all of which involve an effort to simplify the complexity of the real world into a cartoonish battle between good and evil:

- To project an image of strength and victory.
- To excite those with violent tendencies by pairing extreme violence with a moral justification in the form of its alleged utopian society.

- To manipulate the perceptions of ordinary citizens in its enemies' lands to incite demand for military action, while at the same time planting doubt that such action can succeed.
- To place the blame for any conflict that does result on the aggression of Western governments and the incitement of "Zionists."
- To recast any military action against ISIS as an action against Muslims in general, specifically by highlighting civilian casualties.

Each of these goals is vulnerable to a messaging counteroffensive, but some Western messaging reinforces ISIS's goals—such as news stories repeatedly describing ISIS videos as "terrifying" or overstated descriptions of the threat the organization presents. Such statements are an effort to combat ISIS's message with a similarly (not equally) simplified narrative, and they ultimately serve to reinforce ISIS's goal of framing its place in the world as part of a cosmic battle between pure good and pure evil.

Therefore a first step in countering ISIS is to put it in perspective. We should not downplay its threat below a realistic level—that only sets up future hysteria by creating unrealistic expectations. But neither should we inflate it.

ISIS relies on its projection of strength and the illusion of utopian domestic tranquility. Even under the coalition assault, it has labored to maintain its aura of invincibility and defiance. Changing conditions on the ground could cause ISIS to shift its message focus, which would offer a powerful opportunity for countermessaging. But regardless of whether that happens, the West should use every tool available to counter ISIS's stage-managed illusions with the harsh reality.

When Western policy makers discuss "degrading" ISIS, it should be in the context of forcing ISIS to make visible concessions in order to counter military pressure. Strikes designed to degrade the group's real internal strength are good, but our targeting priorities should also aim to expose vulnerabilities.

While we can make some progress amplifying the stories of defectors and refugees from areas ISIS controls, we can make even more by fully exploiting aerial and electronic surveillance and remote imaging to show what really happens in the belly of the beast.

We should pay particular attention to documenting war crimes and atrocities against Sunni Muslims in regions controlled by ISIS. It is patently obvious that ISIS has no qualms about advertising its war crimes against certain classes of people—Shi'a Muslims primarily, and religious minorities such as the Yazidis.

To simply highlight ISIS's barbarity is inadequate to undercut its messaging goals; in many cases, it accomplishes them. There is no doubt that ISIS wants to send a message about its harsh treatment of enemies. Amplifying the very messages the group wishes may resonate with an audience that already opposes ISIS, but it may further energize those who are vulnerable to its radicalizing influence.

While ISIS does not completely suppress information about its massacres against uncooperative Sunni tribes in the region, neither does it highlight them. And such stories have impact. In August, global jihadists on social media were enraged by an ISIS massacre of hundreds of Sunni tribesmen. By documenting such crimes, we can make a significant impact on how ISIS is perceived by those most susceptible to its ideology.[30]

We can also degrade the perception of ISIS's strength and its claims of victory by revealing its failures, particularly within its borders, such as incidents in which local people rise up against its control, failure of infrastructure, corruption, poverty, or other forms of domestic disintegration. The sources-and-methods trade-off will certainly favor disclosure in at least some of these cases.

Finally, we can offset ISIS messaging priorities by refusing to play into its apocalyptic narrative. As seen in Chapter 10, ISIS wants to enact specific prophecies regarding the end times, such as a victorious confrontation with the "crusaders" in the town of Dabiq. Our policies and military actions should not rise to the bait. For both military and messaging purposes, it is foolhardy to show up at the exact place and time that an enemy most desires. Whatever ambush lies in wait at Dabiq, let it rot there unfulfilled.

AGAINST ISIS OR FOR SOMETHING?

Finally, we would raise the question of what we are fighting for.

In the years since September 11, the West in general and the United States in particular have embraced a "war on terrorism" without stated limits. In the name of that war, or as an unintended consequence of its policies, we have vastly increased surveillance authorities, militarized domestic police forces, and used air strikes and drones to dispatch lethal force virtually anywhere that al Qaeda operates. Many of these actions have been taken in response to fear.

Osama bin Laden once said, "All that we have to do is to send two mujahideen to the furthest point east to raise a piece of cloth on which is written al Qaeda, in order to make the generals race there."[31] ISIS has exploited this tendency, in part following the blueprint in *The Management of Savagery* and in part to serve its apocalyptic dream of a confrontation with the "Crusaders" in Dabiq.

We must find better ways to balance our security against common sense and widely accepted ethical principles. That means refusing to rush into war every time we are invited by someone waving a black flag, but it also means taking a closer look at our strategies and tactics, and asking how they can better reflect our values. In the conflict with ISIS, messaging and image are half the battle, and we do ourselves no favors when we refuse to discuss the negative consequences of our actions.

We must be involved in a visible process of continually evaluating and improving the way we conduct war, asking if our responses are not only proportionate and economically responsible, but ethical. For instance, the Foreign Policy Research Institute's Clint Watts has tried to tackle this challenge as it pertains to drones, arguing for a judicial process similar to that currently used by the FISA court, an idea we endorse.[32]

In December 2014, the release of a Senate report on the use of torture by the United States after September 11 provoked a national debate on the morality of our tactics to fight terrorism. Beyond the argument over the results produced by such techniques lies a fundamental question of values and our standing in the world. The use of torture helps validate jihadist claims about the immorality and hypocrisy of the West. We must not fight violent extremism by becoming the brutal enemy that jihadists want. While painful, the process of publicly disclosing and confronting such incidents is, as David Rothkopf argues in *Foreign Policy*, "very American"[33] in its transparency, which, in our view, is something to embrace.

We should be seen, constantly, as balancing the scales of justice and individual freedom rather than letting the weight of groups like al Qaeda and ISIS constantly drag us toward an irrevocable mandate for more action, more compromise, and less concern for innocent people caught in the crossfire.

"The Second Coming," a poem by W. B. Yeats, is often quoted (maybe too often), because it feels so relevant to many modern situations. But its apocalyptic tone and cutting observations could have been written for the challenge of ISIS.

> Things fall apart; the centre cannot hold;
> Mere anarchy is loosed upon the world,
> The blood-dimmed tide is loosed, and everywhere
> The ceremony of innocence is drowned;
> The best lack all conviction, while the worst
> Are full of passionate intensity.

The dilemma of Syria and Iraq finds full-throated expression in the poet's words, written as a comment on wars and politics nearly one hundred years ago.

Perhaps these problems are universal in history, relevant again for each generation. Or perhaps they are iterative, situations repeating and refining until the reality of the world is distilled to the razor-sharp essence that the best poetry provides.

It is hard to imagine a terrible avatar of passionate intensity more purified than ISIS. More than even al Qaeda, the first terror of the twenty-first century, ISIS exists as an outlet for the worst—the most base and horrific impulses of humanity, dressed in fanatic pretexts of religiosity that have been gutted of all nuance and complexity.

And yet, if we lay claim to the role of "best," then Yeats condemns us as well, and rightly so. It is difficult to detect a trace of conviction in the world's attitude toward the Syrian civil war and the events that followed in Iraq. Why do we oppose ISIS and not Assad? There are pragmatic reasons, among them the explicit threat ISIS poses to Western allies and interests in the region, as opposed to the less overt risks to Western allies associated with Assad. But it is difficult to explain the dichotomy between our approaches to each of these villains on the basis of a clear moral imperative. Syria poses a profound dilemma, more so than Rwanda or Bosnia. Our moral impulse is to act on behalf of the Syrian people. But an intervention that simply removes Assad, as the Libyans removed Gadhafi, creates new and different problems for the Syrian people, and these new problems may be even more intractable. Strengthening ISIS would be just one of the possible unintended consequences, but likely the most dangerous—both for the Syrian people and the region.

In the past, the United States has gone terribly awry in its efforts to promote electoral democracy around the world. ISIS is only the latest example of the failure of democracy promotion, although it may be the starkest.

One of the goals for the 2003 invasion of Iraq and the war on terrorism more broadly was to spread democracy, in the belief that "replacing hatred and resentment with democracy and hope" would "deny the militants future recruits," in President Bush's words.[34] Democracy promotion—and the claim that it was a critical component of the war on terrorism—became a theme of his presidency. But people in the Middle East were, and remain, deeply skeptical that this was his goal or a U.S. goal more broadly.[35]

Thomas Carothers, a leading expert on democracy, characterized the policy dilemma this way: The imperative to degrade terrorist capacities tempts policy makers to put aside democratic scruples and seek closer ties with autocracies willing to join the war on terrorism.

On the other hand, some policy makers have come to believe that it is precisely the lack of democracy that breeds Islamic extremism in the first place.[36] But these policy makers are wrong in imagining that promoting electoral democracy is a panacea against terrorism. Many studies have shown that it clearly is not. Economist Alberto Abadie found that countries with intermediate levels of political freedom are even more vulnerable than those with the highest or lowest levels, which suggests that the transition from authoritarian rule is a particularly dangerous period.[37]

Edward Mansfield and Jack Snyder warned in 2007, "When authoritarian regimes collapse and countries begin the process of democratization, politicians of all stripes have an incentive to play the nationalist card."[38] This is precisely what happened in Iraq: After the collapse of Saddam's regime, due to their majority, Shi'a groups had the upper hand. Sunnis felt abandoned and resentful, and were able to mount a fierce insurgency. The elements that led to the violence had not been rectified when U.S. troops left.

Long before the war on terrorism, Fareed Zakaria warned that constitutional liberalism is not about the procedures of *selecting* a government, but the government's goals. "It refers to the tradition, deep in Western history, that seeks to protect an individual's autonomy and dignity against coercion, whatever the source—state, church, or society."[39] Constitutional liberalism argues that human beings have certain "inalienable" rights, and that governments must accept limitations on their own power.[40]

Electoral democracy, which can lead to domination by the most populous ethnic groups, has to be held in check by something like a bill of rights that protects minorities, allows religious freedom, and guarantees freedom of the press. This is the long-term goal for Arab countries, Marwin Muasher argues.[41]

King Abdullah of Jordan, who has shown himself to be extraordinarily courageous, argues that fighting ISIS will require the Muslim world to work together. He calls it a "generational fight" and "a third world war by other means." In the long term, he said, the fight is ideological. As threatening as ISIS is to the West, more than anything else it is an existential threat to Sunni Islam. "This is a Muslim problem. We need to take ownership of this. We need to stand up and say what is right and what is wrong," he said.[42]

Perhaps most important, we must embrace the idea that what we seek is continual progress toward these goals rather than their institution by fiat. Insistence on the latter is the way of dictators, the way of ISIS, of all extremism, and its hypocrisy is self-evident. The West has spent decades trying to impose structures of politics and governance in the Middle East, and the results sadly speak for themselves.

This is work that will never be finished; it is a mission to span generations. It requires patience and attention to detail. It requires humility. We in the West must continually ask if we are living up to our own values of human rights and the importance of self-determination, and we must correct our course if we go astray. Like al Qaeda before it, ISIS derives far more strength from our response to its provocation than from the twisted values it promotes.

Jihadi Salafism is not a monolithic ideology. Despite our sense that movements like al Qaeda and ISIS share a single agenda, there is incredible diversity among such militant groups. On many issues, they simply do not agree: they embrace different religious beliefs and practices, they adopt different standards of conduct in war, and they pursue different strategic objectives. Importantly, these differences often have deep roots and long histories. As a result, making sense of ISIS requires looking at both the past and the present. It requires understanding some of the early history and core components of Islam, tracing the evolution of jihadi Salafism in the twentieth century, and exploring the issues that continue to divide these groups today.

ISLAM: A (VERY) BRIEF HISTORY

MUHAMMAD AS A MESSENGER; HIJRA FROM MECCA TO MEDINA

Islamic tradition holds that Muhammad was born in Mecca around 570 CE. Orphaned as a young child, he was raised by his paternal uncle and belonged to the powerful Quraysh tribe. He had a relatively unremarkable childhood and early adulthood, but around the age of forty he began to have visions in which he received a series of messages from God. Though

---

* Written by Megan K. McBride, a doctoral student, focusing on religious violence and terrorism, in the department of Religious Studies at Brown University. She has an M.A. in Liberal Arts from the Great Books program at St. John's College and an M.A. in Government from John Hopkins University.

initially reluctant to talk about these experiences, he was ultimately persuaded by his wife to share the revelations with his family and community. Muhammad gathered followers slowly, but after a few years he found himself at odds with the people of Mecca since his message—encouraging reform, emphasizing the oneness of God, and declaring polytheism to be sinful—challenged their religious practices and traditions. Growing tensions compelled Muhammad and his followers to leave Mecca and travel to Medina. This move, known as the *hijra* or migration, marked a turning point as the new community transitioned from an oppressed minority movement to a self-governing religious and political community.[1] The years spent in Medina were important ones, and Muhammad used this time to clarify his message, expand his community, and extend his regional influence. He died in 632 having successfully done all three but without having appointed a successor.

CHOOSING A CALIPH; SUNNI AND SHI'A COMMUNITIES EMERGE

Because God's message indicated that Muhammad would be the last prophet, it wasn't clear who should guide the young community after his death. By general consensus, his family and followers decided that the community should be led by a caliph. The caliphs were not seen as replacements for Muhammad or as prophets; they were simply leaders selected to rule in the tradition that he had established.[2] The first four caliphs, who ruled consecutively from 632–661 and were known as the Rightly Guided Caliphs, continued the work that Muhammad had started by overseeing the compilation of the Quran, by consolidating power, and by undertaking a series of conquests. The death of the third caliph, however, precipitated a serious debate and resulted in a fracturing of the Muslim community. One group, whose members came to be known as Sunni Muslims, believed that the leader could be any male member of the Quraysh tribe chosen by the authorities of the Muslim community; thus the term Sunni is derived from the phrase *Ahl al Sunnah wa'l jama'a*, which means "people of the tradition and community."[3] Another group, whose members came to be known as Shi'a Muslims, believed that the leader needed to be a direct male descendant of Muhammad; thus the term Shi'a is an abbreviation of *Shi'at 'Ali*, meaning "followers of Ali" (the son-in-law and cousin of Muhammad). Ali was, in fact, chosen to be the fourth caliph (and the last of the Rightly Guided Caliphs), but he was assassinated after just five years and the caliphs that followed him were not direct descendants of Muhammad and did not have the support of the entire Muslim community.

SOME DIFFERENCES BETWEEN SUNNI ISLAM AND SHI'A ISLAM

Though Sunni and Shi'a Muslims agree on the core tenets of Islam, the two groups have developed unique identities and adopted distinct religious traditions. These differences crystallized not long after the assassination of Ali. Shi'a Muslims objected to the caliphs selected to follow Ali, and questioned the legitimacy of the government. The conflict came to a head when Husayn (Ali's son, Muhammad's grandson, and the individual that the Shi'a community recognized as the rightful leader) directly challenged the reigning caliph. In the ensuing battle at Karbala, Husayn and his family were killed by the caliph's forces. Husayn's death—his martyrdom—became central to the identity of the Shi'a community.

Over the next thousand years, the Shi'a identity was informed by this early experience with "martyrdom, persecution, and suffering."[4] By contrast, the Sunni identity was influenced by the political, military, and cultural successes of the Sunni caliphate.[5] The two groups consequently came to different understandings of what it meant to be Muslim; moreover, their different historical experiences resulted in different religious traditions. While both Sunni and Shi'a Muslims believe that mosques in Mecca, Medina, and Jerusalem are holy sites of great importance, Shi'a Muslims also identify Najaf (where Ali is buried) and Karbala (where Husayn was martyred and is buried) as holy sites. As a result, when ISIS (a Sunni group) threatens to invade Najaf and Karbala (Shi'a holy sites) the objective is more than mere military conquest; it is also a symbolic gesture likely intended to stoke sectarian violence.

Another significant difference between today's Sunni and Shi'a communities lies in their respective approaches to authority. Within the Shi'a community, great emphasis is placed on formalized and institutionalized religious authority. Shi'a clergy are educated at sanctioned seminaries where they study for years and become proficient in subjects such as law, theology, and philosophy. At the end of this period, when a student has completed this course of study in a satisfactory manner, he is permitted to become an official member of the community of religious scholars who protect the legacy of Islam and interpret it to meet the challenges of the modern era.[6] By contrast, religious authority in Sunni Islam is less centralized and hierarchical. Sunni Islam, unlike Shi'a Islam, lacks the formal titles that distinguish the rank of one scholar from another. There is also no clear institutional path to religious authority in Sunni Islam. While many Sunni clergy are highly educated in subjects such as law, theology, and philosophy, this education is not a prerequisite for leading a religious community. As a result, within Sunni Islam it is possible for individuals with little formal religious training to become both prominent and influential religious leaders.[7]

Though the comparison is imperfect for a number of reasons, it can be helpful to think of Shi'a Islam as being analogous to Roman Catholicism, and Sunni Islam as being analogous to Protestantism. Shi'a clergy, like Roman Catholic priests, are educated in a centralized system of seminaries. Additionally, they have formal titles designating rank and creating a clear hierarchy among their leaders. By contrast, both Sunni clergy and their Protestant counterparts are educated in a loose network of institutions. Moreover, neither Sunni Islam nor Protestantism has a formalized system of rank organizing their religious leaders into a unified and recognized hierarchy.

THE CALIPHATE

During the first few centuries of the caliphate (from approximately the seventh to ninth centuries) the Muslim world experienced significant growth and nurtured a civilization that was the most advanced of the era.[8] This period saw a staggering proliferation of intellectual work: "Poetry, grammar, Quranic studies, history, biography, law, theology, philosophy, geography, the natural science—all were elaborated in Arabic and in a form that was distinctively Islamic."[9] At the same time, the Muslim world continued to grow geographically and at its peak extended its reach from Spain to India. Ultimately, a number of factors undermined the strength of the caliphate. The sheer size of the empire made administration from a single seat of power difficult, and internal tensions undermined the stability of the government. At the same time, the Shi'a community continued to challenge the authority of the caliphs. By the middle of the ninth century, the caliphate was a much-weakened institution, and those who believed that it was important were forced to justify its continued existence. In doing so, they offered a rich description of the office. The caliph, they argued, should "maintain orthodoxy, execute legal decisions, protect the frontiers of Islam, fight those who refuse to become Muslims when summoned, raise the canonical taxes, and in general, himself to supervise the administration of affairs without delegating too much authority. He must possess certain qualifications, physical, intellectual, and spiritual, as well as the extraneous qualification of belonging to the same tribe as Muhammad, that of Quraysh." [10]

Despite these efforts to justify and strengthen the office, the caliphate continued to decline. Following the assassination of the reigning caliph in 1258—during the Mongol invasion of Baghdad—the Muslim world was ruled at a more local level with no overarching government uniting what had once been a vast empire. In the fifteenth century, however, a number of powerful Muslim empires emerged from the local governments that had come to control the region. The most important of these, for our purposes, was the Ottoman Empire which revived the office of caliph and lasted for over four hundred years. It was a major economic and military power that at its height controlled territory on three continents. The Ottoman Empire collapsed in the early twentieth century when its remaining territories were parceled out by the British and French following World War I, and the Turkish government that took its place abolished the office of the caliph.

Although the Muslim community was led by a caliph for much of its history—during the Umayyad dynasty (approximately 650 to 750), the Abbasid dynasty (approximately 750 to 1250), and the Ottoman Empire (approximately 1450 to 1923)—the office of the caliphate changed over time. As a result, contemporary calls for a return to the caliphate are unclear about what exactly a revived caliphate would look like. Nevertheless, the office is a potent symbol of Muslim unity and prosperity that many Muslims today hope to restore.

SOME CORE BELIEFS AND PRACTICES OF ISLAM

Though the modern Muslim world is one of staggering diversity—and includes an estimated 1.6 billion people—most Muslims turn to the same sources of authority (the Quran, the sunnah, and the *Hadith*) and embrace a core set of practices (commonly referred to as the five pillars of Islam).[11] Not all Muslims engage with these beliefs and practices in the same way, however. Much as individual Christians understand the Bible and the communion in different ways, individual Muslims come to different understandings of their own scripture and practices. Despite these differences, though, it is possible to identify some central components of the faith:

Quran

The Quran is a full account of the revelations that came to Muhammad. It was collected into a single written volume just one generation after his death, and it contains more than 6,000 verses. It is understood by many Muslims to be a literal transcription of what was relayed to Muhammad in his visions, and consequently the literal word of God.[12] It emphasizes the oneness of God, warns that the apocalypse is approaching, and provides broad guidelines for living a moral and upright life.[13]

Sunnah

The sunnah are the practices, deeds, and words of Muhammad. The Quran does not offer detailed guidance on how Muslims should behave in their daily lives. As a result, many Muslims turn to the sunnah—the example of Muhammad—in order to determine how best to conduct themselves.

Hadith

The sunnah are preserved in the *Hadith*, a collection of the practices, deeds, and words of Muhammad and his companions. The *Hadith* were transmitted orally for the first two centuries following Muhammad's death, but were ultimately collected and standardized. Central to the collection of *Hadith* is the issue authoritative transmission. In the centuries following Muhammad's death, stories of what he had done and said proliferated; in order to identify which anecdotes were reputable, scholars attempted to establish the path via which the stories were transmitted. A *Hadith* that is strong is one that is consistent with other scripture (that is, it doesn't contradict the Quran, and it makes sense along-side other accepted *Hadith*) and well-documented (it originated with a companion of Muhammad, it was transmitted via a relatively small number of people, and there are no breaks in the chain of transmission).

The Five Pillars
The five pillars of Islam are often described as the essential practices endorsed and followed by all Muslims.[14] They include the profession of faith, daily prayer, almsgiving, fasting during the holy month of Ramadan, and pilgrimage to Mecca at least once in a lifetime. Some have suggested that *jihad* is an unofficial sixth pillar of Islam, but this position is not widely held.

SALAFISM AND WAHHABISM

While there is considerable diversity in the Muslim world, the majority of violent jihadist organizations like al Qaeda and ISIS are Salafi. Some familiarity with Salafism is, as a result, critical to understanding a group like ISIS.

DEFINING SALAFISM

Salafism is a loosely organized movement within Sunni Islam; there are no clear requirements for being Salafi and there is no consensus over who should be considered Salafi.[15] But there are core features to the movement. Salafism is a call for a return to the beliefs, practices, and sincerity of early Islam. In fact, the term "Salafism" is a direct reference to these early years, and refers to the first few generations of Muslims, known as the *salaf*. Salafis prefer the Islam of these early Muslims and believe that centuries of human interpretation—influenced by preexisting religious traditions, cultural biases, political agendas, and individual self-interests—have corrupted Islam and led to decline across the Muslim world. They reject this interpretation and maintain that the only sources of authority necessary to be a pious Muslim are the Quran and the sunnah (the example of Muhammad and his companions). In rejecting centuries of scholarship and interpretation, Salafis effectively argue that the sources of authority necessary to being a pious Muslim can be understood without the assistance of intellectual elites.[16] One can, within this model, be a devout Muslim without understanding the intricacies of complex theological arguments.

Despite the modern nature of this movement, Salafis draw inspiration from the scholarship of famed medieval scholar Taqi al-Din Ahmad Ibn Taymiyyah (d. 1328). Ibn Taymiyyah lived in a tumultuous time, and wrote as the Muslim community grappled with the invasion of the Mongol Empire, the destruction of Baghdad, and the assassination of the last Abbasid caliph. These events marked the end of a period of great prosperity, intellectual achievement, military success, and cultural development during which the entire region was politically united under the caliphate. Ibn Taymiyyah argued that the end of this era was the result of a corruption of Islam, and he believed that returning to the beliefs and practice of the early Muslim community would lead to a revival of the Muslim world. Like Ibn Taymiyyah, many Salafis today believe that the misfortunes of the Muslim world have been caused by a corruption of Islam, and that a revival of Islam is an essential corrective.

Another layer can now be added to the comparison of Sunni Islam and Protestantism. Specifically, a helpful comparison can be made between Salafism (a movement within Sunni Islam) and Protestant fundamentalism (a movement within Protestantism). Salafis are, in fact, frequently referred to as "Islamic fundamentalists." Though "fundamentalism" is a term that was originally used to describe an early twentieth century American movement (and we should be careful when applying it to other groups) this label can be useful in helping to foreground a constellation of features shared by distinct religions. As Scott Appleby, a scholar of religion at the University of Notre Dame and co-director of The Fundamentalism Project, has noted, what unites fundamentalists is not a common set of beliefs or religious practices; instead, fundamentalists share an "attitude towards religion itself" in which religion is, among other things, "the best defense against the threatening encroachments of secularism."[17] In other words, both Salafis and Protestant fundamentalists turn to religion in an effort to respond to the destabilizing changes of a rapidly evolving world.

THE ORIGINS OF WAHHABISM

Ibn Taymiyyah wrote centuries before today's jihadi Salafi movement took shape, but remains relevant in no small part due to the eighteenth century rise of Wahhabism. This movement, a type of conservative Salafism, began with Muhammad Ibn Abd al Wahhab (d. 1792). Ibn Abd al Wahhab drew upon the writings of Ibn Taymiyyah and argued for a strict interpretation of Sunni Islam. He believed that Muslims who engaged in practices that he considered idolatrous—practices such as polytheism, venerating the graves of saints, mysticism, and Shi'ism in general—were not Muslims at all. Moreover, he precipitated a series of confrontations by calling on his neighbors to change their practices and embrace his interpretation of Islam. In pursuit of this goal Ibn Abd al Wahhab allied himself with Muhammad bin Saud (d. 1765), the leader of the House of Saud. Over the course of the nineteenth and twentieth centuries the Wahhabis worked with the Saud family to unite the people living on the Arabian Peninsula under a single religious and political authority. This effort culminated in the creation of the modern state of Saudi Arabia, and the Saudi government continues to have a close relationship with Wahhabi religious authorities even today.

SALAFISM: FROM QUIETISM TO *JIHAD*

Though early Salafism and Wahhabism are typically thought of as religious movements, neither was ever apolitical. Moreover, a number of important twentieth and twenty-first century events resulted in the movements' increased involvement with recognizably political issues. These events resulted in what Quintan Wiktorowitz, a former member of the National Security Council and expert on Islamic movements and counterterrorism, has described as three distinct waves of modern Salafism: a quietist faction, a political faction, and a jihadi faction.[18]

QUIETIST SALAFISM

The quietist faction is, in a sense, the strain of Salafism that has responded the least to the world events of the twentieth century. Individuals in this group understand their central project to be the purification of Islam and do not participate in politics.[19] Though there are quietist Salafis across the Muslim world, the center of gravity for this movement is the existing religious establishment in Saudi Arabia. Saudi Arabia is somewhat atypical for a country in the Muslim world, but the very things that make it unique have made it hospitable to the quietists. For most of its existence, the country has been financially independent due to its massive oil reserves, and the ruling family has consequently been insulated from pressures to moderate. Additionally, Saudi Arabia didn't experience colonialism and so its religious scholars were never forced to grapple with the many questions that arise when two political and cultural systems attempt to occupy the same space.[20] In other words, the quietists in Saudi Arabia (the most vocal and powerful of the quietists) have been sheltered. This has allowed the movement to flourish, but it has also laid the groundwork for the rise of the political faction by making the quietists vulnerable to the charge that they are out of touch and incapable of responding to the challenges of the contemporary world.

POLITICAL SALAFISM

The political faction criticizes the quietist faction for its political naiveté and rejects the idea that political activism is un-Islamic. Though this type of Salafism can be found across the Muslim world, the faction was greatly influenced by a subset of the Muslim Brotherhood, a twentieth century Egyptian movement founded by Hasan al Banna (d. 1949). Like Ibn Taymiyyah, al Banna and his contemporaries lived in a tumultuous time. By the end of World War I the Ottoman Empire had collapsed, the office of the caliph had been abolished by the secular Republic of Turkey, and much of the Muslim world was under colonial rule. Al Banna shared the Salafi concern that traditions accumulated over the centuries had corrupted Islam and he worried that the slow Westernization of the Muslim world was having a similar effect. Like his predecessors, he responded to these crises by calling for a return to the religious beliefs, practices, and sincerity of the early Muslim community. Islam, al Banna said, "does not stand helpless before life's problems nor [before] the steps one must take to improve mankind."[21] It is an all-encompassing way of life and the best mechanism for responding to the crises brought on by modernity. Al Banna went on to emphasize the importance of education, and to highlight the ways in which individuals could be knowledgeable about Islam without relying on intellectual elites.[22] At the same time, he offered a justification of militant jihad and articulated a sophisticated political program.[23] He founded the Muslim Brotherhood in an effort to pursue this agenda. The group took its principal task to be a full-scale reformation of society with the utopian hope that this would result in a revitalized Muslim world.

The Muslim Brotherhood was an organization composed of both liberals and ultraconservatives. Its influence over ultraconservatives outside the movement expanded when a subset of its members fled the persecution of the Egyptian government and migrated to Saudi Arabia during the 1960s and 1970s. A number of these refugees became teachers, and injected their political engagement into Saudi Arabia's more quietist Salafism.[24] At the same time, the oil boom of the 1970s ensured that the Saudi Arabian government had the funds to spread Salafism—now influenced by these politically oriented thinkers from Egypt—throughout the region via a far-reaching network of schools and institutions.[25] By the 1980s and 1990s a distinctly conservative and political strand of Salafism had taken root in Saudi Arabia and the greater region; still concerned with ensuring the purity of Islam, this faction believed that doing so required engaging in political action and overthrowing corrupt regimes that threatened Islam.

Political Salafis didn't claim to be as religiously knowledgeable or sophisticated as the quietist Salafis; their authority was based, instead, on their political analysis of the modern world.[26] In articulating their position, they drew on the thinking of Sayyid Qutb (d. 1966), who had been influenced by both Ibn Taymiyyah and Ibn Abd al Wahhab.[27] Debates about whether or not Qutb was really Salafi still persist, but there is no question that he spoke a language understood by both the quietist Salafis from Saudi Arabia and the political Salafis from Egypt. He was, as a result, a central figure in both the growing political Salafi movement and the nascent jihadi Salafi movement that soon followed.[28] Interestingly, Qutb's younger brother was among those who emigrated from Egypt to Saudi Arabia. Like many of his colleagues, he secured a teaching position once in Saudi Arabia and offered lectures that were sometimes attended by a young Osama bin Laden.[29]

JIHADI SALAFISM

The jihadi faction coalesced in large part due to the 1980s war against the Soviets in Afghanistan. The war functioned, unfortunately, as a "dangerous incubator" in which Salafis from across the region came into contact—sometimes on actual battlefields; sometimes in military training camps—with radicalized groups that believed violence could be a solution to some of the problems confronting the Muslim world.[30] Like both the quietist and political Salafis, the jihadi Salafis were concerned about the corruption of Islam and the oppression of the Muslim world. This faction also accepted that the quietists were more knowledgeable about Islam, but they were concerned that the symbiotic relationship between the quietists and the governments (for example, the relationship between the Wahhabi religious establishment and the Saudi Arabian government) had corrupted the religious leadership. They believed, moreover, that it was acceptable to use violence to respond to this crisis.[31]

SALAFISM, WAHHABISM, AND ISIS

By the late twentieth century Salafism had quietist, political, and jihadi factions. Additionally, Wahhabism had managed to "co-opt the language and symbolism of Salafism . . . until the two had become practically indistinguishable."[32] As a result, when analysts, academics, and journalists writing today say that ISIS is following Salafi principles, what they mean is that ISIS's ideology contains elements of both Salafism and Wahhabism. And when they say that the movement is following Wahhabi principles, they mean the same thing.[33] That said, mentioning Wahhabism unquestionably evokes the thought of Saudi Arabia given the long-standing relationship between the Wahhabi religious authorities and the Saudi Arabian government. As a result, describing the movement as Wahhabi is a subtle reference to the fact that Saudi Arabia has been an influential champion of Salafism. It suggests, in a sense, that Saudi Arabia is responsible for movements such as ISIS because of the role that the Saudi Arabian government has played in facilitating the spread of Salafism across the region.[34]

Importantly, while today's Salafis share a set of core beliefs—about monotheism, about the corrupting threat posed by human interpretation, and about the importance of returning to a pure and authentic Islam—the movement is wholly decentralized. Salafism has no official leaders, and individuals are empowered to trust their own understandings of the Quran and sunnah. This simultaneous marginalization of religious scholars and authorization of the individual has, according to scholars like Khaled Abou El Fadl, resulted in a crisis of authority.[35] As a result, there is space within Salafism for both increasingly radical interpretations of Islam and the popular embrace of self-proclaimed experts with little to no training in Islamic law (to include a number of prominent leaders within organizations like al Qaeda and ISIS). There are, consequently, significant differences not only between the major factions of Salafism, but also between individuals and groups within the same faction. While knowing that a group is jihadi (and not quietist or political) is important to understanding the group's commitments, it is still necessary to look closely at the group's specific beliefs and practices.

DECLARING WAR: THE PRACTICE OF JIHADI SALAFISM

Making sense of the disagreements between jihadi Salafi movements requires looking closely at both justifications for engaging in war and accepted practices within war. In doing so, we cannot offer the thinking of "all Salafis" because the movement is diverse and fragmented; nor can we summarize the thinking of "all jihadi Salafis" given that there are clear disputes between groups like al Qaeda and ISIS. We can, however, highlight the issues that separate the non-violent Salafi population from violent Salafi movements like al Qaeda and ISIS. And we can bring to the fore the different interpretations that create conflicts between these jihadi groups.

TAKFIR

One issue that is central to these disagreements is that of *takfir*. To declare a person a nonbeliever is, in Islam, a matter of great significance and the process for doing so is known as *takfir* (the "pronouncement that someone is an unbeliever and no longer Muslim").[36] There is, as a result, considerable debate among Salafis over when invoking *takfir* is appropriate. The quietist and political Salafis typically refrain from using *takfir*, even in the case of dictators ruling corrupt regimes. In fact, the quietists adhere to a "high evidentiary threshold" that makes it quite difficult to use *takfir*.[37] In many instances they do this by differentiating action from belief. They concede that you might accuse a person of engaging in heretical *acts*, but they maintain that this fails to establish that the person is not Muslim because—unless this person claims that the heretical act is Islamic or in some way superior to Islam—there is simply no way to know what the person believes.[38] This high standard functions as a check against rampant accusations of apostasy; because it is difficult to know what a person is thinking, it is difficult to demonstrate that a person is a nonbeliever. Unfortunately, this also makes it difficult to denounce terrorists in a black and white way that might be appealing; the terrorists might be accused of committing *acts* of apostasy, but without a thorough investigation they cannot be labeled as nonbelievers or said to be no longer Muslim.

Unlike the quietist and political factions, the jihadi faction has adopted a more expansive use of *takfir*. Groups in this faction have demonstrated little tolerance for pluralism, and prefer instead to effectively excommunicate those who fail to embrace their interpretation of Islam. They argue that a ruler's refusal to heed the warnings of scholars (that is, warnings that the ruler or government is engaged in un-Islamic practices) is evidence of corrupt belief.[39] In other words, they argue that if the ruler's actions are un-Islamic then his beliefs must also be un-Islamic. One important consequence of this interpretation is that the religious trials required by the quietists play a smaller role for jihadi Salafi groups like ISIS. Actions, it seems, offer sufficient insight to justify declaring an individual an apostate. This doesn't mean, however, that there is no evidentiary requirement. As one scholar noted, most of today's jihadi Salafis believe that "proper evidence must be presented" to sustain the charge of apostasy.[40] The barrier to using *takfir* isn't wholly removed; it is, though, considerably lower for jihadi Salafis than for quietist or political Salafis.

This more radical approach to *takfir* has clear roots in the positions articulated by Ibn Abd al Wahhab. It can also, however, be traced to the early twentieth century Indo-Pakistani Islamic scholar Sayyid Abu'l A'la Mawdudi (d. 1979). Mawdudi argued that the world was experiencing a modern *jahiliyyah*—a period of ignorance; "a government system, ideology, or institution based on values other than those referring to God"— that threatened Islam.[41] He argued that it was the duty of true Muslims to respond to this crisis by fighting against the influence of the heretical individuals that undermined Islam.[42] For Mawdudi, it was critically important to differentiate between believers and nonbelievers. In separating the two, he argued that those whose behavior was not wholly Islamic were nonbelievers.[43] In other words, Mawdudi believed that much of the Muslim world was ruled and inhabited by nonbelievers, and that devout Muslims were obligated to change these circumstances. Mawdudi was read extensively by Qutb who agreed that the world was experiencing a modern *jahiliyyah*, accepted that much of the Muslim world was ruled and inhabited by nonbelievers, and embraced the idea that Muslims were obligated to respond to this crisis.[44] Unlike Mawdudi, though, Qutb concluded that this obligation must take the form of militant *jihad*.

JIHAD

Jihad is an incredibly complex term. In the Quran, it is used to "refer to the act of striving to serve the purposes of God on this earth."[45] In some instances this might mean struggling to be a good person; in other instances this might mean fighting on a battlefield. The word has, as a result, been used to capture a wide spectrum of behaviors ranging from spiritual struggle (sometimes referred to as *greater jihad*) to armed conflict (sometimes referred to as *lesser jihad*). In the context of jihadi Salafism, *jihad* most frequently refers to physical warfare or armed struggle. It is this particular definition of jihad that Mawdudi and Qutb were invoking, and it is the call to this type of jihad that they split over. Mawdudi didn't object to violence on principle and much of his project does sound revolutionary.[46] In fact, though, he advocated for a methodical approach to reform and preferred political solutions to violent ones. He maintained that only a government could declare a jihad, he insisted that it be an option of last resort, and he suggested that it could only be pursued when there was some assurance of victory.[47] His was a decidedly moderate approach to jihad. Qutb, by contrast, adopted a more aggressive approach. He criticized the idea that corrupt governments could be changed from within the system and instead advocated for revolution. However, he understood militant jihad to be merely part of the solution and he insisted that it be coupled with an internal re-education.[48] He did not—notwithstanding his reputation as the founder of modern jihadism—advocate for indiscriminate violence. Qutb's argument was popular and influential as he offered his readers a compelling and articulate call to jihad. He was not, however, alone in this line of thought and a series of increasingly radical thinkers extended this argument (very possibly beyond what Qutb might have accepted).

An important argument was offered, for example, by Mohammad Abd al Salam Faraj (d. 1982), who wrote in his widely read pamphlet, *The Neglected Duty,* that "jihad is second only to belief," a neglected sixth pillar of Islam, and an obligation of every devout Muslim.[49] Faraj rejected al Banna and Qutb's call for education, suggesting that it was no path to change and that militant jihad was the only viable way forward. As Nelly Lahoud, a Senior Associate at the Combating Terrorism Center at West Point, noted: "Faraj's treatise essentially argued that military *jihad* and Islam are one and the same."[50] Faraj argued for a highly deregulated approach to jihad in which individuals acting independently were obligated to attack corrupt regimes. Quietist Salafis have, for some time, rejected the idea of independent jihad and argued that the sanction of a Muslim ruler is necessary to justify jihad; in Saudi Arabia, for example, individual jihad is impermissible and the country's deradicalization programs take pains to emphasize that "only the legitimate rulers of Islamic states, not individuals such as Osama bin Laden, can declare a holy war."[51] Even al Banna had advocated for a regulated approach to jihad, and so Faraj's position was quite radical.[52] The jihadi Salafi faction has, however, followed Faraj and adopted this less centralized approach to jihad that doesn't seem to require the sanction of authority.

DEFENSIVE JIHAD

A similar position—coupled with a gripping call to action—was articulated by Abdullah Azzam (d. 1989). Writing against the backdrop of the 1980s conflict in Afghanistan, Azzam suggested that the non-Muslim invasion of a Muslim territory created an obligation to engage in jihad even if the threat was not local.[53] Azzam essentially shifted the parameters of jihad, transitioning away from Qutb's focus on corrupt Muslim regimes and towards a new focus on the defense of Muslim lands. This argument was particularly powerful because it framed jihad as defensive. A defensive jihad is understood to be a justified response to an external party invading a Muslim state; in such a situation, Muslims are obligated to respond. Because defense is a widely accepted justification for jihad (few question that there is an obligation to respond to invasion), jihadi Salafis frequently argue that the United States is occupying Muslim lands by maintaining military bases in some Muslim-majority countries. By casting the Americans in the role of invading force, the violent Salafis are able to argue that their response is a defensive jihad and thus justified and obligatory.

HIJRA

In addition to framing the jihad in Afghanistan as defensive, Azzam consistently invoked the language of *hijra*. *Hijra* is typically understood to be a reference to Muhammad's migration from Mecca (a city that was in conflict with the new Muslim community) to Medina (a city that welcomed Muhammad and his companions). Using Muhammad as an example, many Muslims have concluded that the only acceptable ways to respond to an un-Islamic environment are jihad (that is, to fight in defense of Islam) and *hijra* (to flee the un-Islamic environment).[54] That said, Fred Donner, a scholar of Islam at the University of Chicago, has noted that a careful examination of the Quran reveals that in some passages *hijra* is invoked in a way that is almost synonymous with jihad and is associated with "leaving home for the purpose of fighting."[55] This is the meaning that Azzam appears to have been gesturing toward when he suggested that Muslim men were obligated to travel to Afghanistan in order to defend Islam. Azzam also argued that these men did not need to obtain the sanction of political leaders before undertaking jihad; they could, in other words, engage in individual jihad anywhere that Islam was under threat.[56] Azzam was, a result of these arguments, quite influential. By invoking both obligations—by explicitly linking jihad and *hijra*—and by authorizing people to act independently, he laid the groundwork for the flood of foreign fighters that have filled the ranks of groups like ISIS.

TAKFIR, JIHAD, AND ISIS

To be clear, the invocation of *takfir* has no necessary relationship with the decision to engage in jihad. An individual or group might invoke *takfir* with respect to a corrupt ruler and yet simultaneously believe that militant jihad is only appropriate on rare occasions. The innovation of these jihadi Salafi theorists was to expand the use of *takfir* while simultaneously describing militant jihad as an individual global obligation. In combination this meant that there were both more justified opportunities for militant jihad and a requirement to participate. Thus this thought—moving through thinkers like Qutb, Faraj, Azzam, and others—significantly influenced jihadi Salafi movements like al Qaeda and ISIS. It influenced Osama bin Laden, who took up the concern with Muslim oppression by non-Muslim parties and radically extended the territory in which jihad was permissible; it was invoked by al Qaeda in Iraq leader Abu Mus'ab al Zarqawi who expanded the range of viable targets by arguing that it was acceptable to kill both Muslims and non-combatants; and it has shaped the thinking of Ayman al Zawahiri (currently the leader of al Qaeda) who has claimed that to renounce jihad is an act that betrays Islam and deserves death.[57] Each of these earlier thinkers, in other words, contributed a small piece to the arguments used to fuel and justify jihad today.

WAGING WAR: THE JIHADI FACTION ON THE BATTLEFIELD

Justifying jihad and engaging in jihad are, however, two very different endeavors and important questions remain even in a situation in which a jihad appears to be justified: Who is the appropriate target of the jihad? Are civilian casualties acceptable? Should it be permissible to kill Muslims? What types of violence can be deployed?

SELECTING A TARGET (NEAR ENEMY AND FAR ENEMY)

Most jihadi Salafis agree that it is appropriate to use violence to challenge corrupt governments in the Middle East. Beyond this consensus, though, a number of strategic, logistical, and moral issues split the faction. To begin, there is the issue of whom to target. The overwhelming military and police power of today's Middle Eastern governments were, prior to the Arab Spring, understood to pose serious challenges to those hoping to launch successful revolutions. As a result, some jihadi Salafis conclude that it makes more sense to undermine these governments by targeting the Western countries that support these regimes. This position found a particularly clear articulation in the writings of Zawahiri, who worked with Bin Laden to launch a global jihad in the late 1990s. This approach was not uncontroversial, though, and a number of jihadi Salafis expressed concern that such a move was strategically unsound because it might provoke an overpowering military response from Western nations, and because it might engender anti-Muslim feelings worldwide.[58] It is, as a result, possible to think of jihadi Salafism as being divided into two camps: a near-enemy group (committed to the use of violence directly against corrupt Middle Eastern governments; Zarqawi's al Qaeda in Iraq is an excellent example of this type) and a far-enemy group (committed to the use of violence against Western governments; Bin Laden and Zawahiri's al Qaeda is an excellent example of this type).

JUSTIFYING THE KILLING OF CIVILIANS AND MUSLIMS

Salafis must also grapple with the question of how to engage in the practice of war. Nonviolent Salafis have, for example, typically accepted that the intentional targeting of civilians is prohibited (though they acknowledge that some civilian casualties are likely to occur as part of any war).[59] The current practice of some jihadi Salafi groups, though, includes the purposeful targeting of civilian populations. In defending these practices, jihadi groups have articulated a number of arguments. First, civilians can be killed if doing so is part of a proportional response.[60] If the Americans are known to be purposefully killing Muslim civilians, the argument goes, then jihadi Salafis are justified in killing American civilians. In making this argument, some have suggested that U.S. war technology is so accurate that every civilian death must be intentional, thus justifying the targeting of any American civilian.[61] Second, civilians can be killed if they betray Islam by assisting the enemy. In making this argument, jihadi Salafi groups have greatly broadened the definition of betrayal to include anyone (journalists, researchers, government workers, etc.) who might be seen as supporting the enemies of Islam.[62] Working in this tradition, bin Laden claimed that all American citizens may be targeted because they live in a democratic nation and are directly responsible for the actions of their government. Similarly, a Muslim civilian can be killed for assisting the enemy since such assistance serves as evidence that the individual is not really Muslim.

Similarly, nonviolent Salafis have typically held that it is impermissible to target Muslims but a number of jihadi Salafi groups embrace this practice and target marketplaces, hotels, and other venues that they know will be filled with Muslim civilians. In defending this practice, the jihadi

Salafis argue that the Muslim casualty is an agent, and not a victim, of the movement itself. In other words, the death can be justified by framing the victim as an (unwitting) martyr and not as a mere casualty or victim of the war. A slightly more complicated argument has been made concerning the targeting of the Shi'a population. Though many reject this tactic on the grounds that the Shi'a are Muslims and should not be killed, there is a long tradition of anti-Shi'a violence among jihadi Salafis. Ibn Taymiyyah identified the Shi'a as a clear enemy of Islam and Abd al Wahhab adopted a number of anti-Shi'a positions that resulted in a spate of violence against Shi'a populations during his 19th century conquest of the Arabian Peninsula.[63] This attitude persists today, and has resulted in significant anti-Shi'a violence in the 20th century. Thus the jihadi Salafi argument for targeting the Shi'a population—to include Zarqawi's declaration of a "full-scale war on Shiites"—has a long and complex history.[64]

BEHEADINGS AND SUICIDE MISSIONS

Even among jihadi Salafis, there is little consensus on the use of tactics such as beheadings and suicide missions. Beheading—a practice embraced by terror groups like ISIS and an accepted method of execution in Saudi Arabia—was actually a preferable mode of execution in the pre-modern era because it was considered to be swift and merciful (and in an era with many trained swordsmen there was no shortage of individuals capable of beheading a man with a single blow).[65] The adoption of this practice by jihadi Salafi groups has little to do with the desire to be humane. Instead, beheading is embraced because it is a powerful means of expressing authority and an effective way for groups like ISIS to intimidate potential enemies. Unfortunately, the lack of a recognized central authority in Salafism makes it difficult to challenge this practice. Beheading might be condemned by religious scholars across the Muslim world, but jihadi Salafis simply ignore these condemnations and turn to their own religious leaders in search of a justification for this tactic.[66]

A similar dynamic makes it difficult to challenge the use of suicide bombings. While religious leaders in Saudi Arabia have explicitly identified suicide bombing as an un-Islamic practice, some movements within the jihadi Salafi faction continue to embrace this tactic.[67] Justifications for suicide bombings vary, but Assaf Moghadam has suggested that we might trace acceptance of this practice to Azzam's argument that martyrs would be rewarded in the afterlife and Zawahiri's fervent embrace of the tactic.[68] Importantly, jihadi Salafis concede that suicide is an impermissible practice but they argue that the act should be judged based on the intent of the perpetrator. That is, they focus on the actor's intention to engage in jihad and reject the idea that the intent was to commit suicide. This effectively recasts the act as one of "legitimate martyrdom" and not as one of suicide.[69]

ISIS'S DEFIANCE: RADICAL AMONG RADICALS

ISIS IN CONFLICT WITH OTHER JIHADI SALAFI MOVEMENTS

ISIS and its predecessors have long been in conflict with mainstream jihadi Salafism and there are major differences of opinion over what we might call the ISIS-approach to jihad. Thus when Zarqawi's al Qaeda in Iraq coordinated suicide bombings at three hotels in Amman (resulting in the deaths of more than sixty mostly Muslim civilians) bin Laden was reportedly "furious."[70] In fact, Zarqawi's excesses concerned al Qaeda so much that it sent him a series of lengthy letters encouraging him to reconsider his strategy. Not quite rebukes, the letters made it clear that Zarqawi's actions were permissible but problematic. A 2005 letter from Zawahiri, for example, explained that al Qaeda's goal of establishing a caliphate couldn't be accomplished without popular support, and that it was important for Zarqawi to avoid actions that wouldn't be understood by the masses. Given this, Zawahiri encouraged Zarqawi to reconsider both his antagonistic engagement with the Shi'a population of Iraq and his habit of publicizing "scenes of slaughter."[71]

This barbaric and excessive approach to jihad—embraced today by groups like ISIS—remains controversial. Sustained criticism of these excesses has come from a number of sources, including the widely influential scholar Abu Muhammad al Maqdisi. Al Maqdisi became prominent in large part because of his role as the spiritual advisor to Zarqawi (the two met in Pakistan and were imprisoned together in Jordan). His influence, however, far transcends the bounds of this relationship; recent analysis of jihadi Salafi literature has shown that al Maqdisi is the "most influential living Jihadi Theorist" and there is little question that he has played a critical role in articulating the ideological foundation of jihadi Salafism.[72]

Like many jihadi Salafis, al Maqdisi has argued that the leaders of the Muslim world are nonbelievers (because they have adopted and applied comprehensive systems of law that are not Islamic) and that this justifies a jihad against them. Al Maqdisi has, however, been considerably more conservative in articulating how this jihad should be practiced. He has cited Ibn Taymiyyah in his argument for a limited use of *takfir*, and he has advocated for a restrained approach to conflict that limits the potential targets of jihad.[73] He suggests that appropriate targets include only the rulers themselves and the government officials who support the regimes.[74] In making his argument, al-Maqdisi broke publicly with Zarqawi and wrote a tract in which he criticized the latter for his indiscriminate use of *takfir*, overbroad targeting, and excessive violence.[75] Al-Maqdisi has also argued against the frequency with which suicide bombers have been deployed suggesting that this practice should be used rarely.[76]

Importantly, al-Maqdisi is no moderate; he has praised the 9/11 hijackers for their actions and he supports jihad against both Americans and the corrupt rulers of Muslim nations.[77] He seems to be concerned, though, that the current generation of jihadi Salafis are engaging in jihad unwisely and that some of their more immoderate choices are undermining their strategic objectives.[78] He is also concerned that these fighters are pursuing jihad without the religious knowledge necessary to do so properly. He is worried, in other words, about the integrity of the movement and he has expressed concern that some practices corrupt jihad itself.[79] His work, according to Joas Wagemakers, might actually be understood as "an effort to take greater scholarly control of a trend that he feels responsible for but has also witnessed becoming more and more the prerogative of fighters instead of scholars."[80] In other words, al-Maqdisi may be attempting to undo some of the decentralization of authority that accompanied the rise of jihadi Salafism.

ISIS'S CLAIM TO THE CALIPHATE

ISIS's approach to the caliphate is similarly controversial. ISIS has established a *de facto* state that currently occupies land in Iraq, Syria, and Libya. It is an organization that controls a third of Syria and a quarter of Iraq, covers an area larger than Great Britain, and has a population larger than Denmark, Finland, or Ireland.[81] Whatever it may have been in the beginning, this now-independent organization "holds territory, provides limited services, dispenses a form of justice (loosely defined), most definitely has an army, and flies its own flag."[82] It also announced, in the summer of 2014, that it was the caliphate.

In making this claim the group effectively demanded allegiance from Muslims far beyond the borders that it currently controls. The group claimed that "the legality of all emirates, groups, states, and organizations becomes null by the expansion of the caliph's authority and arrival of its troops to their areas."[83] In short, the group has suggested that existing nations should quietly defer to its authority. This move, perhaps not surprisingly, was not universally embraced by the Muslim nations in the region. The group is, nonetheless, committed to defending its position and it has made an effort to reshape the public profile of its leader, Abu Bakr al-Baghdadi, in order to strengthen his claim to be the caliph. Sunni tradition—dating to the first years of the Muslim community—dictates that the caliph be a descendant of the Quraysh tribe, and since 2010 al Baghdadi has been increasingly vocal in claiming that he is one such descendant.[84]

ISIS's claim to have restored the caliphate is important for a variety of reasons: It gestures symbolically to a glorious past, it calls for allegiance and cooperation across the Muslim world, and it explicitly rejects Western models of governance and secularism. That said, it would be wrong to interpret this move as purely symbolic; it is unquestionably an attempt to return to an idealized form of government understood to have existed in an era when the Muslim world flourished.

CONCLUSION

Serious debates and deep conflicts have led to significant fractures among different jihadi Salafi groups. Despite the fact that individual movements within this faction often have much in common, the groups clearly disagree on a number of critical issues. In some cases, these divisions are rooted in the disagreements between the thinkers that inspired them. In other cases, today's movements have been influenced by the same thinkers and disagree principally on how to interpret their positions or how these interpretations should be translated into coherent strategies. What is critically important, though, is that the positions that we see taken by groups like al Qaeda and ISIS are not aberrations born of nothing; they are products of long-standing discussions about the authentic practice of Islam, the call to jihad, and the practice of war. Disagreements of this nature will, as a result, likely continue for years to come and will not disappear with the death of a single leader or the dissolution of a single group. Whether or not this means that the jihadi Salafism will continue to grow and flourish, however, is another question entirely.

AFTERWORD

ISIS represents a continually evolving threat. Since the hardcover version of this book went to press, ISIS's strategy has become clearer, as have possible steps to fight it. ISIS's long-term goal is a transnational caliphate. Its strategy for achieving its goals is becoming increasingly clear—both through its actions and statements:[1]

(1) Establish a presence in societies that are riven by political, ethnic, tribal, and/or sectarian tensions, whether just simmering, as in Pakistan, Saudi Arabia, and Tunisia; or in openly bloody civil wars, as in Syria, Nigeria, Libya, and now Yemen.

(2) Accentuate these divisions using calculated terrorist attacks, creating internal conflicts or external conflicts among potential adversaries to erode their morale and effectiveness, and whenever possible, pit them against each other in an atmosphere of escalating violence.

(3) When military control of such territories can be established, quickly extract their resources to fund and fuel additional expansion. This approach allows ISIS to enhance its overall resources by even temporarily controlling a town or region.

(4) Use a carefully calibrated messaging apparatus to project an image of strength and unity that exceeds the reality on the ground, while highlighting the political and military weaknesses of adversaries.

(5) Inspire leaders of other terrorist and insurgent organizations to pledge *bayah* to ISIS's leadership, currently in the person of Abu Bakr Baghdadi.

(6) Indoctrinate recruits using the language and techniques of an

apocalyptic, millenarian cult. This may be a calculated strategy, or it may reflect the actual beliefs of at least some ISIS leaders. Recruits may be local or global, and may be initially drawn to ISIS for pragmatic, political reasons, or they may be attracted primarily by its ideology. But once they are in the crucible, the ideology is liberally reinforced using every available means.

(7) Inspire "lone wolf" attacks and other action by loosely networked supporters throughout the region and the world.

On a tactical level, ISIS uses two tools consistently throughout the fronts it contests:

(1) **Extreme violence.** ISIS's extraordinary violence contributes to all of these goals in fairly quantifiable ways, such as polarizing the politics of its adversaries, intimidating military enemies, and exciting fanaticism.

(2) **Semi-controlled chaos.** Creating a chaotic situation is not the same thing as losing control. Chaos can be strategic, if one is poised to capitalize on it, and it serves the strategic goals of inflaming division and instilling apocalyptic fervor in the rank and file.

EXAMPLES OF ISIS STRATEGY AT WORK

IN IRAQ AND SYRIA

In April 2015, Iraq and the anti-ISIS coalition recaptured Tikrit from ISIS forces. This was a psychologically important victory, in part because Tikrit was Saddam Hussein's birthplace. For a time, it looked as though ISIS was beginning to falter in Iraq amid rumors of significant losses.

Izzat Ibrahim al-Douri, the leader of a former Baathist contingent allied with ISIS and Saddam's former vice president, was reportedly killed. Soon afterward, however, he resurfaced. There were also frequent rumors that Abu Bakr al-Baghdadi had been killed or wounded, and that his deputy, Abu Alaa al-Afri, was dead.

By mid-May 2015, however, ISIS forces were again on the rise. Ramadi—the capital of Iraq's largest province—fell to ISIS. As occurred when ISIS captured Mosul a year earlier, ISIS forces were able to capture military vehicles, weaponry, fuel, and ammunition. The town of Palmyra in Syria—the site of an ancient Roman city—fell to ISIS the same week. On May 14, Baghdadi issued a statement demonstrating he was still alive and in touch.

Military strategist David Kilkullen argued that the group had adapted to the coalition air campaign and had seized the initiative on both sides of the increasingly irrelevant border between Iraq and Syria.[23] As of this writing in November 2015, Kurdish forces were actively contesting ISIS control of Sinjar.[4] We anticipate that each side in the conflict will continue to wrest territory from the other for some time to come. The ISIS strategy of creating chaos and dividing its enemies received a boost in late September 2015, when Russia began air strikes intended to prop up the Assad regime. While characterized as fighting ISIS, most of the strikes targeted other rebel groups, including some supported by the United States. In addition to helping ISIS by putting pressure on its main rivals, Russian planes were involved in close calls with both American and Turkish planes, further ratcheting up tensions and raising the prospect of accidental escalation.[5] The increasingly crowded skies over Syria offer more opportunities every day for the many enemies of ISIS to get involved in distracting conflict among themselves.

IN ARAB COUNTRIES

On March 20, 2015, synchronized suicide bombings targeted two Shi'a mosques in the Yemeni capital of Sanaa and killed more than 140 people. The mosques were targeted specifically as gathering places for members of Yemen's Houthi rebels, a political movement with roots in the minority Zaydi sect of Shi'a Islam (although the coalition it leads in Yemen covers a number of different parties and issues).[6]

The bombing provided a pretext for an already surging Houthi rebellion to advance on the former government's last major stronghold in the port city of Aden. This in turn prompted Saudi Arabia to begin air strikes on Houthi positions and mass forces on its border with Yemen in advance of a possible ground invasion. Saudi Arabia sees the Houthi rebels as Iranian proxies and a direct threat to its national security.

On May 25, 2015, Saudi Arabia assembled a group of nine Arab governments to begin an air campaign against the rebels. While ISIS has not provided incontrovertible evidence it was directly responsible for the attacks on the mosques, no credible competing theory has emerged. Saudi Arabia's decision to begin bombing the Houthi rebels has increased sectarian tensions throughout the region, including within Saudi Arabia itself. ISIS followed on with two attacks at Shi'a mosques in Eastern Saudi Arabia—the first on May 22, 2015, and the second a week later. The first attack killed at least 21 people and injured 120 more; the second killed 4 people. Both attacks were carried out by suicide bombers in the middle of midday Friday prayers. ISIS also claimed credit for attacks in Tunisia on March 28 and May 25, 2015.[7] In the first, three gunmen wearing Tunisian military uniforms attacked a group of visitors to the Bardo National Museum in Tunis, leading to a three-hour siege during which the visitors were held hostage inside the museum. Twenty-two people died, most of them foreign tourists. In the second, a Tunisian soldier opened fire inside an army barracks. ISIS later released video as evidence to support its claim.

IN THE WILAYAT

How much does the ISIS central leadership oversee its wilayat or provinces? It's not entirely clear. It's possible that the provinces are taking their cues from the group's public urging to carry out attacks, but it is also possible that the central ISIS organization sometimes directs attacks. The level of visible coordination between ISIS and its recognized provinces—primarily an exchange of technical capability among their respective media branches—appears to vary from place to place. In Libya, Egypt, and Nigeria (home to Boko Haram, the most recent group to align with ISIS), the local branches showed significant upgrades to their communications apparatus, bringing them roughly in line with the central branch's standards.

The upgrade to Boko Haram's media operations is particularly noteworthy. It began just weeks ahead of its merger with ISIS and proceeded briskly: An audio recording of Boko Haram emir Abubakar Shekau swearing loyalty to the caliphate was released by ISIS's propaganda operatives online, and almost certainly required a sign-off from the group's senior leadership before being released.

It's easy to see how the channels that enable this level of coordination can also be used for command and control. The release of Shekau's pledge clearly involved private communications between Boko Haram and ISIS senior leadership, given that audio of the pledge was transferred to the group prior to its public release. A private message could easily have traveled the same route.

In ISIS's Algeria, Saudi Arabia, Yemen, and Afghanistan "provinces," visible communications improvements have been more modest, and propaganda output has been less prolific, but the evidence of coordination is growing. May saw the debut of ISIS's signature style in propaganda releases from its Yemen branch, and from late May to early June, it began releasing pictures and video relating to attacks in Tunisia and Saudi Arabia.

Each province has a different security posture, and those that control significant territory seem to communicate more easily and more often. Taken together, all of this strongly suggests ISIS's central command has robust communications capabilities, including at least some capacity to transmit instructions from its senior leadership. Meanwhile, its branches outside Syria and Iraq may have individualized limitations as far as their ability to send messages or receive instructions from the central command.

This capacity appears to be significantly more advanced than al Qaeda's equivalent infrastructure, based on the most recent intelligence. Al Qaeda emir Ayman al Zawahiri has gone silent for months at a time, and affiliates have complained about his silence.[8] The lack of timely coordination was thrown into sharp relief in September 2015, when al Qaeda released a series of embarrassing speeches by Zawahiri, which attacked ISIS for breaking its oath of loyalty to Taliban leader Mullah Omar—despite the fact the Taliban publicly admitted in August that Omar had been dead since 2013.[9]

On October 31, a passenger plane traveling from Sharm El-Sheikh, Egypt, to Saint Petersburg, Russia, crashed in the Northern Sinai, killing all 224 people aboard. ISIS's Sinai Wilayat immediately claimed that it had downed the plane. The Russian government confirmed that the crash was caused by a bomb on board, with ISIS as the likely culprit.[10]

IN THE WEST

In Paris, on January 7, 2015, two gunmen forced their way into the offices of the French satirical magazine *Charlie Hebdo*, killing 11 members of the magazine's staff and a police officer at the scene. The gunmen, who were brothers, claimed that they had struck the magazine on behalf of al Qaeda in the Arabian Peninsula, which soon took credit for the attack.

AQAP had earlier urged lone wolves to attack the magazine and its editor in its online magazine, Inspire. In a video recorded before the *Charlie Hebdo* attack, a third jihadist claimed that he was working with the brothers on behalf of ISIS, stating that he had pledged allegiance to ISIS's leader as soon as the caliphate was declared. He carried out two subsequent attacks, murdering a French police officer and holding up a kosher supermarket, where he shot four people.

The incident raised a broader question about how seriously we should take the possibly temporary enmity between al Qaeda and ISIS. Might they sometimes work together, even if they are stated enemies? This kind of competition/cooperation is typical of "chaordic organizations," which are networked, self-governing organizations working toward a common goal. Jihadi groups are typically organized in this way—competing for attention, recruits, and funding while also sometimes working together. We expect to see continuing "chaordic" activities among the various al Qaeda offshoots.

This doesn't preclude or negate open hostility between the parent organizations. Since many ISIS recruiters were previously al Qaeda supporters, their social networks are intermingled and personal bonds may trump leadership disputes. It's also worth noting that since the attribution of a terrorist attack may not be clear until long after the fact, members of one or the other organization may independently carry out a follow-up attack without knowing which team was responsible for the first.

A week after the Paris attacks, Belgian police claimed to have thwarted a major ISIS attack on a city in eastern Belgium. The cell was led by an ISIS operative known as Abu Umar al-Belgiki, who escaped the Belgian raid and left behind a trail of breadcrumbs on social media that stretched from Europe to Syria.[11] European officials voiced concerns about ISIS cells operating inside Europe, some of which could include trained operatives returning from Syria.[12]

Those concerns were well founded. ISIS had everything it needed to carry out external operations outside its geographical domain—men, money, organization, and time. It had already flexed those muscles in Sana'a in March, but few were prepared for its fall offensive.

Shortly after the downing of the Russian airliner, suicide bombers struck Beirut on November 12, killing dozens of civilians in a Shia neighborhood with ties to Hezbollah.[13]

On November 13, ISIS struck Paris with a ruthlessly efficient coordinated attack. Three teams of operatives employing guns and suicide belts killed at least 130 people. There were eight attackers in total, and more waiting in the wings. France was temporarily paralyzed, and President Hollande called for "merciless" strikes on ISIS's strongholds in Syria and Iraq. More than 150 raids were reportedly carried out, until on November 17, another major cell was discovered. Eight more conspirators were arrested, and two were killed, including Abu Umar al-Belgiki (real name Abdelhamid Abaaoud)—the cell leader who had escaped the raid in January. The cell was believed to be close to carrying out a follow-up attack.[14]

ISIS soon claimed all the attacks, publicizing a picture of the IED used to take down the Russian plane and publishing a special issue of *Dabiq* magazine devoted to the attack.[15] The attacks fit with the ISIS strategy of using terrorism to put pressure on existing tensions in society, creating fears that ISIS had infiltrated the massive outflow of Syrian refugees and prompting a xenophobic backlash across Europe and in the United States.[16]

Following ISIS's declaration of the caliphate in June 2014, analysts and policy makers had struggled to pin down the organization's terrorist capabilities. As an insurgency, it had proved both effective and resilient. As a terrorist actor, projecting a threat outside its borders, it was less proven (although Sana'a had been a heavy clue). With this spate of attacks outside Islamic State borders, the debate was settled.

Many experts concluded these attacks were a sign that ISIS had changed its strategy. In our view, the change was only one of tactics and emphasis. ISIS represents a continually evolving threat, but its claims regarding its overarching goals have not wavered. ISIS has stressed again and again that it has two principle (and possibly contradictory) objectives: to expand its caliphate, and to increase sectarian and anti-Muslim sentiment—with the ultimate goal of provoking the West into a final battle in Syria.

With these attacks beyond Islamic State borders, ISIS was deploying the assets that we described in detail in this book—foreign fighters and disenfranchised Muslims living in the West, some of whom had been trained by ISIS. As we warned earlier, given ISIS's resources—both in terms of personnel and capital—we cannot rule out the possibility of more spectacular attacks, even on or near the scale of 9/11.

Lone wolf strikes on behalf of ISIS have also been attempted in the United States. In Garland, Texas, on May 3, 2015, two apparent ISIS supporters were killed attempting to attack an event that involved drawing the Prophet Mohammed. A police officer was wounded. ISIS supporters had openly urged attacks on the event online for more than a week prior. While the attack was thwarted, it was not prevented. A month later, law enforcement officers shot and killed an ISIS-inspired individual in Boston who was threatening them with a large military-style knife. The assailant had recently told an associate that he planned to attack police, a strategy that ISIS has been promoting to its followers.

In 2015 alone, as of this writing in late November, more than 55 Americans have been arrested or killed by law enforcement for activities related to ISIS (including both plots and efforts to travel to Syria)—a staggering increase from the pace of al Qaeda–related arrests in preceding years. Every one of these cases involved significant social media contact with or exposure to ISIS.

SOCIAL MEDIA CHANGES

The social media landscape has changed significantly since the first edition of this book. Although ISIS continues to maintain a strong presence on Twitter, the company has cracked down on extremist networks as the result of considerable public pressure. Twitter has suspended tens of thousands of ISIS-supporting accounts, mostly in response to user abuse reporting, driven in turn by online campaigns by hacktivists such as Anonymous. Because the reporting is tied to organized pushes by hacktivists, certain kinds of high-profile accounts have been heavily targeted, while upward of 10,000 ISIS supporters have been largely untouched. Many suspended accounts belong to the same users, who create new accounts time and again, only have them suspended again. As a result, the number of suspensions far outpaces the reduction in the number of active ISIS accounts.

Nevertheless, the campaign has proven effective. As of this writing, ISIS is unable to game hashtags and search results (as discussed in Chapter 7). It is rare to find an ISIS supporter online with more than 10,000 followers, and a significant majority now have fewer than 150 each. The estimated total number of ISIS-supporting accounts has been reduced from about 46,000 in the fall of 2014 to a level somewhere around 30,000, with some variation from day to day. While the network is down, it is by no means out, as the forty American cases referenced above clearly indicate.

In cases where a clear trajectory could be determined, about one-third of the suspects appear to have been radicalized by al Qaeda–affiliated content prior to the rise of ISIS, and only later shifted allegiance to ISIS. The remainder were reportedly radicalized by ISIS directly. While this points to the growing influence of ISIS among those vulnerable to radicalization, it also highlights the fact that this activity takes place in an evolving context, rather than being an entirely new or different problem.

The main challenge ahead in the messaging war is one of volume. Even at its reduced levels, ISIS deploys thousands of supporters to push its narrative, in a coordinated manner, across multiple languages. In contrast, organized anti-ISIS activity employs numbers of accounts in uncoordinated clusters ranging in size from a handful of accounts to the low dozens, and they tweet far less prolifically than their extremist counterparts. In some ways, the specific content of anti-ISIS messaging is much less important than the volume. Numbers matter in social media, and a few thousand accounts can shape the activity of a much larger social network.

CLOSING THOUGHTS

It will not be possible to fulfill President Obama's goal of defeating ISIS until we recognize that we are fighting a hybrid organization that includes a military organization, a proto-state, and a terrorist group.

As we write these words, ISIS is continuing to expand the territory it controls militarily and to attract new recruits to carry out terrorist strikes in its name, in the region and beyond. The spread of the military organization is a symptom of governments' inability to provide security for Sunnis, state failure, and/or civil war. Curtailing its impact will require not just a military response, but also successful efforts to address these underlying causes.

A critical first step is to develop a clear vision for Syria's future and a strategy for achieving it, an extremely tall order given the conflicting agendas of the members of the anti-ISIS coalition. Finding an acceptable end state for Syria that satisfies the diverse interests of the United States, Turkey, Iran, and Saudi Arabia may not be possible, and this is perhaps the most powerful asset in ISIS's war chest.

ISIS's appeal as a terrorist organization depends in part on its military successes, but also on its ability to seduce young people aiming to reinvent themselves as ruthless fighters for a dark, new interpretation of violent Salafi Islam. The long-term goal must be to find a way to defeat, not only the organization but also the appeal of its twisted ideology.

Here too there may be fundamental limits to our efforts. With its capabilities for self-organizing around content, social media empowers people who hold fringe ideas to discover and connect with each other in ways that were never possible before. While we can defuse some of ISIS's most manipulative tactics to exploit these dynamics, the genie of self-organizing communities cannot be put back in the bottle.

Finally, while a coalition may eventually solve for the territorial problem of Iraq and Syria, ISIS adherents displaced from that territory are likely to take up terrorism with a vengeance. A glimpse of things to come can be seen in Somalia, where al Shabab was substantially defeated on the military front, only to double down on deadly terrorist attacks.[17] The apocalyptic emphasis of ISIS suggests this problem could be even worse if its adherents are faced with the failure of its prophesied victories and choose to "up the ante" with even more violence.

For all these reasons, and more, the problem of ISIS is likely to be with us for a long time to come, whether in its current form or in some future mutation, barring an extraordinary act of self-destruction, which still lies within the realm of possibility. Much of ISIS's messaging is based on distortion, but its slogan and promise—baqiyyah, to remain—continues to be tragically credible.

ACKNOWLEDGMENTS

J.M. BERGER

It was a long road to this book and many people helped along the way. I can't list them all here, but I can make a dent.

First of all, I want to thank Jessica Stern for inviting me to work on this project with her. I am extraordinarily grateful for both the vote of confidence and the opportunity to collaborate with someone whose work I have admired for years. I would not have had this opportunity to work on this important subject at such length without her, and she brought many ideas and resources to the book that would otherwise have been absent. For all this and more, I thank her.

Over the course of many years slogging it out alone, I have been fortunate to encounter others as well who have also placed their confidence in me and allowed me pleasure and privilege of their insights.

Those relationships, and all I have learned from them, are reflected in these pages, often very directly. Among those most represented are Aaron Zelin of the Washington Institute for Near East Policy and Heather Perez, colleagues and friends whom I trust completely for their judgment, knowledge, and skills. I'm lucky to know both of them, and I couldn't have done this without them.

Providing additional crucial input on the manuscript itself were valued friends and esteemed colleagues Will McCants and Charles Lister, both of the Brookings Institution, and Brian Fishman of the New America Foundation. Their invaluable feedback on the book is surpassed only by everything I've learned from their work in the past and expect to learn in the future.

As ISIS was on the rise, before this book was even a glimmer in anyone's eye, I was lucky to share the counsel, collaboration, conversation, feedback, friendship, and good (if occasionally gallows) humor of Clint Watts of the Foreign Policy Research Institute and Aaron Weisburd. It was a pleasure to learn from these pros, and I look forward to learning more.

Many more have generously shared their expertise and camaraderie over the years, and to each I owe an individual debt of knowledge or support, or both. The list includes but is by no means limited to Humera Khan, Daniel Kimmage, Kirsten Fontenrose, Tamar Tesler, Christina Nemr, Thomas Hegghammer, Peter Bergen, J. C. Brisard, Dave Gomez, Don Rassler, Chris Heffelfinger, Rachel Milton, and many others, including some who for various reasons cannot be named.

John Horgan and Mia Bloom deserve an extra-special shout-out here for their professional support, personal friendship, and the role they played in bringing Jessica and me together for this project. (Not to mention Thomas Hegghammer, who first introduced us some months earlier.)

In the world of journalism, I am thankful for those who have had the good grace to take me seriously, especially James Gordon Meek of ABC News, Scott Shane of the *New York Times*, Thanassis Cambanis of the *Boston Globe*, and Josh Meyer of the Medill National Security Journalism Initiative.

Patiently and creatively supporting my education on the technical and social media side of things have been, among others, Daniel Sturtevant, Bill Strathearn, Jonathan Morgan, Justin Seitz, Yasmin Green, and Brendan Ballou.

For giving me platforms from which to write, I thank Foreign Policy, and editors Ben Pauker, Blake Hounshell, Peter Scoblic, Susan Glasser, Uri Friedman, Noah Shactman, Hillary Claggett, and many others. And given my work's focus on social media, both in the book and out, I would be remiss not to mention those who have over the medium of Twitter consistently encouraged me, shared their knowledge, or both, including @NewNarrative, @el_grillo1, @stick631, @hipbonegamer, @ibsiqilli, @gregorydjohnsen, @blogsofwar, @morgfair, and @hlk01.

For the writing and research of this book, many hands helped carry the load. They include Jessica's research assistant Abigail Dusseldorf, who put in many long hours on a host of issues; my research assistant Sam Haas, who provided critical help on ISIS's external operations and a host of other matters; and on Twitter-related issues, Jonathan Morgan, Youssef Ben Ismail, Yasmin Green, and Jana Levene. Without their labors, this wouldn't have come together, and all helped make this book better and more complete.

At Ecco Books, the enthusiastic support and editing of Daniel Halpern was one of the primary factors in making this project happen, with Gabriella Doob sharing editing duties and catching countless fixes. Martha Kaplan of the Martha Kaplan Agency provided tremendous assistance in bringing all the elements together and sage counsel along the way.

Finally, I want to thank my family, most especially my wife, Janet, without whom none of this would have been possible, or even imaginable. Most of the people named herein are part of a complex web of events and capabilities. Subtract any of them, and the course of my career might have veered away from this moment. But no one is more essential than Janet. From the start of my interest in the topic of terrorism, through long years during which that interest seemed of questionable utility, through the writing of my first book and through every step of writing this book, Janet's emotional, logistical, and editing support have been crucial. None of my professional steps forward could or would have happened without her, but my gratitude to her for all the things she has enabled in my work is only a tiny fraction of my gratitude for all the things she has brought to my life.

JESSICA STERN

First, I thank Dan Halpern, whom I am lucky to call my editor, whose idea it was to write this book. I would not have dared to tackle the topic without Dan's encouragement and counsel. I also thank Gabriella Doob, associate editor at Ecco, for her extraordinary efforts on our behalf, including lightning-fast editing and other help.

J.M. Berger is a beloved scholar, and after this experience of working with him, I understand why. I could not have done this without him as coauthor, sometimes editor, sometimes antagonist, sometimes cheerleader. As is perhaps not surprising, I first noticed J.M.'s work on social media, when I started following him to observe his conversations with Omar Hammami, the now deceased member of the terrorist group Al

Shabab. J.M. follows terrorists online as obsessively as I would speak to them, were that still possible. Very few people develop an intuition about terrorists. J.M. is one of them, and I feel lucky to have had the chance to work with him.

My colleague Saida Abdi brought J.M. to my attention. Thank you, Saida. And Thomas Hegghammer, you are the reason I met J.M. in the flesh.

I thank all those who were willing to read the manuscript and provide comments, including, especially, Scott Atran, Dean Atkins, Mia Bloom, Brian Fishman, Martha Kaplan, Will McCants, Charles Lister, and Aaron Zelin. I have learned so much from all of you.

John Horgan and Mia Bloom have been great friends to both of us throughout the project.

We had a team of research assistants who were vitally important, Sam Haas, my doctoral student Megan McBride (who also wrote the appendix), Jennie Spector, and Youssef Ben Ismail. I thank especially Abigail Dusseldorp, who often worked with us from early morning until late at night.

Martha Kaplan, you are the best agent in the world, always seeing the best in me, pushing me further than I imagine to be possible.

The Hoover Institution has provided me an intellectual home away from home and a set of excellent colleagues, the Members of the Task Force on National Security and Law. I thank you all for the conversations we've had over the years, which have taught me a great deal.

Jennifer Leaning and Jacqueline Bhabha provided me an institutional home at the FXB Center for Health and Human Rights at the Harvard School of Public Health. I cherish you both and am in awe of your good works in the world.

Ron Schouten, associate professor of psychiatry at Mass General Hospital, thank you for teaching me about psychopathy and for being such a good ally and collaborator.

Ted Flinter, an outstanding former student, announced out of the blue that he wanted to fund my research. Hassan Abbas, another remarkable former student, came to see me while I was laboring over the end of time. When the student appreciates the teacher, it means the world.

My own professors—Ash Carter, Matthew Meselson, and Richard Zeckhauser—taught me how to think about national security. You continue to inspire me.

Several anonymous Iraqi citizens were kind enough to spend time with me, both in the United States and abroad. They helped me a great deal in my efforts to make sense of the rise of ISIS.

Finally, I thank my family. I thank especially Chet and Evan. Thank you for making love possible.

This book is written with the victims of ISIS's terrorism very much in mind and heart. My thoughts and prayers are with the citizens who are bravely fighting ISIS in their midst.

NOTES

INTRODUCTION

1. "Profile: James Foley, US Journalist Beheaded by Islamic State," BBC World News, August 20, 2014, http://www.bbc.com/news/world-28865508.

2. Rukmini Callimachi, "The Horror Before the Beheadings: Isis Hostages Endured Torture and Dashed Hopes, Freed Cellmates Say," *New York Times*, October 25, 2014.

3. "Bigley's Wife Tells of Her Grief," BBC World News, October 9, 2004, http://news.bbc.co.uk/1/hi/uk/3729158.stm; Joel Roberts, "Report: Japanese Hostage Killed," CBS News, October 30, 2009, http://www.cbsnews.com/stories/2004/10/30/iraq/main652421.shtml; Toby Harnden, "South Korean Hostage Beheaded by al-Qaeda," *The Telegraph*, June 23, 2004, http://www.telegraph.co.uk/news/worldnews/middleeast/iraq/1465300/South-Korean-hostage-beheaded-by-al-Qaeda.html; videos downloaded from al Qaeda in Iraq file servers, 2004 to 2006.

4. David Remnick, "Going The Distance," *New Yorker*, January 2014, http://www.newyorker.com/magazine/2014/01/27/going-the-distance-2?currentPage=all.

5. Conversation with Haidar Alaloom (Senior Policy Analyst and Strategist at Humanize Global), November 2014.

6. George W. Bush, "Speech at the National Endowment for Democracy," October 6, 2005, http://www.presidentialrhetoric.com/speeches/10.06.05.html.

7. Thomas Carothers, "The End of the Transition Paradigm," *Journal of*

Democracy 13, no. 1 (2002): 5–21. See also Fareed Zakaria, "The Rise of Illiberal Democracy," *Foreign Affairs*, November/December 1997. Fareed Zakaria, *The Future of Freedom: Illiberal Democracy at Home and Abroad* (New York: W.W. Norton and Company, 2003); F. Gregory Gause, "Can Democracy Stop Terrorism?" *Foreign Affairs*, September/ October 2005.

8. David Kirkpatrick, "ISIS' Harsh Brand on Islam Is Rooted in Austere Saudi Creed," *New York Times,* September 24, 2014, http://www .nytimes.com/2014/09/25/world/middleeast/isis-abu-bakr-baghdadi -caliph-wahhabi.html?_r=0.

9. Ed Husain, "Saudis Must Stop Exporting Extremism," *New York Times,* August 22, 2014, http://www.nytimes.com/2014/08/23/opinion/isis -atrocities-started-with-saudi-support-for-salafi-hate.html.

10. Section on naming adapted from J.M. Berger, "What's in a Name?" *IntelWire*, August 10, 2014, http://news.intelwire.com/2014/08/whats -in-name.html.

11. Patrick Lyons and Mona El-Naggar, "What to Call Iraq Fighters? Experts Vary on S's and L's," *New York Times,* June 18, 2014, http://www .nytimes.com/2014/06/19/world/middleeast/islamic-state-in-iraq -and-syria-or-islamic-state-in-iraq-and-the-levant.html?_r=0.

12. J.M. Berger, "Gambling on the Caliphate," *IntelWire*, June 29, 2014, http://news.intelwire.com/2014/06/gambling-on-caliphate.html.

13. Liz Peek, "Obama's Use of ISIL, Not ISIS, Tells Another Story," Fox News, August 24, 2014.

14. Matt Apuzzo, Twitter post.

15. Adam Taylor, "France is ditching the 'Islamic State' name," *Washington Post,* September 17, 2014, http://www.washingtonpost.com/blogs /worldviews/wp/2014/09/17/france-is-ditching-the-islamic-state -name-and-replacing-it-with-a-label-the-group-hates/.

16. Portions of this discussion of terminology were adapted from J.M. Berger, *Jihad Joe: Americans Who Go to War in the Name of Islam* (Washington, DC: Potomac Books, 2011).

17. Steve Emerson, "Abdullah Assam: The Man Before Osama Bin Laden," International Association for Counterterrorism and Security Professionals website, undated, http://www.iacsp.com/itobli3.html, accessed August 25, 2010.

18. Just war requires that two conditions be met: just cause (*jus ad bel-*

lum) and just means or justice in war (*jus in bello*). *Jus in bello* requires that the belligerents' methods be proportional to their ends, and that they do not directly target noncombatants. While terrorism may, in principle, meet the requirement of a just cause, it does not meet the second: Terrorists by definition target noncombatants, which is explicitly prohibited by both the Judeo-Christian and Islamic Just War Tradition. The double effects rule modifies this requirement to allow for acts of war that inadvertently result in loss of civilians lives, which military strategists call collateral damage. What matters here, according to philosopher Steven Lee, is intention. There is a morally relevant difference, he argues, "between merely foreseeing the deaths of noncombatants as an effect of military activity and intending to bring about those deaths; the principle of discrimination rules out the activity only in the latter case." Steven Lee, "Is the Just War Tradition Relevant in the Nuclear Age?" *Research in Philosophy and Technology* 9 no. 85. But, what if the adversary knows in advance that many civilian lives will be lost in an attack that is aimed at military targets? This is a moral problem that military strategists still grapple with. The ability to aim at a particular target, with minimal damage, is what makes drones so attractive to those who aim to comply with the law of war. See also, James Turner Johnson, *Modern Contemporary Warfare* (New Haven: Yale University Press, 1999) and John Kelsay, *Arguing the Just War in Islam* (Cambridge: Harvard University Press, 2007).

19. Terrorists' goals can be instrumental (changing the world) and expressive (drawing attention to a cause) or both. For more on this topic, see Jessica Stern and Amit Modi, "Producing Terror: Organizational Dynamics of Survival," in Thomas Biersteker and Sue Eckert, eds., *Countering the Financing of Terrorism* (Rutledge, 2007).

20. Kenneth Anderson writes frequently on this topic. See, for example, "Alan Dershowitz on Degrees of 'Civilianality,'" Kenneth Anderson's Law of War and Just War Theory Blog, July 22, 2006, http://kennethandersonlawofwar.blogspot.com/2006/07/alan-dershowitz-on-degrees-of.html. See also, Monika Hlavkova, "Reconstructing the Civilian/Combatant Divide: A Fresh Look at Targeting in Noninternational Armed Conflict," *Journal of Conflict & Security Law* 19 no. 2 (2014): 251–78.

21. Peter Singer, "Facing Saddam's Child Soldiers," Brookings Institu-

tion, January 14, 2003, http://www.brookings.edu/research/papers/2003/01/14iraq-singer.

22. Tim Arango, "A Boy in ISIS. A Suicide Vest. A Hope to Live," *New York Times*, December 28, 2014, http://www.nytimes.com/2014/12/27/world/middleeast/syria-isis-recruits-teenagers-as-suicide-bombers.html.

23. Article 26 of the Rome Statute states that: "The Court shall have no jurisdiction over any person who was under the age of 18 at the time of the alleged commission of the offence." Matthew Happold, "The Age of Criminal Responsibilty in International Criminal Law," in Karin Arts and Vesselin Popovski, eds., *International Criminal Accountability and Children's Rights* (The Hague: T.M.C. Asser Press, 2007) http://www.asser.nl/Default.aspx?site_id=9&level1=13337&level2=13345. The prosecution of children under 18 is relegated to national courts, and national laws regarding the age of criminal responsibility vary. The Paris Principles, "Principles and Guidelines on Children Associated with Armed Forces or Armed Groups," urges that children accused of war crimes be considered "primarily as victims," and "not only as perpetrators," which would seem to leave room for debate. "The Paris Principles: Principles and Guidelines on Children Associated with Armed Forces or Armed Groups," United Nations International Children's Emergency Fund, February 2007, http://www.unicef.org/emerg/files/ParisPrinciples310107English.pdf. For a case in which a child was treated as a war criminal, see "United States of America vs. Omar Ahmed Khadr," available at http://media.miamiherald.com/smedia/2010/10/26/10/stip.source.prod_affiliate.56.pdf.

24. Arango, "A Boy in ISIS. A Suicide Vest. A Hope to Live."

CHAPTER 1. THE RISE AND FALL OF AL QAEDA IN IRAQ

1. Mary Anne Weaver, "The Short, Violent Life of Abu Musab al-Zarqawi," *Atlantic*, June 8, 2006, http://www.theatlantic.com/magazine/archive/2006/07/the-short-violent-life-of-abu-musab-al-zarqawi/304983/.

2. Jean-Charles Brisard, *Zarqawi: The New Face of Al-Qaeda* (New York: Other Press, 2004).

3. Barbara Metcalf, "Traditionalist Islamic Activism: Deoband, Tablighis and Talibs," Social Service Research Council, November 1, 2004.

4. Jessica Stern, *Terror in the Name of God: Why Religious Militants Kill* (New York: Ecco, 2003).

5. Weaver, "The Short, Violent Life of Abu Musab al-Zarqawi."

6. Ibid.

7. Assaf Moghadam, "The Salafi-Jihad as a Religious Ideology," *CTC Sentinel* 1, no. 3 (February 2008), https://www.ctc.usma.edu/posts/the -salafi-jihad-as-a-religious-ideology.

8. Nibras Kazim, "A Virulent Ideology in Mutation: Zarqawi Upstages Maqdisi," Hudson Institute, http://www.hudson.org/content /researchattachments/attachment/1368/kazimi_vol2.pdf.

9. Weaver, "The Short, Violent Life of Abu Musab al-Zarqawi."

10. Nir Rosen, "Iraq's Jordanian Jihadis," *New York Times Magazine,* February 19, 2006, http://www.nytimes.com/2006/02/19/magazine/iraq .html?pagewanted=all&_r=0.

11. Henry Schuster, "Al-Zarqawi and al Qaeda in Jordan," CNN.com, November 12, 2005, http://www.cnn.com/2005/WORLD/meast/11/11 /zarqawi.jordan/index.html?_s=PM:WORLD.

12. Brisard, *Zarqawi.*

13. Weaver, "The Short, Violent Life of Abu Musab al-Zarqawi."

14. Ibid.; Bruce Reidel, *The Search for Al Qaeda: Its Leadership, Ideology, and Future* (Brookings Institution Press, 2010), 94.

15. Craig Whitlock, "Al-Zarqawi's Biography," *Washington Post,* June 8, 2006, http://www.washingtonpost.com/wpdyn/content/article/2006 /06/08/AR2006060800299_pf.html.

16. Weaver, "The Short, Violent Life of Abu Musab al-Zarqawi."

17. Senate Select Committee on Intelligence, *Postwar Findings About Iraq's WMD Programs and Links to Terrorism and How They Compare with Prewar Assessments* (Washington, DC: U.S. Government Printing Office, 2006), http://www.intelligence.senate.gov/phaseiiaccuracy.pdf, 92; "Ansar al-Islam," Stanford University Mapping Militant Organizations, last updated August 14, 2014.

18. R. Jeffrey Smith, "Hussein's Prewar Ties to Al-Qaeda Discounted: Pentagon Reports Says Contacts Were Limited," *Washington Post,* April 6, 2007.

19. "Full Text of Colin Powell's Speech," *Guardian,* February 5, 2003, http://www.theguardian.com/world/2003/feb/05/iraq.usa.

20. U.S. Senate Select Committee on Intelligence, *Postwar Findings,* 110; National Commission on Terrorist Attacks upon the United States, *The 9/11 Commission Report* (Washington, DC: U.S. Government Printing Office, 2004), 66; Smith, "Hussein's Prewar Ties to Al-Qaeda Discounted."

21. "President Bush Pledges to Rout Terrorism 'Wherever It Exists,'" IIP Digital, February 2003, http://iipdigital.usembassy.gov/st/english /texttrans/2003/02/20030214160346porth@pd.state.gov0.5013697 .html#axzz3KQBs2LRT.

22. Abu Musab al-Suri as summarized in Lawrence Wright, "The Master Plan," *New Yorker,* September 11, 2006.

23. Donna Miles, "Bush Calls Iraq Central Front in Terror War, Vows Victory," U.S. Department of Defense, press release, October 6, 2005, http://www.defense.gov/News/NewsArticle.aspx?ID=18145.

24. Jessica Stern, "How America Created a Terrorist Haven," *New York Times,* August 20, 2003, http://www.nytimes.com/2003/08/20 /opinion/how-america-created-a-terrorist-haven.html.

25. Statistics based on data from the National Consortium for the Study of Terrorism and Responses to Terrorism's Global Terrorism Database and meet the following criteria: each event is an intentional act of violence committed or threatened by a nonstate actor; it is committed with the purpose of attaining a political, economic, religious, or social goal ($crit1 = 1$); it is designed to coerce, intimidate, or convey a message to an audience beyond its immediate victims ($crit2 = 1$); it is committed outside the context of legitimate warfare and violates the conventions outlined in international humanitarian law ($crit3 = 1$); it is unambiguous in meeting this criteria ($doubtterr = 0$) but not necessarily carried out successfully ($success = 0, 1$). See Jessica Stern and Megan McBride, "Terrorism after the 2003 Invasion of Iraq," http://costsofwar .org/sites/default/files/articles/47/attachments/McBride1.pdf.

26. National Consortium for the Study of Terrorism and Responses to Terrorism, Global Terrorism Database, 2012, http://www.start.umd .edu/gtd. See Stern and McBride, "Terrorism after the 2003 Invasion of Iraq," accessed January 31, 2013.

27. Visions of Humanity, "Terrorism Index 2007: Global Rankings," ed-

ited by Institute for Economics and Peace, 2007, accessed November 2014.

28. Brian Fishman, "Redefining the Islamic State," New America Foundation, August 2011, http://security.newamerica.net/sites/newamerica.net/files/policydocs/Fishman_Al_Qaeda_In_Iraq.pdf.

29. Brian Fishman, "Redefining the Islamic State," New America Foundation, August 2011, http://security.newamerica.net/sites/newamerica.net/files/policydocs/Fishman_Al_Qaeda_In_Iraq.pdf; Maamoun Yousef, "Islamic Web Sites Criticize Jordan Bombing," Associated Press, November 15, 2005.

30. Sharon Otterman, "Iraq: Debaathification," Council on Foreign Relations, April 7, 2005, http://www.cfr.org/iraq/iraq-debaathification/p7853#p3.

31. Sarah Childress, Evan Wexler, and Michelle Mizner, "Iraq: How Did We Get Here?," *Frontline*, PBS, July 29, 2014; Lee Hudson Teslik, "Profile: Abu Musab al-Zarqawi," Council on Foreign Relations, June 8, 2006, http://www.cfr.org/iraq/profile-abu-musab-al-zarqawi/p9866#p4.

32. Bruce R. Pirnie, Edward O'Connell, *Counterinsurgency in Iraq (2003–2006): RAND Counterinsurgency Study—Volume 2* (Rand Corporation, 2008), 50–51; email interview with Charles Lister, Visiting Fellow, Foreign Policy, Brookings Institution Doha Center.

33. Email interview with Phillip Smyth, November 21, 2014.

34. Neil MacFarquhar, "After the War: Attack at Shrine; Car Bomb in Iraq Kills 95 at Shiite Mosque," *New York Times*, August 30, 2003, http://www.nytimes.com/2003/08/30/world/after-the-war-attack-at-shrine-car-bomb-in-iraq-kills-95-at-shiite-mosque.html; Elizabeth Farnsworth, "Return of Shiite Leader Ayatollah Mohammad Bakr al Hakim," *PBS Newshour* (radio), May 12, 2003, http://www.pbs.org/newshour/bb/middle_east-jan-june03-basra_05-12/.

35. Murad Batal al-Shishani, "Al-Zarqawi's Rise to Power: Analyzing Tactics and Targets," Terrorism Monitor, Jamestown Foundation, November 17, 2005.

36. Aaron Zelin, "The War Between ISIS and al-Qaeda for Supremacy of the Global Jihadist Movement," Washington Institute, June 2014.

37. Charles Lister, "Profiling the Islamic State," Brookings Doha Center Analysis Papers, November 2014, http://www.brookings.edu

/~/media/Research/Files/Reports/2014/11/profiling%20islamic%20 state%20lister/en_web_lister.pdf; Joas Wagemaker, "Abu Muhammad al-Maqdisi; A Counter-Terrorism Asset?" *CTC Sentinel* 1, no. 6 (May 2008), https://www.ctc.usma.edu/posts/abu-muhammad-al -maqdisi-a-counter-terrorism-asset.

38. Joseph Felter and Brian Fishman, "Al-Qa'ida's Foreign Fighters in Iraq: A First Look at the Sinjar Records," January 2, 2007, The Combating Terrorism Center at West Point, https://www.ctc.usma.edu/posts/al -qaidas-foreign-fighters-in-iraq-a-first-look-at-the-sinjar-records; Matthew Levitt, "Foreign Fighters and Their Economic Impact: a Case Study of Syria and al-Qaeda in Iraq (AQI)," *Perspectives on Terrorism* 3, no. 3, 2009, http://www.terrorismanalysts.com/pt/index.php/pot /article/view/74/html; Phil Sands, "Syria stops insurgents on Iraq border," *The National (UAE)*, November 2, 2008.

39. Robin Simcox, "Al-Qaeda's Global Footprint," Henry Jackson Society, 2013.

40. "Takfir," *Oxford Dictionary of Islam,* ed. John L. Esposito (Oxford: Oxford University Press, 2003).

41. Don Rassler, Gabriel Koehler-Derrick, Liam Collins, Muhammad al-Obaidi, and Nelly Lahoud, *Letters from Abbottabad: Bin Laden Sidelined,* ed. Harmony Program (West Point, NY: Combating Terrorism Center at West Point, 2012).

42. Aaron Zelin, "Al-Qaeda Disaffiliates with the Islamic State of Iraq and al-Sham," Washington Institute, February 4, 2014.

43. Letter from Zawahiri, July 9, 2005, https://www.ctc.usma.edu/wp -content/uploads/2010/08/CTC-Zawahiri-Letter-10-05.pdf.

44. Stephen Ulph, "New Online Book Lays Out Al-Qaeda's Military Strategy," Jamestown Foundation, March 16, 2005, http://www.jamestown .org/programs/tm/single/?tx_ttnews%5Btt_news%5D=27713&tx _ttnews%5BbackPid%5D=238&no_cache=1#.VIhfUmTF_LB.

45. Ibid.

46. Brian Fishman, "Jihadists Target the American Dream," *CTC Sentinel* 1 no. 4 (March 2008), https://www.ctc.usma.edu/posts/jihadists-target -the-american-dream.

47. Abu Bakr Naji, *The Management of Savagery,* translated by William McCants, May 23, 2006, John M. Olin Institute for Strategic Stud-

ies, Harvard University, http://azelin.files.wordpress.com/2010/08/abu-bakr-naji-the-management-of-savagery-the-most-critical-stage-through-which-the-umma-will-pass.pdf.

48. David Cook, "Abu Musa'b al-Suri and Abu Musa'b al Zarqawi: The Apocalyptic Theorist and the Apocalyptic Practitioner," unpublished ms., 9.

49. Lawrence Wright, "The Master Plan," *New Yorker*, September 11, 2006, http://www.newyorker.com/magazine/2006/09/11/the-master-plan.

50. Richard Oppel, "Iraq's New Premier Gains Support in Talks with Shiite Leaders," *New York Times*, April 28, 2006, http://www.nytimes.com/2006/04/28/world/middleeast/28iraq.html.

51. "Iraq's President Appoints Shiite as Prime Minister," *China Daily*, August 8, 2005, http://www.chinadaily.com.cn/english/doc/2005-04/08/content_432343.htm; Oppel, "Iraq's New Premier Gains Support in Talks With Shiite Leaders."

52. David Ignatius, "Iraq's Choice, A Chance for Unity," *Washington Post*, April 26, 2006, http://www.washingtonpost.com/wp-dyn/content/article/2006/04/25/AR2006042501650.html.

53. Ellen Knickmeyer, "Blood on Our Hands," *Foreign Policy*, October 25, 2010; Bobby Ghosh, "An Eye for an Eye," *Time*, February 26, 2006.

54. Nada Bakos, "Humility Now! The Miseducation of Jackson Diehl," *Foreign Policy*, April 2, 2013, http://foreignpolicy.com/2013/02/humility-now/.

55. J.M. Berger, "Analysis: Release of Zarqawi Death Photo Offers Propagandists Opportunity," *IntelWire*, June 8, 2006, http://news.intelwire.com/2006/06/analysis-release-of-zarqawi-death-photo.html; J.M. Berger, "First Zarqawi Tribute Video Appears Online; U.S. Death Photos Prominently Featured," *IntelWire*, June 9, 2006, http://news.intelwire.com/2006/06/first-zarqawi-tribute-video-appears.html.

56. Gilles Kepel, *Beyond Terror and Martyrdom: The Future of the Middle East* (Cambridge, MA: Belknap Press of Harvard University Press, 2008), 133–39.

57. "Pressure Grows on al Qaeda in Iraq," ABC News, January 30, 2006, http://abcnews.go.com/International/story?id=1557349&singlePage=true.

58. Lister, "Profiling the Islamic State"; Brian Fishman, "Redefining the

Islamic State," *New America Foundation*, August 2011, http://security
.newamerica.net/sites/newamerica.net/files/policydocs/Fishman
_Al_Qaeda_In_Iraq.pdf.

59. Fishman, "Redefining the Islamic State."

60. Letter from Zawahiri, July 9, 2005, https://www.ctc.usma.edu/wp
-content/uploads/2010/08/CTC-Zawahiri-Letter-10-05.pdf.

61. "David Petraeus: ISIS's Rise in Iraq Isn't a Surprise," David Petraeus
interview with *Frontline*, PBS, July 29, 2014.

62. Ibid.

63. Ibid.

64. John Hendren, " 'Sunni Awakening': Insurgents Are Now Allies,"
ABC News, December 23, 2007, http://abcnews.go.com/International
/story?id=4045471.

65. "David Petraeus: ISIS's Rise in Iraq Isn't a Surprise."

66. Zaid al-Ali, "How Maliki Ruined Iraq," *Foreign Policy*, June 19, 2014,
http://www.foreignpolicy.com/articles/2014/06/19/how_maliki_ru
ined_iraq_armed_forces_isis.

67. Ibid.

68. "Zalmay Khalilzad: Maliki and the 'Unmaking of Iraq,' " *Frontline*,
PBS, July 29, 2014; Phillip Smyth, email communication with authors,
November 21, 2014.

69. Phillip Smyth, email communication with authors, November 2014.

70. "Zalmay Khalilzad: Maliki and the 'Unmaking of Iraq.' "

71. "Ryan Crocker: 'Our National Security . . . Is At Stake Right Now,' "
Frontline, PBS, July 29, 2014.

72. Ibid.

73. Ibid.

74. "Who Is Nouri al-Maliki?" *Frontline*, PBS, July 29, 2014, http://www
.pbs.org/wgbh/pages/frontline/iraq-war-on-terror/losing-iraq/who
-is-nouri-al-maliki/.

75. Jack Healy, "Arrest Order for Sunni Leader in Iraq Opens New
Rift," *New York Times*, December 19, 2011, http://www.nytimes
.com/2011/12/20/world/middleeast/iraqi-government-accuses-top
-official-in-assassinations.html?pagewanted=all.

76. Hashimi agreed to participate in the Iraqi government, and continued
his participation, at great personal cost. His brother and sister were
murdered by al Qaeda, due to his willingness to participate in the gov-

ernment. "That was worth a lot," General Petraeus says, "and rather than feeling relieved by the opportunity to rid himself of his Sunni Vice-President, as he so clearly was, [Maliki] should have recognized that this was a serious loss for Iraq." "David Petraeus: ISIS's Rise in Iraq Isn't a Surprise."

77. Childress, Wexler, and Mizner, "Iraq: How Did We Get Here?"

78. "Iraq: Investigate Violence at Protest Camp," Human Rights Watch, January 4, 2014, http://www.hrw.org/news/2014/01/03/iraq-investi gate-violence-protest-camp.

79. "2 Wounded as Iraq Protesters Are Disperesed by Security Forces," *New York Times,* December 30, 2012, http://www.nytimes.com/2012/12/31 /world/middleeast/2-wounded-as-iraq-protesters-dispersed.html.

80. Adam Schreck, "Rights Group: Iraq Panel Needs Help Probing Raid," Associated Press, May 4, 2013, http://bigstory.ap.org/article/rights -group-urges-more-backing-iraq-raid-probe.

81. "Iraq: Investigate Deadly Raid on Protest," Human Rights Watch; Martin Smith, *The Rise of ISIS,* video, *Frontline,* PBS, 2014.

82. "Iraq: Investigate Deadly Raid on Protest," Human Rights Watch; Smith, *The Rise of ISIS.*

83. "Iraq: Investigate Violence at Protest Camp," Human Rights Watch.

84. "Iraq: Absolute Impunity: Militia Rule in Iraq," Amnesty Interna- tional, October 14, 2014, http://www.amnesty.org/en/library/info /MDE14/015/2014/en.

85. Schreck, "Rights Group: Iraqi Panel Needs Help Probing Raid."

86. Patrick Cockburn, *The Jihadis Return* (New York: OR Books, 2014), 32.

87. Shashank Bengali and Nabih Bulos, "Influential Iraqi Cleric Calls for Inclusive Government," *Los Angeles Times,* June 20, 2014.

CHAPTER 2. THE RISE OF ISIS

1. Turki al-Binali, "Emir of the Islamic State in Iraq and the Levant," https://archive.org/stream/TheBiographyOfSheikhAbuBakrAlBagh dadi/The%20biography%20of%20Sheikh%20Abu%20Bakr%20Al -Baghdadi_djvu.txt; Cole Bunzel, "The Caliphate's Scholar-in-Arms," *Jihadica.com,* July 9, 2014, http://www.isn.ethz.ch/Digital-Library/Ar ticles/Detail/?id=182205.

2. Ruth Sherlock, "How Baghdadi Went from Shy Scholar to the Most Wanted Jihadist Leader," *Sunday Telegraph,* July 6, 2014.

3. Ibid.

4. Ibid.

5. Aryn Baker, "Jihad's Unholy Ghost," *Time,* January 20, 2014.

6. Aaron Zelin, "Abu Bakr Al-Baghdadi: Islamic State's Driving Force," Washington Institute, 2014.

7. Tim Arango and Eric Schmitt, "U.S. Actions in Iraq Fueled the Rise of a Rebel," *New York Times,* August 10, 2014.

8. Andrew Thompson and Jeremi Suri, "How America Helped ISIS," op-ed, *New York Times,* October 1, 2014, http://www.nytimes.com/2014/10/02/opinion/how-america-helped-isis.html?_r=0.

9. Alissa Rubin, "U.S. Remakes Jails in Iraq, but Gains Are at Risk," *New York Times,* June 2, 2008, http://www.nytimes.com/2008/06/02/world/middleeast/02detain.html?pagewanted=all&_r=0.

10. General Douglas Stone, interviews, September 9–11, 2014.

11. "A Biography of Osama Bin Laden," *Frontline,* PBS, http://www.pbs.org/wgbh/pages/frontline/shows/binladen/who/bio.html.

12. Declan Walsh, "Ayman al-Zawahiri: From Doctor to Osama bin Laden's Successor," *Guardian,* June 16, 2011, http://www.theguardian.com/world/2011/jun/16/ayman-al-zawahiri-osama-bin-laden.

13. Anthony Shadid, "Iraqi Insurgent Group Names New Leaders," *New York Times,* At War blog, May 16, 2010, http://atwar.blogs.nytimes.com/2010/05/16/iraqi-insurgent-group-names-new-leaders/?_php=true&_type=blogs&_r=0.

14. Baathism is a secular movement, originally formed to rescue Arab states from the legacy of colonialism.

15. Anthony Lloyd, "ISIS is deadly revenge of Saddam's henchmen," *The Times of London,* June 14, 2014, http://www.thetimes.co.uk/tto/news/world/middleeast/iraq/article4118901.ece1226954414468?nk=76b7d457b627e2e8644f4018c2679717; Bill Roggio, "ISIS confirms death of senior leader in Syria," February 5, 2014, http://www.longwarjournal.org/archives/2014/02/isis_confirms_death.php#ixzz3N1CN1NRb.

16. Richard Barrett, "The Islamic State," Soufan Group, November 2014, 19.

17. Ibid.

18. Baker, "Jihad's Unholy Ghost."

19. Ibid.

20. Ibid.

21. Rick Gladstone, "Iraq: 1,057 Killed in July, U.N. Says," *New York Times,* August 1, 2013.

22. "Rewards for Justice: Abu Bakr Al-Baghdadi," U.S. Department of State, press briefing, July 9, 2014.

23. Islamic State of Iraq, "But Allah Will Not Allow but That His Light Should Be Perfected," audio, Abu Bakr al-Baghdadi, posted online July 12, 2012.

24. http://www.understandingwar.org/sites/default/files/JessVBIED _PartII_3Oct.pdf; "Over 500 'Al-Qaeda militants' Escape Iraq's Abu Ghraib in Violent Break-out," Reuters, July 22, 2013; "David Petraeus: ISIS's Rise in Iraq Isn't a Surprise."

25. "The Arab Spring," Wikipedia, http://en. wikipedia.org/wiki/Arab _Spring, accessed December 1, 2014.

26. The Second Arab Awakening took form as "popular movements" against despotic rule; Marwan Muasher, *The Second Arab Awakening* (New Haven, CT: Yale University Press, 2014), 2–3, 160.

27. "Syria," U.S. Department of State, September 8, 2010, http://www .state.gov/outofdate/bgn/syria/158703.htm.

28. "'We've Never Seen Such Horror': Crimes Against Humanity by Syrian Security Forces," Human Rights Watch, http://www.hrw.org /sites/default/files/reports/syria0611webwcover.pdf.

29. "Syria: Crimes Against Humanity in Daraa," Human Rights Watch, June 1, 2011, http://www.hrw.org/news/2011/06/01/syria-crimes -against-humanity-daraa.

30. "Syrian Death Toll Surpasses 1000," Al Jazeera, May 24, 2011, http:// www.aljazeera.com/news/middleeast/2011/05/2011524182251952727 .html.

31. Liam Stack, "Video of Tortured Boy's Corpse Deepens Anger in Syira," *New York Times,* May 30, 2011, http://www.nytimes.com/2011/05/31 /world/middleeast/31syria.html.

32. Rosemary D'Amour, "Syria Utilizes 'Kill Switch' as Internet Freedom Debate Heats Up," BroadbandBreakfast.com, June 17, 2011.

33. "Syria: Political Detainees Tortured, Killed," Human Rights Watch,

October 3, 2013, http://www.hrw.org/news/2013/10/03/syria-politi cal-detainees-tortured-killed.

34. "Syria: Sexual Assult in Detention," Human Rights Watch, June 15, 2012, http://www.hrw.org/news/2012/06/15/syria-sexual-assault-de tention.

35. Email interview with Charles Lister.

36. Charles Lister, "Profiling the Islamic State," Brookings Doha Center Analysis Papers, November 2014, http://www.brookings.edu /~/media/Research/Files/Reports/2014/11/profiling%20islamic%20 state%20lister/en_web_lister.pdf; Aaron Zelin, "Al-Qaeda in Syria: A Closer Look at ISIS (Part 1)", Washington Institute, September 10, 2013, http://www.washingtoninstitute.org/policy-analysis/view/al -qaeda-in-syria-a-closer-look-at-isis-part-i.

37. Zelin, "Al-Qaeda in Syria."

38. Matt Levitt, Statement to the House Financial Services Committee, Terrorist Financing and the Islamic State, Hearing, November 13, 2014.

39. Charles Lister, "The 'Real' Jabhat al-Nusra Appears to Be Emerging," Brookings Institution, August 7, 2014, http://www.brookings.edu /research/opinions/2014/08/07-the-real-jabhat-al-nusra.

40. Lister, "The 'Real' Jabhat al-Nusra Appears to Be Emerging."

41. Zelin, "Al-Qaeda in Syria."

42. Aaron Zelin, "Al-Qaeda Announces an Islamic State in Syria," Washington Institute, April 9, 2013, http://www.washingtoninstitute.org /policy-analysis/view/al-qaeda-announces-an-islamic-state-in-syria.

43. Bessma Atassi, "Al Qaeda Chief Annuls Syrian-Iraqi Jihad Merger," Al Jazeera, June 9, 2013, http://www.aljazeera.com/news/middle east/2013/06/2013699425657882.html.

44. "Iraqi al-Qaeda Chief Rejects Zawahiri Orders," Al Jazeera, June 15, 2013, http://www.aljazeera.com/news/middleeast/2013/06 /2013615172217827810.html.

45. Lister, "Profiling the Islamic State"; Rania Abouzeid, "Syria's Uprising Within an Uprising," European Council on Foreign Relations, January 16, 2014, http://www.ecfr.eu/content/entry/commentary_syrias _uprising_within_an_uprising238.

46. Zelin, "Al-Qaeda Disaffiliates with the Islamic State of Iraq and al-Sham."

47. Aaron Lund, "Who and What Was Abu Khalid al-Suri? Part 1," Carnegie Endowment for International Peace, February 24, 2014, http://carnegieendowment.org/syriaincrisis/?fa=54618.

48. Abu Muhammad al-Adnani, "Apologies, Amir of al-Qaidah," translated by Aaron Zelin, https://azelin.files.wordpress.com/2014/05/shaykh-abc5ab-mue1b8a5ammad-al-e28098adnc481nc4ab-al-shc481mc4ab-22sorry-amc4abr-of-al-qc481_idah22-en1.pdf.

49. Alessandria Masi, "ISIS Youth Recruitment: Life in ISIS-Ruled Raqqa So Grim, Parents Are Giving Up Their Children," *International Business Times*, November 6, 2014, http://www.ibtimes.com/isis-youth-recruitment-life-isis-ruled-raqqa-so-grim-parents-are-giving-their-1719124.

50. Charles Lister, *Profiling the Islamic State* (The Brookings Institution/Brookings Doha Center, 2014), 19.

51. Liz Sly, "Insurgents Seize Iraqi City of Mosul as Security Forces Flee," *Washington Post*, June 10, 2014, http://www.washingtonpost.com/world/insurgents-seize-iraqi-city-of-mosul-as-troops-flee/2014/06/10/21061e87-8fcd-4ed3-bc94-0e309af0a674_story.html.

52. Melissa Block, "In New Iraqi Conflict, 'Sunni Awakening Stays Dormant,'" NPR, *All Things Considered* (radio), June 26, 2014, http://www.npr.org/2014/06/26/325909167/in-new-iraqi-conflict-sunni-awakening-stays-dormant; Mushreq Abbas, "The False 'Awakening' Model to Deal with ISIS in Iraq," al-Monitor, July 3, 2014, http://www.al-monitor.com/pulse/originals/2014/07/iraq-separating-sunni-armed-militants-isis-difficulties.html#.

53. Phillip Smyth, "Iranian Proxies Step Up Their Role in Iraq," Washington Institute, June 13, 2014, http://www.washingtoninstitute.org/policy-analysis/view/iranian-proxies-step-up-their-role-in-iraq.

54. Hassan Hassan, "More than ISIS, Iraq's Sunni Insurgency," Carnegie Endowment for International Peace, June 17, 2014, http://carnegieendowment.org/sada/2014/06/17/more-than-isis-iraq-s-sunni-insurgency/hdvi.

55. Jacob Siegel, "With Friends Like These, ISIS Is Doomed," Daily Beast, July 24, 2014, http://www.thedailybeast.com/articles/2014/07/24/with-friends-like-these-isis-is-doomed.html; Mohammad Ballout, "Invasion of Mosul May Backfire on ISIS in Syria," al-Monitor, June 13, 2014, http://www.al-monitor.com/pulse/security/2014/06/iraq-fall-mosul-isis-regional-repercussions.html.

56. Liz Sly and Ahmed Ramadan, "Insurgents Seize Iraqi city of Mosul as Security Forces Free," *Washington Post,* June 10, 2014, http://www .washingtonpost.com/world/insurgents-seize-iraqi-city-of-mosul-as -troops-flee/2014/06/10/21061e87-8fcd-4ed3-bc94-0e309af0a674_ story.html.

57. David Zucchino, "Why Iraqi Army Can't Fight, Despite $25 Billion in U.S. Air, Training," *Los Angeles Times,* November 3, 2014, http:// www.latimes.com/world/middleeast/la-fg-iraq-army-20141103-story .html#page=1.

58. Ibid.

59. Ned Parker, Isabel Coles and Raheem Salman, "Special Report: How Mosul fell—An Iraqi general disputes Baghdad's story," Reuters, October 14, 2014, http://www.reuters.com/article/2014/10/14/us-mideast -crisis-gharawi-special-report-idUSKCN0I30Z820141014.

60. Jeremy Bender, "Iraqi Bankers Say ISIS Never Stole $430 Million from Mosul Banks," *Business Insider,* July 17, 2014, http://www.businessin sider.com/isis-never-stole-430-million-from-banks-2014-7.

61. Patrick B. Johnston, "Countering ISIL's Financing," Congressional Hearing Before the Committee on Financial Services, U.S. House of Representatives, November 13, 2014, http://www.rand.org/content /dam/rand/pubs/testimonies/CT400/CT419/RAND_CT419.pdf.

62. Levitt, Statement to the House, Financial Services Committee.

63. David Cohen, "Attacking ISIL's Financial Foundation" (remarks presented at Carnegie Endowment for International Peace, October 23, 2014). Juan Zarate, who served in the U.S. government from 2005–2009, helped to develop laws and regulations to target sources of funding for al Qaeda and other terrorist groups. See Juan Zarate, *Treasury's War* (New York: Public Affairs, 2013). According to Zarate, Cohen, and Levitt, these instruments are far less useful in regard to ISIS because its main sources of funding are criminal operations, and there is less money laundering to target.

64. David Cohen, Statement to the House Financial Services Committee, Terrorist Financing and the Islamic State, Hearing, November 13, 2014.

65. Peter Bergen, "Should Western Nations Just Pay ISIS Ranson?" CNN Opinion, August 22, 2014, http://www.cnn.com/2014/08/22/opinion /bergen-schneider-isis-ransom/.

66. Bill Roggio, "ISIS Takes Control of Bayji, Tikrit in Lightning South-ward Advance," *Long War Journal*, July 11, 2014, http://www.longwar journal.org/archives/2014/06/isis_take_control_of_1.php; Mehmet Kemal Firik, "ISIS's Weapon Inventory Grows," *Daly Sabah*, July 3, 2014, http://www.dailysabah.com/mideast/2014/07/03/isiss-weapon -inventory-grows.

67. "This Is the Promise of Allah," Al Hayat Media Center, 2014, https:// azelin.files.wordpress.com/2014/06/shaykh-abc5ab-mue1b8a5ammad -al-e28098adnc481nc4ab-al-shc481mc4ab-22this-is-the-promise-of -god22-en.pdf; J.M. Berger, "Gambling on the Caliphate," *IntelWire*, June 29, 2014, http://news.intelwire.com/2014/06/gambling-on-ca liphate.html.

68. Loveday Morris, "Islamic State Seizes Town of Sinjar, Pushing out Kurds and Sending Yazidis Fleeing," *Washington Post*, August 3, 2014, http://www.washingtonpost.com/world/islamic-state-seize-town-of -sinjar-pushing-out-kurds-and-sending-yazidis-fleeing/2014/08/03/52 ab53f1-48de-4ae1-9e1d-e241a15f580e_story.html.

69. "Who, What, Why: Who Are the Yazidis?" *BBC Magazine Monitor*, August 7, 2014, http://www.bbc.com/news/blogs-magazine-monitor -28686607.

70. Jane Arraf, "Islamic State Persecution of Yazidi Miniority Amounts to Genocide, UN Says," *Christian Science Monitor*, August 7, 2014, http:// www.csmonitor.com/World/Middle-East/2014/0807/Islamic-State -persecution-of-Yazidi-minority-amounts-to-genocide-UN-says-video.

71. Ivan Watson, "'Treated Like Cattle,' Yazidi Women Sold, Raped, Enslaved by ISIS," CNN World, November 7, 2014, http://www.cnn .com/2014/10/30/world/meast/isis-female-slaves/.

72. Dan Wilkofsky and Osama Abu Zeid, "US-Backed SRF 'No Longer' in South Idlib After Nusra Victory," Syria: Direct, November 4, 2014, http://syriadirect.org/main/30-reports/1653-us-backed-srf-no-longer -in-south-idlib-after-nusra-victory.

73. Aron Lund, "Syria's Ahrar al-Sham Leadership Wiped Out in Bomb-ing," Carnegie Endowment for International Peace, September 9, 2014, http://carnegieendowment.org/syriaincrisis/?fa=56581; Email inter-view with Charles Lister.

74. Office of the Press Secretary, The White House, "Statement by the

President on ISIL," September 10, 2014, http://www.whitehouse.gov /the-press-office/2014/09/10/statement-president-isil-1.

75. Ashley Fantz, "Who Is Doing What in the Coalition Battle Against ISIS?" CNN World, September 17, 2014, http://www.cnn.com/2014/09/14 /world/meast/isis-coalition-nations/.

76. Martha Raddatz and Luis Martinez, "Airstrikes in Syria That Targeted Khorasan Group Disrupted Plots Against US, Gen. Dempsy Says," ABC News, October 7, 2014, http://abcnews.go.com/Interna tional/airstrikes-syria-targeted-khorasan-group-disrupted-plots-us /story?id=26030142.

77. Wilkofsky and Abu Zeid, "US-Backed SRF 'No Longer' in South Idlib After Nusra Victory."

78. Charles Lister, "In Syria, a Last Gasp Warning for U.S. Influence," Brookings Institution, December 5, 2014, http://www.brookings.edu /blogs/markaz/posts/2014/12/05-syria-united-states-losing-last-gasp -at-leverage

79. Aron Lund, "The Levant Front: Can Aleppo's Rebels Unite?," *Carnegie Endowment for International Peace*, December 26, 2014 http://carnegie endowment.org/syriaincrisis/?fa=57605.

80. "Islamic State: Fighting Intensifies in Syrian Town of Kobane," BBC News Middle East, November 29, 2014, http://www.bbc.com/news /world-middle-east-30260834; Ben Hubbard, "ISIS Wave of Might Is Turning into a Ripple," *New York Times,* November 5, 2014, http:// www.nytimes.com/2014/11/06/world/middleeast/isis-wave-of-might -is-turning-into-ripple.html?_r=0.

81. Brian Katulis, "Good News from Iraq: ISIS Setbacks," *Wall Street Journal,* November 17, 2014, http://blogs.wsj.com/washwire/2014/11/17/good -news-from-iraq-isis-setbacks/; "Thirty U.S.-Led Strikes Hit Islamic State in Syria's Raqqa: Monitoring Group," Reuters, November 30, 2014, http://www.reuters.com/article/2014/11/30/us-mideast-crisis -syria-raqqa-idUSKCN0JE07020141130.

82. Charles Lister, "Assessing Syria's Jihad," *Survival: Global Politics and Strategy* 56, no. 6 (2014): 87–112.

83. Patrick Cockburn, "ISIS Consolidates," *London Review* 36, no. 16 (August 21, 2014), http://www.lrb.co.uk/v36/n16/patrick-cockburn/isis -consolidates.

84. Lister, "Assessing Syria's Jihad."

CHAPTER 3. FROM VANGUARD TO SMART MOB

1. Descriptions of al Qaeda's founding in this chapter derived from al Qaeda internal records obtained by J.M. Berger. Translations of most of the documents can be found in *Beatings and Bureaucracy: The Founding Memos of Al Qaeda,* ed. J.M. Berger (Intelwire Press, 2012).

2. Paul Klebnikov, "Who Is Osama Bin Laden?" Forbes, September 14, 2001, http://www.forbes.com/2001/09/14/0914whoisobl.html.

3. For a fuller discussion of Omar Abdel Rahman and al Qaeda see J.M. Berger, "The Death Dealers (1990–1993)," ch. 3 in *Jihad Joe: Americans Who Go to War in the Name of Islam* (Washington, DC: Potomac Books, 2011).

4. Brynjar Lia, *Architect of Global Jihad: The Life of Al-Qaida Strategist Abu Mus'ab Al-Suri* (New York: Columbia University Press, 2008), 370.

5. For a fuller discussion see J.M. Berger, "War on America (1991–1999)," ch. 6 in *Jihad Joe.*

6. *The State of the Ummah*, online video, As-Sahab Media Productions, 2001.

7. "Timeline of al-Qaida Statements," NBC News, http://www.nbcnews .com/id/4686034/ns/world_news-hunt_for_al_qaida/t/timeline-al -qaida-statements/.

8. Alan Cullison, "Inside Al-Qaeda's Hard Drive," *Atlantic*, September 1, 2004, http://www.theatlantic.com/magazine/archive/2004/09 /inside-al-qaeda-s-hard-drive/303428/?single_page=true.

9. "Bin Laden on Tape: Attacks 'Benefited Islam Greatly,'" CNN, http:// edition.cnn.com/2001/US/12/13/ret.bin.laden.videotape/, last modified December 14, 2001.

10. Jane Novak, "Arabian Peninsula Al-Qaeda Groups Merge," *Long War Journal,* http://www.longwarjournal.org/archives/2009/01/arabian _peninsula_al.php, last modified January 26, 2009.

11. Gregory D. Johnson, *The Last Refuge: Yemen, Al-Qaeda, and America's War in Arabia,* 1st ed. (New York: Norton, 2013).

12. "Al-Qa'ida in Iraq (AQI)," National Counterterrorism Center, http:// www.nctc.gov/site/groups/aqi.html, accessed September 20, 2014.

13. "Al-Qa'Ida in the Lands of the Islamic Maghreb (AQIM)," National Counterterrorism Center, http://www.nctc.gov/site/groups/aqim .html, accessed September 20, 2014.

14. "Al-Qa'Ida in the Arabian Peninsula (AQAP)," National Counterter-

rorism Center, http://www.nctc.gov/site/groups/aqap.html, accessed September 20, 2014.

15. "Al-Shabaab," National Counterterrorism Center, http://www.nctc.gov/site/groups/al_shabaab.html, accessed September 20, 2014.

16. "Al-Nusrah Front," National Counterterrorism Center, http://www.nctc.gov/site/groups/al_nusrah.html, accessed September 20, 2014.

17. Ayman al-Zawahiri, "Announcing a New Branch of Al Qaeda Al Jihad," As-Sahab Media, September 3, 2014.

18. Zachary Abuza, "Balik-Terrorism: The Return of the Abu Sayyaf," Army War College Strategic Studies Institute, 2005, 14–15.

19. For a fuller discussion see J.M. Berger, "Project Bosnia (1992–1995)," ch. 4 in *Jihad Joe*.

20. For example, House Committee on Homeland Security, Subcommittee on Counterterrorism and Intelligence, "Jihadist Safe Havens: Efforts to Detect and Deter Terrorist Travel," July 24, 2014. A July 24, 2014, Nexis search for AQAP and "most dangerous" produced more than 1,600 results.

21. Alexander Meleagrou-Hitchens and Peter Neumann, "Al Qaeda's Most Dangerous Franchise," *Wall Street Journal*, May 10, 2012, http://www.wsj.com/articles/SB10001424052702304203604577395611014467928.

22. Special Agent Theodore James Peisig, "United States of America v. Umar Farouk Abdulmutallab," Huffington Post, Big Assets, December 25, 2009.

23. "$4,200," *Inspire*, November 2010.

24. Lia, *Architect of Global Jihad*, 422.

25. Louis Beam, "Leaderless Resistance," http://www.louisbeam.com/leaderless.htm; Daveed Gartenstein-Ross, Leadership vs. Leaderless Resistance: The Militant White Separatist Movement's Operating Model," *Foundation for the Defense of Democracies*, February 18, 2010, http://www.defenddemocracy.org/media-hit/leadership-vs-leaderless-resistance-the-militant-white-separatist-movement/.

26. Shane Dennan, "From 'Abdullah 'Azzam to Djamel Zitouni," *CTC Sentinel* 1, no. 7 (2008), https://www.ctc.usma.edu/posts/constructing-takfir-from-abdullah-azzam-to-djamel-zitouni, accessed September 5, 2014.

27. Ayman Zawahiri, "Zawahiri's Letter to Zarqawi," trans. Center for

Combating Terrorism at West Point, July 9, 2005, https://www.ctc
.usma.edu/posts/zawahiris-letter-to-zarqawi-english-translation-2.

28. Nasir al Wuhayshi to Osama bin Laden, August 27, 2010, in Don
Rassler et al., "Letters from Abbottabad: Bin Laden Sidelined," Combating Terrorism Center at West Point, 2012, https://www.ctc.usma
.edu/posts/letters-from-abbottabad-bin-ladin-sidelined.

29. Bill Roggio, "African Al-Qaeda Leader Side with Zawahiri in Syrian Dispute," *Long War Journal*, http://www.longwarjournal.org/ar
chives/2014/05/african_al_qaeda_lea.php., last modified May 1, 2014.

30. Aaron Zelin, "The War Between ISIS and Al-Qaeda for Supremacy of
the Global Jihadist Movement," *Research Notes*, Washington Institute,
2014.

31. Farouk Chothia, "Ahmed Abdi Godane: Somalia's Killed Al-Shabab
Leader," BBC News Africa, 2014, http://www.bbc.com/news/world
-africa-29034409.

32. Berger, *Jihad Joe*, 170–74.

33. J.M. Berger, "Me Against the World," *Foreign Policy*, May 2012, http://
www.foreignpolicy.com/articles/2012/05/25/me_against_the_
world; J. M Berger, "Omar and Me," *Foreign Policy*, September 2013,
http://www.foreignpolicy.com/articles/2013/09/16/omar_and_me.

34. Berger, "Omar and Me."

35. Aaron Zelin, "The State of Global Jihad Online: A Qualitative, Quantitative, and Cross-Lingual Analysis," New America Foundation, 2013,
http://www.newamerica.net/sites/newamerica.net/files/policydocs
/Zelin_Global%20Jihad%20Online_NAF.pdf.

36. Philipp Holtmann, "The Different Functions of IS Online and Offline Pledges," *Jihadology*, November 15, 2015, http://jihadology
.net/2014/11/15/guest-post-the-different-functions-of-is-online-and
-offline-plegdes-bayat-creating-a-multifaceted-nexus-of-authority/.

37. Berger, "Omar and Me."

38. Tony Busch, "How Twitter Is Messing with Al-Qaeda's Careful
PR Machine," *Atlantic*, http://www.theatlantic.com/international
/archive/2013/05/how-twitter-is-messing-with-al-qaedas-careful-pr
-machine/275697/, last modified May 14, 2013.

39. Abdullah bin Mohammed, "Strategy Affairs," Twitter account, https://
twitter.com/Strategyaffairs.

40. Omar Hammami to J.M. Berger, Twitter message.

41. Hamsa Omar and Sarah McGregor, "U.S. Al-Qaeda Member Rebuked by Somalia Militia over Criticism," Bloomberg, December 18, 2012, http://www.bloomberg.com/news/2012-12-18/u-s-al-qaeda-member-rebuked-by-somalia-militia-over-criticism.html; hundreds of tweets to Hammami by Shabab members archived by J.M. Berger; many of these accounts are now suspended.

42. Majid Ahmed, "Open Letter to Al-Zawahiri Rocks Foundations of Al-Shabaab," Sabahi, April 12, 2013, http://sabahionline.com/en_GB/articles/hoa/articles/features/2013/04/12/feature-01.

43. Sa'eed Bin Jubayr, Twitter post, May 18, 2013, https://twitter.com/Ibn Jubayr/status/335786202542268417, accessed September 20, 2014.

44. Berger, "Omar and Me."

45. Zelin, "The State of Global Jihad Online."

46. Forum activity tracked over time by J.M. Berger; additional information was provided by two analysts who track online jihadist activity but spoke under conditions of anonymity.

47. Zelin, "The War Between ISIS and Al-Qaeda."

48. Ayman Zawahiri. "A Testimony to Save/Preserve Blood of the Mujahideen in Sham," Jihadology, Aaron Zelin, 2005, https://azelin.files.wordpress.com/2014/05/dr-ayman-al-e1ba93awc481hirc4ab-22testimony-to-preserve-the-blood-of-the-mujc481hidc4abn-in-al-shc481m22-en.pdf; see also, Ayman Zawahiri, "Eulogy for the Martyr of the Fitnah Shaykh Abū Khālid Al-Sūrī," Jihadology, Aaron Zelin, 2014.

49. Hashtag activity tracked and analyzed by J.M. Berger, March 25–27, 2014.

50. Aaron Zelin, "Member of Islamic State of Iraq and al-Shām Leaks Unpublished Video Message from Adam Gadahan: 'Islamic State of Iraq and al-Shām: Extremists,'" Jihadology, March 29, 2014, http://jihadology.net/2014/03/29/member-of-islamic-state-of-iraq-and-al-sham-leaks-unpublished-video-message-from-adam-gadahan-islamic-state-of-iraq-and-al-sham-extremists/. On the forum infiltration, members of al Qaeda's media team were identified by their activity uploading videos to popular file-sharing forums, including account information that could be correlated back to both Twitter and to the names used by prominent forum members; analysis of forum Twitter feeds over time pointed to shifting allegiances. Additional information

was provided by two analysts who track online jihadist activity but spoke under conditions of anonymity.

51. Zelin, "Member of Islamic State"; Bill Roggio, "Al Qaeda's American Propagandist Notes Death of Terror Group's Representative in Syria," *Long War Journal*, http://www.longwarjournal.org/archives/2014/03/al_qaedas_american_p.php, last modified March 30, 2014.

52. For example, SayfulAdl, Twitter post, April 22, 2014, 1:28 p.m., https://twitter.com/SayfulAdl/status/458658509434146816.

53. "Amir of the Islamic State of Iraq and Al-Sham," translated by Pieter Van Ostaeyen, http://pietervanostaeyen.wordpress.com/2013/07/15/abu-bakr-al-baghdadi-a-short-biography-of-the-isis-sheikh/, last modified July 7, 2013; Associated Press, "Mosul Video Purports to Show Abu Bakr Al-Baghdadi, Head of Islamic State," *Guardian*, July 5, 2014, http://www.theguardian.com/world/2014/jul/06/mosul-video-purports-abu-bakr-al-baghdadi-islamic-state; Martin Chulov, "Abu Bakr Al-Baghdadi Emerges from Shadows to Rally Islamist Followers," *Guardian*, July 6, 2014, http://www.theguardian.com/world/2014/jul/06/abu-bakr-al-baghdadi-isis.

54. Howard Rheingold, *Smart Mobs: The Next Social Revolution* (Cambridge, MA: Basic Books, 2002), xii.

55. "Iraq PM Calls Emergency after Mosul Seized," Al Jazeera, June 10, 2014, http://www.aljazeera.com/news/middleeast/2014/06/iraq-calls-emergency-after-rebels-seize-mosul-2014610121410596821.html.

56. A more detailed discussion of content can be found in Chapter 5.

57. Patrick Kingsley, "Who Is Behind Isis's Terrifying Online Propaganda Operation?" *Guardian*, June 23, 2014, http://www.theguardian.com/world/2014/jun/23/who-behind-isis-propaganda-operation-iraq.

58. Mosnen, JustPaste File Upload, justpaste.it/Mosnen, accessed September 20, 2014; "Nursing Homes in the State of Nineveh," forum post, http://www.hanein.info/vb/showthread.php?t=383739, accessed September 20, 2014.

59. Abu Naji, *The Management of Savagery,* translated by Will McCants, http://azelin.files.wordpress.com/2010/08/abu-bakr-naji-the-management-of-savagery-the-most-critical-stage-through-which-the-umma-will-pass.pdf, 51.

60. Anwar al-`Awlaqi, "Constants on the Path of Jihad," audio lecture,

2005; J.M. Berger, "The Enduring Appeal of Al-ʿAwlaqi's 'Constants on the Path of Jihad,'" *CTC Sentinel* 4, no. 10 (2011): 12–14.

CHAPTER 4. THE FOREIGN FIGHTERS

1. "Eid Greetings from the Land of Khilafah," video, August 2014.
2. William McCants, "Militant Ideology Atlas," Combating Terrorism Center at West Point, November 1, 2006, https://www.ctc.usma.edu /posts/militant-ideology-atlas, 44, 47.
3. The gun, a World War II–era MP40, was likely a toy or a replica. Aside from its antiquity, the weight would probably have been too great for a small child. Other children in the video brandished guns that were clearly toys.
4. Megan Specia, "Westerners Among Knife-Wielding Militants in ISIS Beheading Video," Mashable.com, November 17, 2014, http://mash able.com/2014/11/17/foreign-fighters-isis-video/; "Danish Jihadist 'Possibly' in ISIS Beheading Video," *Local Denmark*, November 18, 2014, http://www.thelocal.dk/20141118/danish-jihadist-possibly-in-new -isis-video; Jenny Booth and Charles Bremner, "Security Services Race to Identify ISIS Killers," *Times* (London), November 17, 2014, http:// www.thetimes.co.uk/tto/news/world/middleeast/article4270040 .ece.
5. Joshua Holland, "Why Have a Record Number of Westerners Joined the Islamic State?" BillMoyers.com, October 10, 2014, http://billmoy ers.com/2014/10/10/record-number-westerners-joined-islamic-state -great-threat/.
6. Radio Free Europe Radio Liberty, "Foreign Fighter in Iraq and Syria: Where Do They Come From?" infograph, http://www.rferl.org/con tentinfographics/infographics/26584940.html, accessed December 11, 2014.
7. Mohanad Hage Ali, "Meet ISIS's New Breed of Chechen Militants," Al Arabiya, August 31, 2014.
8. Scott Pelly, "FBI Director: We Know the Americans Fighting in Syria," video, produced by Robert Anderson and Pat Milton, October 5, 2014, http://www.cbsnews.com/news/fbi-director-james-comey-we-know -the-americans-fighting-in-syria/; "FBI Tracking 150 Americans Who

Traveled to Syria, Perhaps to Fight," Reuters, November 18, 2014, http://www.reuters.com/article/2014/11/18/us-usa-fbi-syria-idUSKC N0J22HG20141118; "Intel Believes 300 Americans Fighting with Islamic State, Posing Threat to US," *Washington Times,* August 26, 2014, http://www.washingtontimes.com/news/2014/aug/26/us-citizens -joining-islamic-state-pose-major-threa/?page=all.

9. "FBI Director: Americans Fighting in Syria Hard to Track," CBS News, August 11, 2014, http://www.cbsnews.com/news/fbi-director -americans-fighting-in-syria-hard-to-track/.

10. Data collected from Facebook and Twitter, December 2013–November 2014.

11. Harriet Alexander, "Jihadists in Syria Write Home to France: My iPod is Broken, I Want to Come Back," *Telegraph,* December 2, 2014, http:// www.telegraph.co.uk/news/worldnews/islamic-state/11268208/ Jihadists-in-Syria-write-home-to-France-My-iPod-is-broken.-I-want- to-come-back.html.

12. "Two Thousand Britons Fighting with Terrorists in Iraq and Syria," *Sunday Telegraph,* November 23, 2014, http://www.telegraph.co.uk /news/worldnews/islamic-state/11248114/Muslim-MP-2000-Britons -fighting-for-Islamic-State.html.

13. Patrick Wintour, "Jihadis Who Travel to Syria Could Be Barred from UK Return for Two Years," *Guardian,* November 14, 2014, http:// www.theguardian.com/uk-news/2014/nov/13/british-jihadis-syria -bar-from-uk-two-years; Martin Robinson, "The Homegrown Jihadists Fighting for ISIS: How one in four foreigners who have signed up for Islamic State is British—and how half of them are ALREADY back in the UK," *Daily Mail,* August 21, 2014, http://www.dailymail.co.uk /news/article-2730602/The-homegrown-jihadists-fighting-ISIS-How -one-four-foreigners-signed-Islamic-State-British-half-ALREADY-UK .html.

14. David Fitzpatrick and Drew Griffin, "Canadians Have Joined ISIS to Fight—and Die—in Syria," CNN World, September 10, 2014, http:// www.cnn.com/2014/09/10/world/canada-isis-jihadists/; "Germany's Intelligence Chief Says at Least 550 Germans in IS Ranks," Radio Free Europe, November 25, 2014, http://www.rferl.org/content/islamic -state-550-germans/26706500.html.

15. Richard Barrett, "Foreign Fighters in Syria," Soufan Group, 2014, http://soufangroup.com/wp-content/uploads/2014/06/TSG-Foreign -Fighters-in-Syria.pdf.

16. Audrey Hamilton, "Speaking Psychology: Getting into a Terrorist's Mind," American Psychological Association, video, http://www.apa .org/research/action/speaking-of-psychology/terrorist-mind.aspx.

17. For widely cited examples, see the Combating Terrorism Center at West Point, https://www.ctc.usma.edu/v2/wp-content/up loads/2012/04/Atlas-ResearchCompendium1.pdf. Hal al-umma alislamiyyawa'lirhaba l-mafqud by Abu `Umar al-Sayf; Abdullah Azzam, the Defense of Muslim Lands; The State of the Ummah (lecture by Anwar Awlaki not the AQ video).

18. Erin Banco, "Why Do People Join ISIS? The Psychology of a Terrorist," *International Business Times,* September 5, 2014, http://www.ibtimes .com/why-do-people-join-isis-psychology-terrorist-1680444.

19. Scott Atran with Lydia Wilson, Richard Davis, and Hammad Sheikh, "The Devoted Actor, Sacred Values and Willingness to Fight— Preliminary Studies with ISIL Volunteers and Kurdish Frontline Fighters," Executive Summary prepared for Senate Armed Services Committee, unpublished manuscript, November 2014.

20. Scott Atran with Lydia Wilson, Richard Davis, Hammad Sheikh, *The Devoted Actor, Sacred Values, and Willingness to Fight: Preliminary Studies with ISIL Volunteers and Kurdish Frontline Fighters,* ARTIS Research, The Gerald R. Ford School of Public Policy at the University of Michigan & The Centre for the Resolution of Intractable Conflict, University of Oxford, 2014), 3, http://artisresearch.com/wp-content /uploads/2014/11/Atran_Soccnet_MINERVA_ISIS.pdf.

21. Scott Atran, Hammad Sheikh, and Angel Gomez, "Devoted Actors Sacrifice for Close Comrades and Sacred Cause," commentary, *PNAS Early Edition;* Charlotte Kathe, "A Summary of ICSR Event in Parliament with Professor Peter Neumann," October 21, 2014, http://www .quilliamfoundation.org/blog/charlotte-kathe-a-summary-of-icsr -event-in-parliament-with-professor-peter-neumann/, accessed December 9, 2014.

22. "Up to 30 British Jihadists Now Dead in Syria but Toll Will Rise with ISIL Lure," Syria HR, October 16, 2014, http://syriahr.com/en

/2014/10/up-to-30-british-jihadists-now-dead-in-syria-but-toll-will
-rise-with-isil-lure/.

23. Miriam Elder and Adrian Carrasquillo, "This Foreign Fighter in Syria IS Terrifyingly Good at the Internet," Buzzfeed, November 22, 2013, http://www.buzzfeed.com/miriamelder/this-foreign-fighter-in-syria -is-terrifyingly-good-at-the-in.

24. http://ask.fm/ijaman08, accessed December 2013.

25. http://ask.fm/abumuhajir1, accessed December 2013 and January 2014.

26. Tweets collected January 2014.

27. al-Ghuraba, *The Chosen Few of Different Lands: Ab? Muslim from Canada*, video, July 2014.

28. Michael S. Schmidt, "Canadian Killed in Syria Lives On as Pitchman for Jihadis," *New York Times*, July 15, 2014, http://www.nytimes .com/2014/07/16/world/middleeast/isis-uses-andre-poulin-a-cana dian-convert-to-islam-in-recruitment-video.html?_r=0.

29. Robin Young and Jeremy Hobson, "ISIS Advertises for Oil Industry Managers," *Here & Now*, NPR Boston, (radio), November 19, 2014, http://hereandnow.wbur.org/2014/11/19/isis-oil-jobs.

30. William McCants, "Islamic State Invokes Prophecy to Justify Its Claim to Caliphate," Brookings Institution, November 5, 2014, http://www .brookings.edu/blogs/iran-at-saban/posts/2014/11/05-islamic-state -prophecy-justify-caliphate-mccants.

31. Data collected from Twitter, August and September 2014.

32. Suzanne Evans, *Mothers of Heroes, Mothers of Martyrs: World War I and the Politics of Grief* (Monreal: McGill-Queen's University Press, 2007), 33–34.

33. Data collected from Twitter, September 2013–November 2014.

34. Kathy Gilsinan, "The ISIS Crackdown on Women, by Women," *Atlantic*, July 25, 2014, http://www.theatlantic.com/international/archive /2014/07/the-women-of-isis/375047/.

35. Aqsa blog, March 7, 2013.

36. Ibid., March 10, 2013.

37. Ibid., March 12, 2013.

38. "Scottish Women Who Married ISIS Fighter 'Wants to Become a Martyr,'" *Guardian*, September 6, 2014, http://www.theguardian

.com/world/2014/sep/06/scottish-woman-married-isis-fighter-martyr
-aqsa-mahmood?CMP=twt_gu.

39. Khaleda Rahman, "'I Will Only Come Back to Britain to Raise the
Black Flag," *Daily Mail,* September 11, 2014, http://www.dailymail
.co.uk/news/article-2752957/I-come-Britain-raise-black-flag-Jihadist
-warning-Wests-Muslims-rejects-appeal-return.html.

40. Mia Bloom, "How the Islamic State Is Recruiting Western Teen
Girls," *Washington Post,* PostEverything blog, October 9, 2014, http://
www.washingtonpost.com/posteverything/wp/2014/10/09/how-the
-islamic-state-is-recruiting-western-teen-girls/.

41. Corey Charlton, "We Made a Big Mistake," *Daily Mail,* October 10,
2014, http://www.dailymail.co.uk/news/article-2788605/Teenage
-Austrian-poster-girls-ISIS-moved-Syria-live-jihadis-pregnant-want
-come-home-officials-say-impossible.html.

42. Arwa Damon and Gul Tuysuz, "How She Went from a Schoolteacher
to an ISIS Member," CNN World, October 6, 2014, http://www.cnn
.com/2014/10/06/world/meast/isis-female-fighter/.

43. "Race Toward Good," video, November 22, 2014.

44. Rita Katz, "The Overlooked Goal of the Islamic State's Beheading
Video: Recruitment," Insite Blog on Terrorism & Extremism, August
2014, http://news.siteintelgroup.com/blog/index.php/about-us/21
-jihad/4418-the-overlooked-goal-of-the-islamic-state-s-beheading
-video-recruitment?start=120.

45. Zenia Karam and Vivian Salama, "Islamic State Group Recruits, Ex-
ploits Children, Yahoo News, November 23, 2014, http://news.yahoo
.com/islamic-state-group-recruits-exploits-children-110950193.html.

46. Liz Sly, "Despite boasts, ISIS failing in attempt at state-building," *Washington Post,* December 26, 2014, http://www.bostonglobe.com/news/
world/2014/12/26/despite-boasts-isis-failing-attempt-state-building/
zr1jz0NHRbViNMW7esFNWI/story.html; Kevin Sullivan and Karla
Adams, "Hoping to Create a New Society, the Islamic State Recruits
Entire Families," *Washington Post,* December 24, 2014, http://www
.washingtonpost.com/world/national-security/hoping-to-create-a-
new-homeland-the-islamic-state-recruits-entire-families/2014/12/24/
dbffceec-8917-11e4-8ff4-fb93129c9c8b_story.html.

47. "Three Iraqis Arrested in Switzerland, Suspected of Planning Attack,"
Reuters, October 31, 2014, http://uk.reuters.com/article/2014/10/31

/uk-mideast-crisis-switzerland-idUKKBN0IK19I20141031; Gianluca Mezzofiore, "Switzerland: Iraqi ISIS Cell Arrested for Planning Terror Attack," *International Business Times,* September 24, 2014, http://www.ibtimes.co.uk/switzerland-iraqi-isis-cell-arrested-planning-terror-attack-1466994.

48. Sylvie Corbet and Lori Hinnant, "French Ex-Hostage Says Tortured in Syria by Brussels Shooting Suspect Mehdi Nemmouche," *World Post,* http://www.huffingtonpost.com/2014/09/06/mehdi-nemmouche-syria_n_5777064.html, last updated November 6, 2014.

49. "Malaysia 'Foiled' Attack Plots by ISIS-Inspired Militants," *Daily Star Lebanon,* August 20, 2014, http://www.dailystar.com.lb/News/Middle-East/2014/Aug-20/267773-malaysia-foiled-attack-plots-by-isis-inspired-militants.ashx#axzz3AtRUMqbA.

50. Ben Quinn, "London Teenager Charged with Preparing to Commit Acts of Terrorism," *Guardian,* August 21, 2014, http://www.theguardian.com/uk-news/2014/aug/21/teenager-remanded-terrorism-charge-brustchom-ziamani.

51. Madeline Grant, "Two Teenage Girls Arrested over French Synagogue Suicide Bomb Plot," *Newsweek,* August 29, 2014, http://www.newsweek.com/two-teenage-girls-arrested-over-french-synagogue-suicide-bomb-plot-267523.

52. "Raids Fail to Deter Terrorist Recruiter Mohammad Ali Baryalei," *Australian,* http://www.theaustralian.com.au/national-affairs/defence/raids-fail-to-deter-terrorist-recruiter-mohammad-ali-baryalei/story-e6frg8yo-1227099143028.

53. Abu Muhammad al Adnani, "Indeed Your Lord Is Ever Watchful," speech, September 21, 2014.

54. Leela Jacinto, "Jund al-Khalifa: The IS-linked Group That Shot into the Spotlight," France 24, September 25, 2014, http://www.france24.com/en/20140923-terrorism-france-algeria-jund-al-khalifa-kidnapping-islamic-state/; "Jund al-Khilafah in Algeria Beheads French Hostage," Insite Blog on Terrorism and Extremism, September 24, 2014, http://news.siteintelgroup.com/blog/index.php/entry/289-jund-al-khilafah-in-algeria-beheads-french-hostage-in-video.

55. "Terror Suspect Abdul Numan Haider Continued to Stab Fallen Officer: Inquest," ABC News, October 3, 2014, http://www.abc.net.au/news/2014-10-03/abdul-numan-haider-continued-to-stab-police

-officer-inquest/5787788; Hilary Whiteman, "Lone Wolf? Australian Police Shoot Dead Teen 'Terror Suspect,'" CNN World, September 25, 2014, http://www.cnn.com/2014/09/24/world/asia/australia-terror -shooting-laws/index.html.

56. Mark Gollom and Tracey Linderman, "Who Is Martin Couture-Rouleau," CBC News Canada, http://www.cbc.ca/news/canada /who-is-martin-couture-rouleau-1.2807285, last updated October 22, 2014. Data collected from Twitter, October 2014.

57. Saeed Ahmed and Greg Botelho, "Who Is Michael Zehaf-Bibeau, the Man Behind the Deadly Ottawa Attack?" CNN World, October 23, 2014, http://www.cnn.com/2014/10/22/world/canada-shooter/.

58. Kevin Fasick, Amber Jamieson, and Laura Italiano, "Ax Attacker Wanted 'White People to Pay' for Slavery," New York Post, October 25, 2014, http://nypost.com/2014/10/25/pro-jihadi-ax-attacker-wanted -white-people-to-pay-for-slavery/; James Gordon Meek and Josh Margolin, "NYC Ax Attacker Was Consumed by Desire to Strike U.S. Authority Figures, Police Say," ABC News, November 3, 2014, http:// abcnews.go.com/US/nyc-ax-attacker-consumed-desire-strike-us-au thority/story?id=26664787.

59. The term "lone wolf" connotes someone acting without direct connections to a terrorist group. However the phrase is commonly, if misleadingly, used to refer to people who act individually or in small cells with minimal support from a terrorist group.

60. "Inspire Magazine: A Staple of Domestic Terror," Anti-Defamation League, http://blog.adl.org/extremism/inspire-magazine-a-staple-of -domestic-terror.

61. Dabiq no. 5 (November 2014): 537.

62. J.M. Berger, "The Islamic State's Irregulars," Foreign Policy, December 23, 2014, http://foreignpolicy.com/2014/12/23/the-islamic-states -deranged-irregulars-lone-wolf-terrorists-isis/.

63. "Norwegian Syria Fighter Seized on Terror Charge," Local, February 14, 2014, http://www.thelocal.no/20140214/norway-jihadi-ar rested-for-terror-on-return-from-syria; "Belgain Terrorist Suspect Held in Luxembourg," Flanders News, September 7, 2014, http://dere dactie.be/cm/vrtnieuws.english/News/1.2027303; "Alleged Terrorist Arrested in Bekasi," Jakarta Post, August 10, 2014, http://www.the

jakartapost.com/news/2014/08/10/alleged-terrorist-arrested-bekasi.html.

64. J.M. Berger, "The Islamic State's Irregulars," *Foreign Policy*, December 23, 2014; Alison Smale and James Kanterjan, "As Europe Moves Aggressively Against Terrorism, New Challenges Emerge," *New York Times*, January 16, 2015.

65. Melissa Eddy, "Nations Ponder How to Handle European Fighters Returning from Jihad," *New York Times*, November 23, 2014, http://www.nytimes.com/2014/11/24/world/nations-ponder-how-to-handle-european-fighters-returning-from-jihad.html?_r=0.

66. Bharati Naik, Atlka Shubert, and Nick Thompson, "Denmark Offers Some Foreign Fighters Rehab Without Jail Time—but Will It Work?" CNN World, October 28, 2014 http://www.cnn.com/2014/10/28/world/europe/denmark-syria-deradicalization-program/.

67. John Horgan and Max Taylor, "Disengagement, De-radicalization and the Arc of Terrorism: Future Directions for Research," ch. 13 in *Jihadi Terrorism and the Radicalisation Challenge: European and American Experiences* (Farnham, England, and Burlington, VT: Ashgate, 2011).

68. Thomas Hegghammer, "Should I Stay or Should I Go? Explaining Variation in Western Jihadists' Choice Between Domestic and Foreign Fighting," *American Political Science Review* 107, no. 1 (February 2013), http://hegghammer.com/_files/Hegghammer_-_Should_I_stay_or_should_I_go.pdf.

CHAPTER 5. THE MESSAGE

1. "The Martyrs of Bosnia," video, 1995.

2. "The State of the Ummah," in *Al Qaeda Propaganda Video*, 2001.

3. Raffi Khatchadourian, "Azzam the American: The Making of an Al Qaeda Homegrown," *New Yorker*, January 22, 2007, http://www.newyorker.com/magazine/2007/01/22/azzam-the-american; J.M. Berger, *Jihad Joe: Americans Who Go to War in the Name of Islam* (Washington, DC: Potomac Books, 2011), 152–56.

4. George Michael, "Adam Gadahn and Al Qaeda's Internet Strategy," *Middle East Policy* (Fall 2009); *The 19 Martyrs*, Al Qaeda propaganda video, 2002; *Voice of the Caliphate*, Al Qaeda propaganda video, 2005;

Daniel Williams, *Voice of Caliphate* Web Broadcast Speaks of Joy over US Hurricane," *Washington Post*, September 25, 2005.

5. Spencer Ackerman, "Watch: Osama's Blooper Reel, Courtesy of the Navy Seals." *Wired*, May 2011, http://www.wired.com/danger room/2011/05/watch-osamas-blooper-reel-courtesy-of-the-navy-seals /?pid=425.

6. David Ensor, Octavia Nasr, Cal Perry, Auday Sadik, and Mohammad Tawfeeq, "Video Apparently Shows Public Beheadings: 12 Killed, Dozens Injured as Wedding Party Attacked," CNN International, January 22, 2005, http://edition.cnn.com/2005/WORLD/meast/01/21 /iraq.main/index.html?iref=newssearch.

7. Ayman al-Zawahiri, letter to Abu Musab Zarqawi, 2005, https:// www.ctc.usma.edu/posts/zawahiris-letter-to-zarqawi-english-trans lation-2.

8. Abu Mohammad al Adnani, "The Islamic State Will Remain," audio, August 2011.

9. Nelly Lahoud, with Muhammad al-Ubaydi, "Jihadi Discourse in the Wake of the Arab Spring," Harmony Program, Combating Terrorism Center at West Point, 2013, https://www.ctc.usma.edu/v2 /wp-content/uploads/2013/12/CTC-Jihadi-Discourse-in-the-Wake -of-the-Arab-Spring-December2013.pdf.

10. Islamic State of Iraq, *Clanging of the Swords, Part 1*, video, Al-Furqan Media, http://jihadology.net/2012/06/30/al-furqan-media-presents -a-new-video-message-from-the-islamic-state-of-iraq-clanging-of-the -swords-part-1/.

11. Islamic State of Iraq, "But Allah Will Not Allow but That His Light Should Be Perfected," audio, Abu Bakr al-Baghdadi, posted online July 12, 2012.

12. Islamic State of Iraq, *Clanging of the Swords, Part 2*, Al-Furqan Media, http://jihadology.net/2012/08/16/al-furqan-media-presents-a-new -video-message-from-the-islamic-state-of-iraq-clanging-of-the -swords-part-2%E2%80%B3/.

13. The Martyrs of Bosnia and the State of the Ummah both follow this model, as did many print outlets such as the *Al Hussam* newsletter of the 1990s. For more discussion of this issue, see Thomas Hegghammer, "Global Jihadism after the Iraq War," *Middle East Journal* 60,

no. 1 (Winter 2006), http://hegghammer.com/_files/Global_Jihad ism_after_the_Iraq_War.pdf.

14. Fawaz A. Gerges, *The Far Enemy: Why Jihad Went Global,* 2nd ed. (New York: Cambridge University Press, 2009), ch. 1; Phillip Mudd, "Are Jihadist Groups Shifting Their Focus from the Far Enemy?" *CTC Sentinel* 5, no. 5 (May 2012): 12–15, https://www.ctc.usma.edu/posts/are -jihadist-groups-shifting-their-focus-from-the-far-enemy; National Commission on Terrorist Attacks upon the United States, *The 9/11 Commission Report: Final Report of the National Commission on Terrorist Attacks upon the United States* (Washington, DC: U.S. Government Printing Office, 2004), 2–4.

15. Islamic State of Iraq, *Clanging of the Swords, Part 3,* video, Al-Furqan Media, http://jihadology.net/2013/01/11/al-furqan-media-presents -a-new-video-message-from-the-islamic-state-of-iraq-clanging-of-the -swords-part-3/.

16. Data collected from Twitter, 2013 and 2014.

17. Data collected from Twitter, May–July 2014.

18. *Islamic State News* no. 1 (May 27, 2014).

19. *Islamic State Report* no. 1 (June 2014).

20. *Islamic State News* nos. 2–4 (June 2014); *Islamic State Report* nos. 2–3 (June 2014).

21. William McCants, "How Zawahiri Lost Al Qaeda: Global Jihad Turns On Itself," *Foreign Affairs,* November 2013, http://www.foreignaffairs .com/articles/140273/william-mccants/how-zawahiri-lost-al-qaeda.

22. Rukmini Callimachi, "Yemen Terror Boss Left Blueprint for Waging Jiahd," Associated Press, The Big Story, August 9, 2013, http://bigstory .ap.org/article/yemen-terror-boss-left-blueprint-waging-jihad; "Al-Qaida Papers," Associated Press, http://hosted.ap.org/specials/inter actives/_international/_pdfs/al-qaida-papers-how-to-run-a-state.pdf.

23. *Islamic State Report* no. 1; social media postings, May–November 2014.

24. *Islamic State News* no. 3, June 2014; *Dabiq* no. 4 (October 2014).

25. Rod McGuirk, "Child Holding Severed Head Photograph Is Proof of Threat from ISIS: John Kerry," *National Post,* August 12, 2014; Tom Wyke, "Fanatics Sink to New Low: Shocking Photograph Shows ISIS supporter getting a BABY to Kick the Severed Head of a Syrian Soldier," *Daily Mail Online,* November 1, 2014, http://www.dailymail

.co.uk/news/article-2816766/WARNING-GRAPHIC-CONTENT
-Shocking-ISIS-photos-shows-toddler-kicking-severed-head-cruci
fixion-ex-regime-policeman-Syria.html; Jack Moore, "Former ISIS
Bodyguard: We Beheaded Children in Front of Parents," *International Business Times*, November 11, 2014, http://www.ibtimes.co.uk
/former-isis-bodyguard-reveals-horrific-secrets-terror-groups-leader
ship-1474296; https://pbs.twimg.com/media/B2Q6OOMCYAE84Ta
.jpg:large (graphic image), accessed from an ISIS supporter's social
media account, November 13, 2014.

26. Brian Fishman, "Jihadists Target the American Dream," *CTC Sentinel*,
March 15, 2008, https://www.ctc.usma.edu/posts/jihadists-target-the-
american-dream; Alastair Crooke, "The ISIS' Management of Savagery in Iraq," Huffington Post, August 30, 2014, http://www.huff
ingtonpost.com/alastair-crooke/iraq-isis-alqaeda_b_5542575.html.

27. Abu Bakr al Naji, *The Management of Savagery*, translated by William
McCants, May 23, 2006, http://azelin.files.wordpress.com/2010/08
/abu-bakr-naji-the-management-of-savagery-the-most-critical-stage
-through-which-the-umma-will-pass.pdf, 109.

28. For instance, Islamic State of Iraq, *Clanging of the Swords, Parts 3–4*.

29. "This Is the Promise of Allah," Al Hayat Media Center, 2014, https://
azelin.files.wordpress.com/2014/06/shaykh-abc5ab-mue1b8a5ammad
-al-e28098adnc481nc4ab-al-shc481mc4ab-22this-is-the-promise-of-god
22-en.pdf; J.M. Berger, "Gambling on the Caliphate," *IntelWire*, June
29, 2014, http://news.intelwire.com/2014/06/gambling-on-caliphate
.html.

30. Abu Bakr al Naji, *The Management of Savagery*, 19.

31. Data collected from Twitter, July 2014.

32. Abu Bakr Al-Hussayni al Qurashi al Baghdadi, "Khutbah and Jum'ah
Pray in the Grand Mosque of Mosul," YouTube video, July 27, 2014, http://
jihadology.net/2014/07/05/al-furqan-media-presents-a-new-video
-message-from-the-islamic-states-abu-bakr-al-%E1%B8%A5ussayni
-al-qurayshi-al-baghdadi-khu%E1%B9%ADbah-and-jumah-prayer-in
-the-grand-mosque-of-mu/.

33. Will McCants, Twitter post, July 5, 2014, https://twitter.com
/will_mccants/status/485463489503195136.

34. For instance, Muhammad Abd Al-Salam Faraj, *The Neglected Duty*,
trans. Mark Juergensmeyer, http://www.juergensmeyer.com/files

/Faraj_The_Neglected_Duty.pdf, and Nosair, referenced at House Committee on Financial Services Subcommittee on Oversight and Investigations, Testimony of Matthew Epstein: Arabian Gulf Financial Sponsorship of Al-Qaida Via U.S. Based Banks, Corporations and Charities (2003).

35. Yusuf al-'Uyayri, "Constants on the Path of Jihad," translated and expanded by Anwar al-'Awlaqi, "Constants on the Path of Jihad," audio lecture, 2005.

36. *Dabiq* no. 1 (July 2014).

37. William McCants, "ISIS Fantasies of an Apocalyptic Showdown in Northern Syria," Brookings Institution, October 3, 2014, http://www .brookings.edu/blogs/iran-at-saban/posts/2014/10/03-isis-apocalyptic -showdown-syria-mccants.

38. *Dabiq* no. 1 (July 2014): 3–4.

39. "A Message to America," August 19, 2014.

40. Tom Vanden Brook, "Foley Rescue Attempt in Syria Failed, Officials Say," *USA Today*, August 21, 2014, http://www.usatoday.com/story /news/usanow/2014/08/20/syria-isis-hostages/14360787/.

41. Abu Bakr al Naji, *The Management of Savagery*, 78.

42. Jamie Orme, "British ISIS Fighter Calls David Cameron a 'Despicable Swine' in Online Video," October 3, 2014, http://www.theguardian .com/world/2014/oct/03/isis-david-cameron-despicable-swine-video; Lizzie Dearden, "Islamic State: David Cameron's Head Will Be on a Spike, Says British Women in Syria," *Independent*, September 9, 2014, http://www.independent.co.uk/news/world/middle-east/islamic -state-david-camerons-head-will-be-on-a-spike-says-british-woman -in-syria-9721757.html; data collected August and September 2014 from Twitter and Facebook.

43. "Although the Disbelievers Dislike It," November 16, 2014: Peter Allen and John Hall, "ISIS' International Death Squad: Jihadist executioners in beheading video include two Britons, two Frenchmen and a German," *Daily Mail*, November 17, 2014, http://www.dailymail.co.uk /news/article-2837742/Now-two-Frenchmen-German-jihadist-identi fied-ISIS-executioners-sick-beheading-video-Extremist-killers-came -Europe.html; Megan Spacia, "Westerners Among Knife-Wielding Militants in ISIS Beheading Video," Mashable, November 17, 2014, http://mashable.com/2014/11/17/foreign-fighters-isis-video/.

44. Greg Botelho, "ISIS Executes British Aid Worker David Haines; Cameron Vows Justice," CNN World, September 14, 2014, http://www.cnn
.com/2014/09/13/world/meast/isis-haines-family-message/; David
Williams, Ian Drury, Jason Groves, and Richard Spillett, "'Happy 47th
Birthday to my Little Bro—Can't Wait to Have a Party on Your Return':
British Hostage Alan Henning's Sister Posts Moving Message to Him Just
Weeks Before He Appeared in Chilling ISIS Video," Daily Mail Online,
September 14 2014, http://www.dailymail.co.uk/news/article-2755705
/Find-s-late-He-s-Manchester-taxi-driver-went-Syria-help-refugees
-Now-s-facing-death-hands-Jihadi-John.html; Rukmini Callimachi,
"Obama Calls Islamic State's Killing of Peter Kassig 'Pure Evil,'" New
York Times, November 16, 2014, http://www.nytimes.com/2014/11/17
/world/middleeast/peter-kassig-isis-video-execution.html?_r=0.

45. Alice Speri, "Islamic State Video Shows Hostage John Cantlie Allegedly Reporting from Ground in Kobane," Vice News, October 27, 2014,
https://news.vice.com/article/new-islamic-state-video-shows-hostage
-john-cantlie-allegedly-reporting-from-the-ground-in-kobane.

46. Jamie Dettmer, "An ISIS Hostage on the Dark Side," Daily Beast, October 27, 2014, http://www.thedailybeast.com/articles/2014/10/27/isis
-hostage-cantlie-in-a-propaganda-video-from-besieged-kobani.html.
Note: A subsequent "Lend Me Your Ears" release by ISIS showed
Cantlie once again in the orange jumpsuit, but other ISIS propaganda
indicated that the series had been filmed all in one sitting, prior to the
Kobane release.

CHAPTER 6. JIHAD GOES SOCIAL

1. Portions of this chapter were adapted from J.M. Berger, "Internet Provides Terrorists with Tools—Just Like Everyone Else," IntelWire, July
31, 2011, http://news.intelwire.com/2011/07/internet-provides-terrorists-with-tools.html.

2. J.M. Berger, "Terrorist Propaganda: Past And Present," Huffington
Post, August 1, 2011, http://www.huffingtonpost.com/jm-berger
/terrorist-propaganda_b_913684.html.

3. U.S. v. Muhamed Mubayyid, Emadeddin Muntasser, and Samir Al Monla,
Criminal Action No. 05-40026-FDS (2007), Exhibit 514A, transcript of
phone conversation of January 27, 1996.

4. Joseph Cox, "How Jihadists Use the Internet," *Vice News*, January 14, 2014, http://www.vice.com/read/how-jihadists-use-the-internet.

5. Jarret Brachman and Alix Levine, "The World of Holy Warcraft," *Foreign Policy*, April 13, 2011, http://www.foreignpolicy.com/articles /2011/04/13/the_world_of_holy_warcraft. The authors here overstate the evidence linking gamification to violent activity or progressive radicalization, but the concept is part of engagement. For a more measured assessment, see Will McCants, "The Limits of Gamification," Jihadica, April 25, 2011, http://www.jihadica.com/the-limits-of -gamification/, as well as a more detailed debate over the premise, J.M. Berger, "Some Thoughts on 'Gamification': Where the Loser Takes It All," *IntelWire*, April 26, 2011, http://news.intelwire.com/2011/04 /some-thoughts-on-gamification.html; Alex Levine, "My Analysis of the Analysis on Gamification," blog post, April 26, 2011, http:// alixlevine.com/Alix_Levine/Blog/Entries/2011/4/27_My_Analysis _of_the_Analysis_on_Gamification_Article.html; and J.M. Berger, "More Gamification: Alix Levine Responds," *IntelWire*, April 27, 2011, http://news.intelwire.com/2011/04/more-gamification-alix-levine -responds.html.

6. "Terrorist on the Internet," narrated by Dick Gordon, *The Connection*, WBUR Boston (radio), June 8, 2005, http://theconnection.wbur .org/2005/06/08/terrorist-on-the-internet.

7. Paul Cruickshank and Tim Lister, "Al Qaeda Calling?" CNN, Security Clearance blog, August 8, 2013, http://security.blogs.cnn.com /2013/08/08/al-qaeda-calling/.

8. "Adam Gadahn Lashes Out Against Al-Qaeda Affiliates, Jihadist Forums," *Central Asia Online*, May 9, 2012, http://centralasiaonline .com/en_GB/articles/caii/features/pakistan/main/2012/05/09 /feature-02.

9. Aaron Zelin, "The State of Global Jihad Online: A Qualitative, Quantitative, and Cross-Lingual Analysis," New American Foundation, January 2013, http://www.newamerica.net/sites/newamerica.net/files /policydocs/Zelin_Global%20Jihad%20Online_NAF.pdf.

10. J.M. Berger, "Gone but Not Forgotten," *Foreign Policy*, September 30, 2011, http://www.foreignpolicy.com/articles/2011/09/30/Anwar_al _Awlaki_dead_but_not_forgotten.

11. Aamer Madhani, "Cleric al-Awlaki Dubbed 'bin Laden of the Inter-

net,'" *USA Today*, August 24, 2010, http://usatoday30.usatoday.com /news/nation/2010-08-25-1A_Awlaki25_CV_N.htm.

12. "Anwar Alawlaki," http://www.anwar-alawlaki.com/, accessed 2009.

13. FBI Letterhead Memorandum, "Anwar Nasser Aulaqi," September 26, 2001, http://intelfiles.egoplex.com/2001-09-26-fbi-lhm-anwar-nasser -aulaqi.pdf; "Terror Leader Awlaki Paid Thousands for Prostitutes in DC Area, Documents Show," Fox News, July 2, 2013, http://www .foxnews.com/politics/2013/07/02/terror-leader-awlaki-paid-thou sands-for-prostitutes-in-dc-area-documents-show/#ixzz2XzwccxWq; J.M. Berger, *Jihad Joe: Americans Who Go to War in the Name of Islam* (Washington, DC: Potomac Books, 2011), 116–26.

14. Brianna Lee, "Under Pressure, YouTube Removes Awlaki Jihadi Videos," PBS, Daily Need blog, November 5, 2010, http://www.pbs .org/wnet/need-to-know/the-daily-need/under-pressure-youtube -removes-awlaki-jihadi-videos/4872/attachment/anwar-al-awlaki/; Spencer Akerman, "Qaida's YouTube Preacher Is Killed In Yemen," *Wired* blog, September 30, 2011, http://www.wired.com/2011/09/aw laki-dead-yemen/.

15. "Lieberman Calls on Google to Take Down Terrorist Content," Senate Committee on Homeland Security and Government Affairs, press release, May 19, 2008, https://web.archive.org/web/20080731055347 /http://hsgac.senate.gov/public/index.cfm?Fuseaction=PressReleases .Detail&PressRelease_id=8093d5b2-c882-4d12-883d5c670d43d269& Month=5&Year=2008&Affiliation=C; "Lieberman to YouTube: Remove al Qaeda Videos," CNN, May 20, 2008, http://edition.cnn.com /2008/POLITICS/05/20/youtube.lieberman/.

16. Craig Kanalley, "YouTube Gives Users Ability to Flag Content That Promotes Terrorism," Huffington Post, May 25, 2011, http://www.huff ingtonpost.com/2010/12/13/youtube-terrorism-flag_n_796128.html.

17. "Dialogue with Sen. Lieberman on Terrorism Videos," YouTube Official Blog, May 19, 2008 http://youtube-global.blogspot.com/2008/05 /dialogue-with-sen-lieberman-on.html.

18. J.M. Berger, "It's Too Easy for Wanna-be Terrorists to Use Google, YouTube for Promotion of Their Extremist Causes," New York *Daily News*, blog, November 21, 2011, http://www.nydailynews.com/opin ion/easy-wanna-be-terrorists-google-youtube-promotion-extremist -article-1.980586?localLinksEnabled=false.

19. J.M. Berger, "Anwar Awlaki E-Mail Exchange with Fort Hood Shooter Nadal Hasan," *IntelWire,* July 19, 2012, http://news.intelwire.com/2012/07/the-following-e-mails-between-maj.html.

20. David Ariosto and Deborah Feyerick, "Christmas Day Bomber Sentenced to Life in Prison," CNN, February 17, 2012, http://www.cnn.com/2012/02/16/justice/michigan-underwear-bomber-sentencing/.

21. "$4,200." *Inspire,* November 2010. See also Thomas Hegghammer, "The Case for Chasing al-Awlaki," *Foreign Policy,* Middle East Channel, November 24, 2010, http://mideastafrica.foreignpolicy.com/posts/2010/11/24/the_case_for_chasing_al_awlaki.

22. John Burns and Miguel Helft, "YouTube Withdraws Cleric's Videos," *New York Times,* November 4, 2010, http://www.nytimes.com/2010/11/05/world/05britain.html.

23. Tom Warren, "Microsoft, like Google, Tips Off Police for Child Porn Arrest," Verge, blog, August 7, 2014, http://www.theverge.com/2014/8/7/5977827/microsoft-tips-off-police-for-child-porn-arrest.

24. For instance, Wisconsin Sikh temple shooter Wade Page played in a "hatecore" band called Definite Hate, whose albums were mainly distributed on Myspace. Eliza Shapiro, "Shooting Suspect Wade Michael Page's White-Power Past," Daily Beast, August 6, 2012, http://www.thedailybeast.com/articles/2012/08/06/shooting-suspect-wade-michael-page-s-white-power-past.html.

25. "Al-Qaeda, Jihadis Infest the San Francisco, California–Based 'Internet Archive' Library," Cyber and Jihad Lab, Middle East Media Research Institute, August 17, 2014, http://cjlab.memri.org/uncategorized/al-qaeda-jihadis-infest-the-san-francisco-california-based-internet-archive-library/.

26. The Twitter rules governing content were one page consisting of 1,123 words as of March 8, 2011. By October 27, 2014, they had expanded to more than nine pages and well over six thousand words, http://support.twitter.com/articles/18311-the-twitter-rules and https://support.twitter.com/articles/18311-the-twitter-rules, both accessed October 27, 2014.

27. For examples, Google "Nu Wexler," "suspension," and "statement" and read the news stories that come up.

28. J.M. Berger, "Zero Degrees of Al Qaeda," *Foreign Policy,* August 14,

2014, http://www.foreignpolicy.com/articles/2013/08/14/zero_de grees_of_al_qaeda_twitter.

29. Ahmad Khan, Al-Emara Twitter account, https://twitter.com/alema rahstudio.

30. Abdualqahar Balkhi, Twitter account, https://twitter.com/ABalkhi.

31. Ali Latifi, "Afghanistan's Online War Of Words," Al Jazeera, October 17, 2012, http://www.aljazeera.com/indepth/features/2012/10 /2012101510373939539.html.

32. "Al Shabaab Launches Apparent Twitter Campaign," Anti-Defamation League, December 20, 2011, http://www.adl.org/combating-hate /international-extremism-terrorism/c/shabaab-launches-twitter .html; J.M. Berger, "#Unfollow: The Case for Kicking Terrorists off Twitter," *Foreign Policy*, February 20, 2013, http://www.foreignpolicy .com/articles/2013/02/20/unfollow.

33. J.M. Berger, "Omar and Me," *Foreign Policy*, September 16, 2014, http:// www.foreignpolicy.com/articles/2013/09/16/omar_and_me.

34. Confidential interviews with U.S. counterterrorism officials working on social media issues; interview with Jeff Weyers, University of Liverpool, researcher/consultant at iBRABO.

35. First discussed during a congressional hearing in 1993 by Undersecretary of State Tim Wirth speaking to Federal News Service, U.S. House Committee on Foreign Affairs Subcommittee on International Security, International Organization and Human Rights, US Anti Terrorism Policy: Capitol Hill Hearing About the Middle-East, July 13 1993, H. Res. 118 (Washington, DC: Government Printing Office, 1993).

36. Central Intelligence Agency, "The Predicament of the Terrorism Analyst," *Studies in Intelligence* 29 (Winter 1985), http://www.foia.cia.gov /sites/default/files/DOC_0000620573.pdf.

37. U.S. House, Committee on Foreign Affairs Subcommittee on Terrorism, Nonproliferation and Trade, U.S Strategy for Countering Jihadist Web Sites, Hearing, September 29, 2010, Serial No. 111-130 (Washington, DC: U.S. Government Printing Office, 2010), http://www.gpo .gov/fdsys/pkg/CHRG-111hhrg61516/html/CHRG-111hhrg61516 .htm.

38. "Terrorist on the Internet," narrated by Dick Gordon, *The Connection*, WBUR Boston (radio), June 8, 2005, http://theconnection.wbur.org /2005/06/08/terrorist-on-the-internet.

39. Pablo Barberá, "How Social Media Reduces Mass Political Polarization: Evidence from Germany, Spain, and the U.S.," October 18, 2014, https://files.nyu.edu/pba220/public/barbera-polarization-social-me dia.pdf.

40. Interview with Abu Suleiman al Nasser by J.M. Berger, 2011.

41. Correspondence with Jeff Weyers, analyst with iBRABO, 2014.

42. Jenna McLaughlin, "Twitter Is Not at War with ISIS. Here's Why," *Mother Jones*, November 18, 2014, http://www.motherjones.com /politics/2014/11/twitter-isis-war-ban-speech; Caleb Garling, "Twitter C.E.O. Dick Costolo on Receiving Death Threats from ISIS," *Vanity Fair*, October 9, 2014, http://www.vanityfair.com/online/daily /2014/10/twitter-ceo-death-threats-isis.

43. Neil Ungerleider, "Despite Ban, YouTube Is Still a Hotbed of Terrorist Group Video Propaganda," *FastCompany*, blog, November 12 2010, http://www.fastcompany.com/1701383/despite-ban-youtube-still -hotbed-terrorist-group-video-propaganda; David Gardner, "Facebook Used by Al Qaeda to Recruit Terrorists and Swap Bomb Recipes, Says US Homeland Security Report," *Daily Mail Online*, December 10, 2010, http://www.dailymail.co.uk/news/article-1337344/Facebook -used-Al-Qaeda-recruit-terrorists-swap-bomb-recipes.html.

44. Confidential interviews with Google employees; confidential interviews with U.S. counterterrorism officials working on social media issues.

45. Confidential interviews with U.S. counterterrorism officials working on social media issues; email interview with Jeff Weyers.

46. Bob Griffin, "Intelligence Analysis to Take Down a Terrorist," army -technology.com, July 12, 2011, http://www.army-technology.com /features/feature123872/.

47. Somini Sengupta, "Twitter Yields to Pressure in Hate Case in France," *New York Times*, July 12, 2013, http://www.nytimes.com/2013/07/13 /technology/twitter-yields-to-pressure-in-hate-case-in-france.html.

48. Meyukh Sen, "'Blasphemy' and Social Media in Pakistan," Internet Monitor blog, Harvard Law School, June 27, 2014, https://blogs.law .harvard.edu/internetmonitor/2014/06/27/blasphemy-and-social-me dia-in-pakistan/.

49. Berger, "#Unfollow."

50. Coauthor J.M. Berger reported the account shortly before it was suspended.

51. "Somalia's al-Shabab Opens New Twitter Account," BBC News Africa, February 4, 2013, http://www.bbc.com/news/world-africa-21321687.

52. Twitter data collected December 2012–February 2013.

53. Berger, "#Unfollow."

54. Associated Press, "Kenya Marks 1 Year Since Westgate Mall Attack," Fox News, September 21, 2014, http://www.foxnews.com/world/2014/09/21/kenya-marks-1-year-since-westgate-mall-attack-left-67-dead-on-high-alert-for/; Jason Straziuso, "NYPD Report on Kenya Attack Isn't US Gov't Review," Yahoo News, December 13, 2013, http://news.yahoo.com/nypd-report-kenya-attack-isn-39-t-us-151711825.html.

55. Lori Hinnant, "Kenya Attack Unfolded in Up and Down Twitter Feeds," Associated Press, The Big Story, http://bigstory.ap.org/article/kenya-attack-unfolded-and-down-twitter-feeds.

56. Coauthor J.M. Berger reported several iterations of the account during this period, roughly corresponding to the accounts' suspensions.

57. Peter Bergen, "Are Mass Murders Using Twitter as a Tool?" CNN, September 27, 2013, http://www.cnn.com/2013/09/26/opinion/bergen-twitter-terrorism/.

CHAPTER 7. THE ELECTRONIC BRIGADES

1. Mark Townsend and Toby Helm, "Jihad in a Social Media Age: How can the west win an online war?" *Guardian*, August 23, 2014, http://www.theguardian.com/world/2014/aug/23/jihad-social-media-age-west-win-online-war; quote taken from image provided by Jeff Weyers, ibRABO, specializing in terrorist's use of social media.

2. Simon Rogers, "Insights into the #WorldCup Conversation on Twitter," Twitter blog, July 14, 2014, https://blog.twitter.com/2014/insights-into-the-worldcup-conversation-on-twitter.

3. J.M. Berger, Twitter post, June 14, 2014, https://twitter.com/intelwire/status/478029497317724160.

4. J.M. Berger, Twitter post, June 15, 2014, https://twitter.com/intelwire/status/478310946385850370.

5. Mohamed J. was identified via the Twitter account he used to design the ISIS app discussed in this chapter. The website registrations associated with the app and additional pseudonymous social media accounts were identified using the email addresses associated with the app and

the websites, and through social network analysis confirmed via the content of the social media accounts. No school matching the name "Los Angeles School of Arts" could be identified.

6. All apps by the company were terminated on Google's app store, but some were still available via Amazon.com as of October 28, 2014. For example, "AKSSWA," http://www.amazon.com/x627-x643-x633-x648 -AKSSWAR/dp/B00IT07W4U/ref=sr_1_2?s=mobile-apps&ie=UTF 8&qid=1414528220&sr=1-2 and "Quran Graphics," http://www.ama zon.com/MAWLAH-Company-Quran-Graphics/dp/B00JV18GBY /ref=sr_1_5?s=mobile-apps&ie=UTF8&qid=1414528220&sr=1-5.

7. The app was available from Google Play, https://play.google.com /store/apps/details?id=com.pashaeer.myapp, prior to its suspension on June 20, 2014.

8. Analysis of the Android app performed by Justin Seitz, author of *Black Hat Python: Python Programming for Hackers and Pentesters* (n.p.: No Starch Press, 2014).

9. An account created by J.M. Berger was used to sign up for the app and track its behavior.

10. "New Pro-Islamic State of Iraq And Syria (ISIS) 'News' App for Android, Available at Google Play Store," Cyber and Jihad Lab, Middle East Media Research Institute, April 24, 2014, http://cjlab.memri.org /lab-projects/tracking-jihadi-terrorist-use-of-social-media/new-pro -islamic-state-of-iraq-and-syria-isis-news-app-for-android-available-at -google-play-store/.

11. Analysis of the app and its behavior is based on more than 100,000 tweets specifically identified as originating with the app by approximately more than 3,000 accounts retrieved from Twitter by J.M. Berger in June 2014 and in October 2014. These samples are considerably less than the total output of the app, and the number of accounts is also likely a conservative estimate.

12. "New Pro-Islamic State of Iraq and Syria (ISIS) 'News' App for Android, Available at Google Play Store."

13. J.M. Berger, "How ISIS Games Twitter," *Atlantic*, June 16, 2014, http:// www.theatlantic.com/international/archive/2014/06/isis-iraq-twitter -social-media-strategy/372856/.

14. http://t.co/fh8ZMgPFQn, accessed November 2, 2014.

15. Data collected in June 2014.

16. Shalom Bear, "IDF's Operation 'Protective Edge' Begins Against Gaza," *Jewish Press*, July 8, 2014, http://www.jewishpress.com/news/breaking-news/idfs-operation-protective-edge-begins-against-gaza/2014/07/08/.

17. Congress-Edits, Twitter account, https://twitter.com/congressedits.

18. Jovi Umawing, "Twitter's Bane: Bad Bots," Malwarebytes Unpacked blog, June 10, 2014, https://blog.malwarebytes.org/social-engineering/2014/06/twitters-bane-bad-bots/.

19. The "ghost" bots were first brought to the attention of coauthor J.M. Berger by Twitter user @hlk01. Subsequent analysis of the ghostbots revealed the existence of additional bot networks.

20. Catherine O'Donnell, "New Study Quantifies Use of Social Media in Arab Spring," University of Washington, September 12, 2011, www.washington.edu/news/2011/09/12/new-study-quantifies-use-of-social-media-in-arab-spring/.

21. Jack Shenker, "Egyptian Army Retakes Tahrir Square," *Guardian*, August 1, 2011, http://www.theguardian.com/world/2011/aug/01/egypt-army-tahrir-square.

22. Blake Hounshell, "The Revolution Will Be Tweeted," *Foreign Policy*, June 20, 2011, http://www.foreignpolicy.com/articles/2011/06/20/the_revolution_will_be_tweeted.

23. "Tunisia: President Zine Al-Abidine Ben Ali Forced Out," BBC Africa, January 14, 2011, http://www.bbc.com/news/world-africa-12195025; David Kirkpatrick, "Egypt Erupts in Jubilation as Mubarak Steps Down," *New York Times*, February 11, 2011, http://www.nytimes.com/2011/02/12/world/middleeast/12egypt.html?pagewanted=all.

24. "Tethered by History," *Economist*, July 5, 2014, http://www.economist.com/news/briefing/21606286-failures-arab-spring-were-long-time-making-tethered-history.

25. Marc Lynch, Deen Freelon, and Sean Aday, "Blogs and Bullets: Syria's Socially Mediated Civil War," USIP Peaceworks Report no. 91, 2014; Adam Rawnsley, "Be Very Skeptical—A Lot of Your Open-Source Intel Is Fake," Medium blog, November 1, 2014, https://medium.com/war-is-boring/be-very-skeptical-a-lot-of-your-open-source-intel-is-fake-5e4a5d5a9195; J.M. Berger, "Syria's Socially Mediated Civil War," *IntelWire*, January 22, 2014, http://news.intelwire.com/2014/01/syrias-socially-mediated-civil-war.html.

26. Brian Fishman and Joseph Felter, "Al-Qa'ida's Foreign Fighters in Iraq," Combating Terrorism Center at West Point, January 2, 2007, https://www.ctc.usma.edu/posts/al-qaidas-foreign-fighters-in-iraq-a-first-look-atthesinjar-records; Lieutenant Colonel Joel Ray, " 'Blowback'—Iraq Comes to Syria," The Caravan Hoover Institution, February 23, 2012, http://www.hoover.org/research/blowback-iraq-comes-syria.

27. For an example of regime-sponsored activities, see Noah Shachtman and Michael Kennedy, "The Kardashian Look-Alike Trolling for Assad," Daily Beast, October 17, 2014, http://www.thedailybeast.com/articles/2014/10/17/the-kardashian-look-a-like-trolling-for-assad.html. The account mentioned in the story is part of a large and highly organized network.

28. Jeffery Carr, "In OSINT, All Sources Aren't Created Equal," Digital Dao blog, November 24, 2014, http://jeffreycarr.blogspot.com/2013/11/in-osint-all-sources-arent-created-equal.html; Charlemagne, "Analyst Beware! The Perils of OSINT," IMSL blog, December 18, 2012, http://intelmsl.com/insights/in-the-news/analyst-beware-the-perils-of-osint/.

29. Alice Speri, "Now Even ISIS Has Its Very Own Whistleblower," *Vice News*, June 19, 2014, https://news.vice.com/article/now-even-isis-has-its-very-own-whistleblower; Jacob Siegel, "Someone Is Spilling ISIS's Secrets on Twitter," Daily Beast, http://www.thedailybeast.com/articles/2014/06/18/someone-is-spilling-isis-s-secrets-on-twitter.html.

30. For instance, Teresa Salvadoretti, "The Role of Social Media in the Syrian Crisis," *Asfar,* 2014, http://www.asfar.org.uk/the-role-of-the-social-media-in-the-syrian-crisis.html, and Laure Nouraout, "Gaza, Syria, Ukraine: Hot to Debunk Fake Images," GEN: Global Editors Network, August 27, 2014, http://www.globaleditorsnetwork.org/news/2014/08/gaza,-syria,-ukraine-how-to-debunk-fake-images/.

31. Data collected on @e3tasimo in December 2013.

32. Jawa Report blog, February 28, 2014, http://mypetjawa.mu.nu/archives/217785.php; J.M. Berger, Twitter post, February 28, 2014, https://twitter.com/intelwire/status/439495997706993664; data collected on @reyadiraq's followers on February 21, 2013.

33. Data on @dawlh_i_sh was last collected on August 26, 2014. The suspension happened soon after.

34. J.M. Berger, Twitter post, January 10, 2014, https://twitter.com/intelwire/status/421707838021316608.

35. Data collected in February and March 2014.

36. Data collected via Topsy.com for thirty days following February 20, 2014.

37. Charles Arthur, "What Is the 1% Rule?" *Guardian,* July 19, 2006, http://www.theguardian.com/technology/2006/jul/20/guardian weeklytechnologysection2; J.M. Berger and Bill Strathearn, "Who Matters Online: Measuring influence, evaluating content and countering violent extremism in online social networks," International Centre for the Study of Radicalisation and Political Violence, March 2013, http://icsr.info/wp-content/uploads/2013/03/ICSR_Berger-and-Strathearn.pdf.

38. Data collected by and analyzed J.M. Berger in March and April 2014 using a methodology derived from "Who Matters Online," ibid.

39. ISIS Twitter strategy document, September 12, 2014; Shayba al-Hamdi; http://justpaste.it/h26t.

40. Ibid.

41. Based on analysis of data collected from February 2014 to November 2014.

42. Active Hashtags, Twitter account: https://twitter.com/activehash tags.

43. Based on data collected in March and April 2014.

44. Ibid.

45. Data collected by J.M. Berger in June 2014; see also Berger, "How ISIS Games Twitter."

46. J.M. Berger, Twitter post, August 8, 2014, https://twitter.com/intel wire/status/497954495109009409.

47. Sophie Evans, "The Dutch Jihadi Fighter (and artist) Training British Teenagers How to Kill in Syria (and when he's not doing that he's taking pictures of cats)," *Daily Mail Online,* July 9, 2014, http://www.dailymail.co.uk/news/article-2685648/Syria-fighter.html; "ISIS Goes Nuts for Nutella: Brutal Jihadists Reveal Bizarre Soft Spot for the Chocolate Hazelnut Spread," *Daily Mail Online,* August 14, 2014, http://www.dailymail.co.uk/news/article-2724889/ISIS-goes-nuts-Nutella-Brutal-Jihadists-reveal-bizarre-soft-spot-chocolate-hazelnut-spread.html; Jon Stone, "Islamist Fighters in Iraq and Syria Keep Tweeting Pictures of Cats," BuzzFeed, June 16, 2014, http://www.buzzfeed.com/jonstone/foreign-jihadi-fighters-in-iraq-and-syria-keep

-tweeting-pict; "Chilling New ISIS Threat Against Americans on Twitter," AOL, August 21, 2014, http://www.aol.com/article/2014/08/21/chilling-new-isis-threat-against-americans-on-twitter-group-be headed-journalist-james-foley/20950677/; Meg Wagner, "'We Are in Your Cities': Alleged ISIS Terrorists Threaten U.S. Via Twitter," New York *Daily News,* August 23, 2014, http://www.nydailynews.com/news/national/alleged-isis-terrorists-threaten-u-s-twitter-article-1.1914363; Taylor Wofford, "ISIS Paranoia Reaches America," *Newsweek,* September 24, 2014, http://www.newsweek.com/where-isis-isnt-273030; Jennifer Smith, "Islamic State has a 'dirty bomb' says British jihadi, amid claims 40kg of URANIUM was taken from Iraqi university," *Daily Mail,* November 30, 2014, http://www.dailymail.co.uk/news/article-2854729/Islamic-State-dirty-bomb-says-British-jihadi-amid-claims-40kg-uranium-taken-Iraqi-university.html; Sarah Kaufman, "Boneheaded ISIS Threat: We Will Infect U.S. With Ebola," Vocativ, September 16, 2014, http://www.vocativ.com/world/isis-2/isis-ebola-threat/.

48. Google Trends, accessed October 31, 2014.

49. Data on financier networks on Twitter was analyzed over several months in 2014 with assistance from Aaron Zelin, the Richard Borow Fellow at the Washington Institute for Near East Policy and Rena and Sami David Fellow at the International Centre for the Study of Radicalisation and Political Violence.

50. For detailed descriptions of how these networks worked, see Aukai Collins, *My Jihad: The True Story of an American Mujahid's Amazing Journey from Usama Bin Laden's Training Camps to Counterterrorism with the FBI and CIA* (Guilford, CT: Lyons Press, 2002); John Miller, Michael Stone, and Chris Mitchell, *The Cell: Inside the 9/11 Plot, Why the FBI and CIA Failed to Stop It* (New York: Hyperion, 2002); and J.M. Berger, *Jihad Joe: Americans Who Go to War in the Name of Islam* (Washington, DC: Potomac Books, 2011).

51. Shaarik Zafar, "Western Foreign Fighters in Syria: Implications for the U.S. CVE Efforts," Washington Institute, March 14, 2014, http://www.washingtoninstitute.org/policy-analysis/view/western-foreign-fighters-in-syria-implications-for-u.s.-cve-efforts.

52. Ibid.

53. Vicky Collins, "Denver Teens Chatted Online with ISIS Recruiter,

Report Says," NBC News, October 30, 2014, http://www.nbcnews
.com/storyline/isis-terror/denver-teens-chatted-online-isis-recruiter
-report-says-n237666.

54. Oren Dorell, "FBI Probes Islamic State Recruiting in Minneapolis,"
USA Today, August 29, 2014, http://www.usatoday.com/story/news
/world/2014/08/29/islamic-state-minneapolis-recruiting/14794307/;
Jamie Yuccas, "Minneapolis Has Become Recruiting Ground for Islamic Extremists," CBS News, August 27, 2014, http://www.cbsnews
.com/news/minneapolis-has-become-recruiting-ground-for-islamic
-extremists/; Philip Ross, "ISIS Recruiters in US: Jihadists Target
Minneapolis Youth for 'Indoctrination,'" *International Business Times*,
October 2, 2014, http://www.ibtimes.com/isis-recruiters-us-jihadists
-target-minneapolis-youth-indoctrination-1698359.

55. Analysis of Twitter networks by J.M. Berger.

56. Dr. Jon Cole et al., "Guidance For Identifying People Vulnerable to Recruitment Into Violent Extremism," Counterrorism.org, 2012, https://
www.counterextremism.org/resources/details/id/224/guidance
-for-identifying-people-vulnerable-to-recruitment-into-violent-ex
tremism.

57. Adam Serwer, "Accused Student Terrorist Was a Social Media Over-
sharer," MSNBC, March 25, 2014, http://www.msnbc.com/msnbc
/instagram-social-media-terrorism-teausant; "Wake Terrorism Investigation Built Over Weeks," WRAL, March 20, 2014, http://www
.wral.com/wake-terrorism-investigation-built-over-weeks/13496093/;
United States of America v. Avin Marsalis Brown (aka Musa Brown),
5:14-MJ-1181-WW, March 20, 2014, http://www.investigativeproject
.org/documents/case_docs/2331.pdf; "Nicolas Teausant Charged with
Attempting to Aid Al Qaeda Fighters from California," Associated
Press, March 27, 2014, http://www.huffingtonpost.com/2014/03/27
/nicholas-teausant-student-al-qaeda_n_5041718.html; analysis of
Facebook networks by J.M. Berger.

58. Jeff Weyers, "The Newest Advertising and Recruiting Mogul: ISIS,"
IBRABO, last updated June 26, 2014, http://ibrabo.wordpress.com
/2014/06/26/the-newest-advertising-and-recruiting-mogul-isis/; interview with Jeff Weyers, 2014.

59. Facebook data collected by Jeff Weyers; additional Facebook data
collected by J.M. Berger in 2013 and 2014; Jeff Weyers, "While the

Ceasefire Holds on the Ground the Battle to Win Hearts and Minds on Social Media Heats Up for al Qassam," IBRABO, last updated September 9, 2014, http://ibrabo.wordpress.com/2014/09/11/while-the-cease fire-holds-on-the-ground-the-battle-to-win-hearts-and-minds-on -social-media-heats-up-for-al-qassam/.

60. "Reporting Abusive Behavior," Twitter Help Center blog, accessed October 31, 2014.

61. "Blocking Users On Twitter," Twitter Help Center blog, accessed October 31, 2014.

62. Peter Bergen, "Are Mass Murders Using Twitter as a Tool?" CNN Opinion, September 27, 2013, http://www.cnn.com/2013/09/26/opin ion/bergen-twitter-terrorism/.

63. Data collected by J.M. Berger, September 2013 to present.

64. Ibid.

65. Jeremy Kessel, "Fighting for More #transparency," Twitter blog, February 6, 2014, https://blog.twitter.com/2014/fighting-for-more -transparency; Jeremy Kessel, "Continuing Our Fight for More #Transparency," Twitter blog, July 31, 2014, https://blog.twitter .com/2014/continuing-our-fight-for-more-transparency; "Removal Requests," Twitter Transparency, https://transparency.twitter.com /removal-requests/2014/jan-jun.

66. Caleb Garling, "Twitter C.E.O. Dick Costolo on Receiving Death Threats from ISIS," *Vanity Fair,* October 9, 2014, http://www.vanity fair.com/online/daily/2014/10/twitter-ceo-death-threats-isis.

67. Entities with legal jurisdiction, a category not clearly defined by Twitter, but likely related to copyright and intellectual property violations, as well as security issues such as phishing and malware.

68. "Removal Requests," Twitter Transparency, https://transparency .twitter.com/removal-requests/2014/jan-jun.

69. Dawn app tweets promoting the video, collected by J.M. Berger in May, June, and October 2014.

70. A note from J.M. Berger: There are established methods for inflating YouTube view counts (see http://youtubecreator.blogspot .com/2014/02/keeping-youtube-views-authentic.html), including commercial spam services of a type known to be employed by ISIS, but not yet observed in the wild. In addition, users can simply view the video over and over again, in part or in whole. The video was rapidly

suspended as it was posted, limiting historical access to statistics, but in light of ISIS's known tactics, I believe it's extremely likely that the views for most ISIS video releases are similarly inflated.

71. Nico Prucha, "Is this the most successful release of a jihadist video ever?," Jihadica blog, May 19, 2014, http://www.jihadica.com/is-this -the-most-successful-release-of-a-jihadist-video-ever/.

72. "India Arrests Man Behind Pro-ISIS Twitter Account @shamiwitness," *NBC News*, December 13, 2014, http://www.nbcnews.com/storyline /isis-terror/india-arrests-man-behind-pro-isis-twitter-account-shami witness-n267701.

73. "ISIS Jihadists Put Out Hollywood-Style Propaganda Film," France 24, June 13, 2014, http://observers.france24.com/content/20140613-holly wood-fim-jihadist-propaganda-isis; Brown Moses, Twitter post, June 11, 2014, https://twitter.com/Brown_Moses/status/476728953856884738; Paul Szoldra, "Militants in Iraq Are Surprisingly Brilliant on Social Media," *Business Insider,* June 25, 2014, www.businessinsider.com/isis -propaganda-2014-6.

74. Miriam Berger, "Twitter Has Suspended an ISIS Account That Live-Tweeted Its Advance in Iraq," Buzzfeed, June 14, 2014, http://www .buzzfeed.com/miriamberger/twitter-has-suspended-an-isis-account -that-live-tweeted-its; data collected from Twitter in June 2014.

75. Data collected June 2014.

76. J.M. Berger, "The State of the 'Caliphate' is . . . Meh," *IntelWire,* July 2, 2014, http://news.intelwire.com/2014/07/the-caliphate-so-far-flatlin ing.html.

77. Data collected from Twitter in August 2014 and thereafter.

78. Miriam Berger, "Where Is ISIS Moving to Online as Twitter Clamps Down?" BuzzFeed, August 20, 2014, http://www.buzzfeed.com /miriamberger/where-is-isis-moving-to-online-as-twitter-clamps -down.

79. Amar Toor, "How Putin's Cronies Seized Control of Russia's Face-book," Verge, January 31, 2014, http://www.theverge.com/2014 /1/31/5363990/how-putins-cronies-seized-control-over-russias-face book-pavel-durov-vk.

80. Rukmini Callimachi, "Militant Group Says It Killed American Jour-nalist in Syria," *New York Times,* August 19, 2014, http://www.nytimes

.com/2014/08/20/world/middleeast/isis-james-foley-syria-execution
.html.

81. Nancy Scola, "Foley Video, Photos Being Scrubbed from Twitter," *Washington Post,* August 19, 2014, http://www.washingtonpost
.com/blogs/the-switch/wp/2014/08/19/foley-video-photos-being
-scrubbed-from-twitter/.

82. "Social Media Pushes Back at Miliant Propaganda," Asharq Al-Awsat,
August 21, 2014, http://www.aawsat.net/2014/08/article55335722.

83. Shirley Li, "Twitter Will Crack Down on Trolls After Robin William'
Daughter Was Bullied off Network," *Wire,* August 14, 2014, http://
www.thewire.com/technology/2014/08/twitter-to-crack-down-on
-trolls-after-robin-williams-daughter-bullied-off-network/376059/.

84. Alex Ben Block, "Robin Williams' Death Spurs New Twitter Policy,
Magazine Sales Spikes," *Hollywood Reporter,* August 28, 2014, http://
www.hollywoodreporter.com/news/robin-williams-death-spurs
-new-728722.

85. "Contacting Twitter About a Deceased User or Media Concerning
a Deceased Family Member," Twitter Help Center, https://support
.twitter.com/articles/87894-contacting-twitter-about-a-deceased-user
-or-media-concerning-a-deceased-family-member.

86. Hayley Tsukayama, "Inside How Twitter Decides Which Images to
Block," *Washington Post,* August 21, 2014, http://www.washington
post.com/blogs/the-switch/wp/2014/08/21/inside-how-twitter-de
cides-which-images-to-block/.

87. Lance Whitney, "Twitter Staff Received Death Threats from ISIS,
Admits CEO," *CNET,* October 10, 2014, http://www.cnet.com/news
/twitter-staff-received-death-threats-from-isis-admits-ceo/.

88. Suspended accounts were logged from a list of approximately 3,000
ISIS-supporting Twitter users monitored by J.M. Berger. The count of
suspensions, which is periodically refreshed, did not include multiple
iterations of some major accounts that returned and were immediately
suspended in between refresh periods.

89. Garling, "Twitter C.E.O. Dick Costolo on Receiving Death Threats
from ISIS"; Whitney, "Twitter Staff Received Death Threats from ISIS,
Admits CEO."

90. Caitlin Dewey, "Twitter Just Overhauled Its Abuse-Reporting Policy.

But Is It Enough to Stop Harassment?" *Washington Post,* December 2, 2014, http://www.washingtonpost.com/news/the-intersect/wp/2014/12/02/twitter-just-overhauled-its-abuse-reporting-policy-but-is-it-enough-to-stop-harassment/.

91. Based on the analysis of 76,747 tweets posted in August 2014, collected on August 29, 2014, against 105,245 tweets posted in October, collected October 31, 2014. On each date, the last 200 tweets from a list of ISIS-supporting accounts were collected. This methodology has the effect of exaggerating the number of retweets (but in a roughly consistent manner between the two periods). The true average number of retweets is lower in both months. Comparisons are for tweets posted in August versus tweets posted in October.

92. ISIS Twitter strategy document, September 12, 2014; Shayba al-Hamdi; http://justpaste.it/h26t.

93. ISIS Twitter strategy document, November 25, 2014, http://justpaste.it/jpabout.

94. J.M. Berger, Twitter post, September 30, 2014, https://twitter.com/intelwire/status/517125173175390209.

95. Jasper Hamill, "Anonymous Hacktivists Prepare for Strike Against ISIS 'Supporters,'" *Forbes,* June 27, 2014, www.forbes.com/sites/jasperhamill/2014/06/27/anonymous-hacktivists-prepare-for-strike-against-isis-supporters/.

96. http://alplatformmedia.com/vb/showthread.php?t=64480, accessed November 3, 2014.

97. Mujahideen Twitter Handbook; http://justpaste.it/h1sg.

98. Data collected September 15 and November 3, 2014.

99. Ali Fisher and Nico Prucha, "ISIS Is Winning the Online Jihad Against the West," Daily Beast, October 1, 2014, http://www.thedailybeast.com/articles/2014/10/01/isis-is-winning-the-online-jihad-against-the-west.html.

100. J.M. Berger, "Resistible Force Meets Moveable Object," *IntelWire,* October 2, 2014, http://news.intelwire.com/2014/10/resistable-force-meets-movable-object.html.

CHAPTER 8. THE AQ-ISIS WAR

1. Brynjar Lia, "Abu Mus'ab al-Suri's Critique of Hard Lined Salafists in the Jihadist Current," *CTC Sentinel,* December 15, 2007, https://www .ctc.usma.edu/posts/abu-musab-al-suri%E2%80%99s-critique-of -hard-line-salafists-in-the-jihadist-current.

2. Portions of this section were adapted from J.M. Berger, "The Islamic State vs. Al Qaeda: Who's Winning the War to Become the Jihadi Superpower?" *Foreign Policy,* September 2, 2014, http://www.foreignpol icy.com/articles/2014/09/02/islamic_state_vs_al_qaeda_next_jihadi _super_power.

3. World Islamic Front, "Jihad against Jews and Crusaders," statement, February 23, 1998, http://fas.org/irp/world/para/docs/980223-fatwa .htm. Al Qaeda did formally change its name to "Qa'idat al-Jihad to commemorate the merger, but the vernacular never embraced the new name.

4. "Jama'a al-Islamiya rejects Assem Abdel Magued," *Egypt Independent,* May 12, 2013, http://www.egyptindependent.com/news/jama-al-is lamiya-rejects-assem-abdel-magued.

5. Michael Smith, "Analysis: Al-Qa'ida Renews Its Allegiance to Mullah Omar," Downrange blog, July 21, 2014.

6. Vahid Brown, "The Façade of Allegiance: Bin Ladin's Dubious Pledge to Mullah Omar," *CTC Sentinel* 3, no. 1 (January 2010), https://www .ctc.usma.edu/posts/the-facade-of-allegiance-bin-ladin%E2%80%99s -dubious-pledge-to-mullah-omar.

7. Aaron Zelin, "The War Between ISIS and al-Qaeda for Supremacy of the Global Jihadist Movement," *Research Notes,* Washington Institute, no. 20 (June 2014), http://www.washingtoninstitute.org/uploads /Documents/pubs/ResearchNote_20_Zelin.pdf.

8. Liz Sly, "Al-Qaeda Disavows Any Ties with Radical Islamist ISIS Group in Syria, Iraq," *Washington Post,* February 3, 2014, http://www.wash ingtonpost.com/world/middle_east/al-qaeda-disavows-any-ties-with -radical-islamist-isis-group-in-syria-iraq/2014/02/03/2c9afc3a-8cef -11e3-98ab-fe5228217bd1_story.html.

9. "This Is the Promise of Allah," Al Hayat Media Center, 2014, https:// azelin.files.wordpress.com/2014/06/shaykh-abc5ab-mue1b8a5ammad -al-e28098adnc481nc4ab-al-shc481mc4ab-22this-is-the-promise-of

-god22-en.pdf; J.M. Berger, "Gambling on the Caliphate," *IntelWire*, June 29, 2014, http://news.intelwire.com/2014/06/gambling-on-ca liphate.html.

10. J.M. Berger, "The State of the 'Caliphate' Is . . . Meh," *IntelWire*, July 2, 2014.

11. J.M. Berger, "IS Backlash Spills Over on Jabhat al Nusra," *IntelWire*, July 12, 2014.

12. Greg Miller, "Fighters Abandoning Al-Qaeda Affiliates to Join Islamic State, U.S. Officials Say," *Washington Post*, August 9, 2014, http://www .washingtonpost.com/world/national-security/fighters-abandoning -al-qaeda-affiliates-to-join-islamic-state-us-officials-say/2014/08/09 /c5321d10-1f08-11e4-ae54-0cfe1f974f8a_story.html.

13. "Senior Al-Qaeda Leader Calls for Followers to Support ISIS," *National News Yemen*, July 5, 2014, http://nationalyemen.com/2014/07/05 /senior-al-qaeda-leader-calls-for-followers-to-support-isis/.

14. Walid Ramzi, "ISIS Caliphate Spilts AQIM," *Magharebia*, July 18, 2014, http://magharebia.com/en_GB/articles/awi/features/2014/07/18/fea ture-01; Mawassi Lahcen, "Islamic State Tempts Morocco Jihadists," *Magharebia*, November 14, 2014, http://magharebia.com/en_GB/ar ticles/awi/reportage/2014/11/14/reportage-01.

15. "Jailed Indonesian Terrorist Abu Bakar Bashir Has Been Funding ISIS: Anti-Terrorism Chief," *Straits Times*, November 18, 2014.

16. "Abu Bakar Bashir's Sons Reject ISIS, Form New Group," *Khabar Southeast Asia*, August 16, 2014.

17. Michelle FlorCruz, "Philippine Terror Group Abu Sayyaf May Be Using ISIS Link for Own Agenda," *International Business Times*, September 25, 2014, http://www.ibtimes.com/philippine-terror-group -abu-sayyaf-may-be-using-isis-link-own-agenda-1695156.

18. John Simpson, "Afghan Militant Fighters 'May Join Islamic State,'" BBC Asia, September 2, 2014, http://www.bbc.com/news/world-asia -29009125.

19. Jon Boone, "ISIS Ascent in Syria and Iraq Weakening Pakistani Taliban," *Guardian*, October 22, 2014, http://www.theguardian.com /world/2014/oct/22/pakistani-taliban-spokesman-isis-pledge.

20. "Nigeria Rejects Boko Haram 'Caliphate' Claim," Al Jazeera, August 25, 2014.

21. "Local Support for Dreaded Islamic State Growing in Pakistan: Re-

port," *Times of India*, November 14, 2014, http://timesofindia.india
times.com/world/pakistan/Local-support-for-dreaded-Islamic-State
-growing-in-Pakistan-Report/articleshow/45149421.cms.

22. Simon Speakman Cordall, "8,000 Young Tunisian Men Are Eager to
Join Islamic State," *Newsweek*, August 28, 2014.

23. Kevin Sullivan, "Tunisia, After Igniting Arab Spring, Sends Most
Fighters to Islamic State in Syria," *Washington Post*, October 28, 2014,
http://www.washingtonpost.com/world/national-security/tunisia
-after-igniting-arab-spring-sends-the-most-fighters-to-islamic-state-in
-syria/2014/10/28/b5db4faa-5971-11e4-8264-deed989ae9a2_story.html.

24. "Revolution Muslim Leader Sentenced to 2.5 Years for Threatening
Jews," ADL blog, April 25, 2014.

25. Shaikh Abdullah Faisal, "Is the New Caliphate Valid?" audio recording, July 21, 2014, http://www.authentictauheed.com/2014/07/audio
-is-caliphate-valid.html.

26. Murray Wardrop, "Anjem Choudary: Profile," *Telegraph*, January 4,
2010, http://www.telegraph.co.uk/news/uknews/terrorism-in-the
-uk/6930447/Anjem-Choudary-profile.html; Scott Kaufman, "CNN
Host Stunned When Radical Muslim Cleric Makes 9/11 Joke During Soundcheck," RawStory, August 31, 2014; "Anjem Choudary held
in London terror raids," BBC, September 25, 2015, http://www.bbc
.com/news/uk-england-29358758; Griff Witte, "In Britain, Islamist extremist Anjem Choudary proves elusive," *Washington Post*, October
11, 2014, http://www.washingtonpost.com/world/europe/in-britain
-islamist-extremist-anjem-choudary-proves-elusive/2014/10/11/eb
731514-4e43-11e4-8c24-487e92bc997b_story.html; Andrew Anthony,
"Anjem Choudary: the British extremist who backs the caliphate," *Guardian*, September 6, 2015, http://www.theguardian.com
/world/2014/sep/07/anjem-choudary-islamic-state-isis.

27. Michael Safi, "Radical Preacher Back in Melbourne After Deportation
from Philippines," *Guardian*, July 22, 2014.

28. Niraj Warikoo, "U.S.: Dearborn Cleric Popular with ISIS Fighters Owes
$250K for Fraud," *Detroit Free Press*, August 21, 2014; Shiv Malik, and
Michael Safi, "Revealed: the radical clerics using social media to back
British jihadists in Syria," *Guardian*, April 15, 2014, http://www.the
guardian.com/world/2014/apr/15/preachers-spiritual-cheerleaders-
social-media-syria-london-university; Joseph A. Carter, Shiraz Maher

and Peter R. Neumann, "Greenbirds: Measuring Importance and Influence in Syrian Foreign Fighter Networks," International Centre for the Study of Radicalization, April 2014, http://icsr.info/2014/04/icsr-report-inspires-syrian-foreign-fighters/; Clarissa Ward, "US Cleric Inspiring Jihadists in Syria?", CBS News, April 15, 2014, http://www.cbsnews.com/news/fiery-speeches-lure-americans-to-fight-in-syria/.

29. Although Twitter had shut down ISIS's official accounts, individual accounts remained on Twitter and on file-sharing services that the authors have determined are operated by members of ISIS's official media team. For the occasion, ISIS also launched a new website for uploading its press releases.

30. *Dabiq* no. 5, November 2014, 24.

31. Shadi Bushra, "Egypt's Ansar Bayt al-Maqdis Swears Allegiance to ISIS: Statement," Al Arabiya, November 4, 2014, http://english.alarabiya.net/en/News/middle-east/2014/11/04/Egypt-s-Ansar-Bayt-al-Maqdis-swears-allegiance-to-ISIS.html; "Egypt Militant Group Denies Pledging Loyalty to Islamic State: Twitter," Reuters, November 4, 2014, http://www.reuters.com/article/2014/11/04/us-egypt-militants-islamicstate-denial-idUSKBN0IO19F20141104.

32. "The Charge of the *Ansar* (supporters)," video, November 15, 2014; http://uk.reuters.com/article/2014/11/14/uk-egypt-security-claim-idUKKCN0IY2HL20141114.

33. Data collected from Twitter, November 2014.

34. Pledge from Libya to the Islamic State, November 10, 2014; "Libya's Islamist Militant Parade with ISIS Flags," video, Al Arabiya, October 6, 2014, http://english.alarabiya.net/en/News/africa/2014/10/06/Video-Libya-s-Islamist-militants-parade-with-ISIS-flags.html; Aaron Zelin, "The Islamic State's First Colony in Libya," Washington Institute, October 10, 2014, http://www.washingtoninstitute.org/policy-analysis/view/the-islamic-states-first-colony-in-libya.

35. Leela Jacinto, "Jund al-Khalifa: The IS-linked Group That Shot into the Spotlight," France 24, September 25, 2014, http://www.france24.com/en/20140923-terrorism-france-algeria-jund-al-khalifa-kidnapping-islamic-state/.

36. "A Statement About What Was Contained in the Speech of Sheikh Abu

Bakr al-Baghdadi 'Even If the Disbelievers Dislike It,'" video, Al Qaeda in the Arabian Peninsula, November 21, 2014.

37. Aaron Zelin, "The Islamic State's Archipelago of Provinces," Washington Institute, November 14, 2014, http://www.washingtonin stitute.org/policy-analysis/view/the-islamic-states-archipelago-of -provinces.

38. *Dabiq* no. 5, 24.

39. Declan Walsh, "Allure of ISIS for Pakistanis Is on the Rise," *New York Times*, November 21, 2014, http://www.nytimes.com/2014/11/22 /world/asia/isis-pakistan-militants-taliban-jihad.html.

40. J.M. Berger, "IS Closes In on JN in Hastag Battle, but Attention Is Divided," *IntelWire*, October 22, 2014, http://news.intelwire .com/2014/10/is-closes-in-on-jn-in-hashtag-battle.html; J.M. Berger, "Gaza Dominates Talk in Jihadist Finance Networks; IS Still Struggles for Acceptance," *IntelWire*, August 9, 2014, http://news.intelwire .com/2014/08/gaza-dominates-talk-in-jihadist-finance.html; J.M. Berger, "IS Backlash Spills Over on Jabhat Al Nusra," *IntelWire*, July 12, 2014, http://news.intelwire.com/2014/07/is-backlash-spills-over-on -jabhat-al.html; Ruth Sherlock, "'Moderate' Syrian Rebels Defecting to ISIS, Blaming Lack of U.S. Support and Weapons," *National Post*, November 11, 2014, http://news.nationalpost.com/2014/11/11/mod erate-syrian-rebels-defecting-to-isis-blaming-lack-of-u-s-support-and -weapons/.

41. Jamie Dettmer, "ISIS and Al Qaeda Ready to Gang Up on Obama's Rebels," Daily Beast, November 11, 2014, www.thedailybeast .com/articles/2014/11/11/al-qaeda-s-killer-new-alliance-with-isis .html?via=mobile&source=email; Deb Riechmann, "AP Sources: IS, Al-Qaida Reach Accord in Syria," Associated Press, The Big Story, November 13, 2014, http://bigstory.ap.org/article/c71e3be959414e 69bde25f792e18aad6/ap-sources-al-qaida-reach-accord-syria; "Did IS Refuse Truce with Syrian Islamist Factions?" Radio Free Europe/ Radio Liberty, November 13, 2014, http://www.rferl.mobi/a/islamic -state-truce-syria-factions-islamist/26690340.html; Nick Paton Walsh, "Al-Nusra Advances in Syria's Idlib Area, Pushing Back Moderates, Activists Say," CNN World, last update November 3, 2014, http://www .cnn.com/2014/11/03/world/meast/syria-al-nusra-front/.

42. Thomas Joscelyn, "Al Qaeda in the Islamic Maghreb Calls for Reconciliation Between Jihadist Groups," *Long War Journal*, July 2, 2014, http://www.longwarjournal.org/archives/2014/07/al_qaeda_in_the_isla.php; Thomas Joscelyn, "Ansar al Sharia Tunisia Leader Says Gains in Iraq Should Be Cause for Jihadist Reconciliation," *Long War Journal*, June 14, 2014, http://www.longwarjournal.org/archives/2014/06/ansar_al_sharia_tuni_8.php#; Thomas Joscelyn, "Saudi Cleric's Reconciliation Initiative for Jihadist Draws Wide Support, then a Rejection," *Long War Journal*, January 27, 2014, http://www.longwarjournal.org/archives/2014/01/saudi_clerics_reconc.php#; Thomas Joscelyn and Bill Roggio, "Shabaab Leader Calls for Mediation in Syria, Says Zawahiri is 'Our Shiekh and Emir,'" *Hiiran Online*, http://hiiraan.com/news4/2014/May/54709/shabaab_leader_calls_for_mediation_in_syria_says_zawahiri_is_our_sheikh_and_emir.aspx#sthash.ldvgpZxe.dpbs, last updated May 17, 2014.

43. Kristina Wong, "al Qaeda in Yemen Declares Support for ISIS," *Hill*, August 19, 2014, http://thehill.com/business-a-lobbying/215480-al-qaeda-in-yemen-declares-support-for-isis; Joscelyn, "Al Qaeda in the Islamic Maghreb Calls for Reconciliation Between Jihadist Groups"; Sam Jones, "AQ Group Offers Support to ISIS," *Financial Times*, September 17, 2014, http://www.ft.com/intl/cms/s/0/d964dcb6-3e82-11e4-adef-00144feabdc0.html.

44. Joanna Paraszczuk, "Which Dagestani Caucasus Emirate Groups Are Switching To IS?", *From Chechnya to Syria*, http://www.chechensinsyria.com/?p=23223.

45. J.M. Berger, "Zawahiri Falls off the Map, Is Rebuked by Top al Nusra Figure," *IntelWire*, August 18, 2014, http://news.intelwire.com/2014/08/zawahiri-falls-off-map-gets-rebuked-by.html.

46. Andres Perez, "With ISIS Stealing Its Thunder, al Qaeda Declares Jihad on India," *Week*, September 4, 2014, http://theweek.com/speedreads/index/267515/speedreads-with-isis-stealing-its-thunder-al-qaeda-declares-jihad-on-india; "Overshadowed by ISIS, al Qaeda Claims Expansion," CBS News, September 4, 2014, http://www.cbsnews.com/news/overshadowed-by-isis-al-qaeda-leader-zawahri-expansion-india/.

47. Arif Rafiq, "The New Al Qaeda Group in South Asia Has Nothing to Do with ISIS," *New Republic*, September 5, 2014, http://www.new

republic.com/article/119333/al-qaeda-indian-subcontinent-not-re
sponse-islamic-state.

48. Data collected from Twitter, September and October 2014.

49. Faith Karimi, Omar Nor, and Jason Hanna, "Al-Shabaab Names Successor to Slain Leader; Somalia on High Alert," CNN, September 7, 2014, http://www.cnn.com/2014/09/06/world/africa/somalia-go dane-high-alert/; "Al-Shabab Names New Leader After Godane Death in US Strike," BBC Africa, September 6, 2014, http://www.bbc.com /news/world-africa-29093200; Thomas Joscelyn, "Shabaab Names New Emir, Reaffirms Allegiance to al Qaeda," *Long War Journal,* September 6, 2014, http://www.longwarjournal.org/archives/2014/09 /shabaab_names_new_em.php.

50. "Some Ruminations on the Syrian Foreign Fighters Problem," Haganah Forum, http://forum.internet-haganah.com/ipb/index .php?/topic/2642-some-ruminations-on-the-syrian-foreign-fighter -problem/, accessed November 24, 2013.

51. For example, Fort Hood shooter Nidal Hasan, attempted Fort Hood assailant Jason Abdo, Seattle plotter Khalid Abdul Latif, the Fort Dix Six, and many others. See J.M. Berger, "Why U.S. Terrorists Reject the Al Qaeda Playbook," *Atlantic,* July 19, 2011, http://www.theatlantic.com /international/archive/2011/07/why-us-terrorists-reject-the-al-qaeda -playbook/242019/.

52. Katherine Zimmerman and Alexis Knutsen, "Warning: AQAP's Looming Threat in Yemen," American Enterprise Institute, August 14, 2014, http://www.aei.org/publication/warning-aqaps-looming-threat-in -yemen/.

53. Jack Jenkins, "The Book That Really Explains ISIS (Hint: It's Not the Quran," Think Progress, September 10, 2014, http://thinkprogress .org/world/2014/09/10/3565635/the-book-that-really-explains-isis -hint-its-not-the-quran/.

54. "Libya, Calvinball, Solidarity: Reflections on Refelctions," Abu Aardvark blog, November 19, 2014, http://abuaardvark.typepad.com /abuaardvark/2014/11/reflections-on-reflections.html.

55. "Oklahoma Beheading: FBI Probing Suspect's Recent Conversion to Islam," Fox News, September 27, 2014, http://www.foxnews.com/us /2014/09/27/woman-beheaded-at-oklahoma-food-distribution-center -police-say/; "Oklahoma Beheading Suspect Described as a 'Little

Odd,'" CBS News, September 27, 2014, http://www.cbsnews.com /news/alton-nolen-oklahoma-beheading-suspect-described-as-a-little -odd/.

56. Henry Lee, "Guns Seized from Man Who Wanted 'We Love ISIS' Hat," *SFGate*, November 13, 2014, http://www.sfgate.com/crime /article/Guns-seized-from-man-who-wanted-We-Love-ISIS-5890600 .php.

57. James Gordon Meek and Josh Margolin, "NYC Ax Attacker Was Consumed by Desire to Strike U.S. Authority Figures, Police Say," ABC News, November 3, 2014, http://abcnews.go.com/US/nyc-ax-attacker -consumed-desire-strike-us-authority/story?id=26664787.

58. Daniel LeBlanc and Steven Chase, "Two Soldiers Struck in Quebec Hit-And-Run," *Globe and Mail*, October 20, 2014, http://www.theglobe andmail.com/news/politics/two-soldiers-injured-in-quebec-hit-and -run/article21177035/. Data collected from Twitter, October 2014.

59. Dan Wilkofsky and Osama Abu Zeid, "US-Backed SRF 'No Longer' in South Idlib After Nusra Victory," Syria: Direct, November 4, 2014, http://syriadirect.org/main/30-reports/1653-us-backed-srf-no-longer -in-south-idlib-after-nusra-victory.

60. Michael Noonan, "15,000-Plus for Fighting: The Return of the Foreign Fighters," War on the Rocks, October 8, 2014, http://waron therocks.com/2014/10/15000-plus-for-fighting-the-return-of-the-for eign-fighters/.

CHAPTER 9: ISIS'S PSYCHOLOGICAL WARFARE

1. Radio Free Europe Radio Liberty, "Foreign Fighter in Iraq and Syria: Where Do They Come From?" infograph, http://www.rferl.org/con tentinfographics/infographics/26584940.html, accessed December 11, 2014.

2. Alessandria Masi, "Where to Find ISIS Supporters: A Map of Militant Groups Aligned with the Islamic State Group," *International Business Times*, October 9, 2014, http://www.ibtimes.com/where-find-isis-sup porters-map-militant-groups-aligned-islamic-state-group-1701878.

3. "Remaining and Expanding," *Dabiq* no. 5 (November 2014): 30.

4. "Brussels Jewish Museum Killings: Suspect 'Admitted Attack,'" BBC

News (2014), June 1, 2014, http://www.bbc.com/news/world-europe
-27654505.

5. Justin Huggler, "ISIL Jihadists 'Offered Teenager $25,000 to Carry
out Bombings in Vienna,'" *The Telegraph*, October 30, 2014, http://
www.telegraph.co.uk/news/worldnews/islamic-state/11199628/Boy
-14-who-planned-Vienna-bombings-was-recruited-on-internet-by
-Isil.html.

6. "Gunman in Ottawa Attack Prepared Video of Himself," *The Telegraph*, October 26, 2014, http://www.telegraph.co.uk/news/world
news/northamerica/canada/11189257/Gunman-in-Ottawa-attack
-prepared-video-of-himself.html.

7. Thomas Hegghammer, "Should I Stay or Should I Go?" *American Politi-
cal Science Review* 107, no. 1 (February 2013); Daniel Byman and Jeremy
Shapiro, "Homeward Bound? Don't Hype the Threat of Returning
Jihadists," *Foreign Affairs* (November/December 2014); Arwa Damon
and Gul Tuysuz, "How She Went from a Schoolteacher to an ISIS
Member," CNN, October 6, 2014.

8. These statistics are based on figures from the United States: Chris Jag-
ger, "The 25 Most Common Causes of Death," MedHelp, http://www
.medhelp.org/general-health/articles/The-25-Most-Common-Causes
-of-Death/193.

9. Paul Slovic, Baruch Fischoff and Sarah Lichtenstein, "Facts and Fears:
Understanding Perceived Risk," in Richard Schwing and Walter Al-
bers, eds., *Societal Risk Assessment: How Safe Is Safe Enough?* (New York:
Plenum Press, 1980), 181–216.

10. Ayman al-Zawahiri, *Letter to Abu Musa al-Zarqawi*, October 11, 2005.

11. We use the term "risk" to denote the possibility of an adverse outcome whose probability is between zero and one. It is important to
point out at the outset that one school of thought, more prevalent
in Europe, rejects many of the assumptions appealed to here. The
alternative school questions the following assertions: that risk (other
than actuarial risk) exists and can be quantified or that risk trade-off
analysis can be accomplished; that experts and laypeople are different from one another; that "dread" is a property of risks; and that
publics are anxious and irrational and that policy should compensate
for these qualities. Sheila Jasanoff, email communication, August 18,

2002. We do not enter into this debate here, except to highlight that Daniel Kahneman, one of the foremost experts, is describing his reaction as a layperson in the text.

12. Amos Tversky and Daniel Kahneman, "Judgment Under Uncertainty: Heuristics and Biases," *Science* 185 (1974): 1124–31; Paul Slovic, Baruch Fischoff, and Sarah Lichtenstein, "Facts and Fears: Understanding Perceived Risk," in Richard Schwing and Walter Albers, eds., *Societal Risk Assessment: How Safe Is Safe Enough?* (New York: Plenum Press, 1980), 181–216. Other biases include that people tend to be overconfident in the accuracy of their assessments, even when those assessments are based on nothing more than guessing. And people seem to desire certainty: they respond to the anxiety of uncertainty by blithely ignoring uncertain risks; and a belief that while others may be vulnerable (for example to driving accidents), they themselves are not.

13. Amos Tversky and Daniel Kahneman, "Rational Choice and the Framing of Decisions," in David E. Bell, Howard Raiffa, and Amos Tversky, eds., *Decision Making* (New York: Cambridge University Press, 1988): 167–92. Originally published in *Journal of Business* 59, no. 4 (1986): 5251–78.

14. For problems in the applicability of prospect theory to group decision making, see Jack S. Levy, "Prospect Theory, Rational Choice, and International Relations," *International Studies Quarterly* 41, no. 1 (1997): 87–112.

15. Daniel Kahneman, *Thinking, Fast and Slow* (New York: Farrar, Straus & Giroux, 2011): 322–23.

16. This paragraph summarizes Susan Nieman, *Evil in Modern Thought: An Alternative History of Philosophy* (Princeton, NJ: Princeton University Press: 2002). For an updated understanding of Eichmann, see Bettina Stangneth, *Eichmann Before Jerusalem* (New York: Random House, 2014).

17. William Miller, *The Anatomy of Disgust* (Cambridge, MA: Harvard University Press), 26, citing Susan Miller, "Disgust: Conceptualization, Development and Dynamics," *International Review of Psychoanalysis*, 13 (1986): 295–307.

18. Steven Pinker, *The Better Angels of Our Nature: Why Violence Has Declined* (New York: Viking, 2011): 143.

19. Ibid., 175.

20. Simon Baron-Cohen, *The Science of Evil* (New York: Basic Books, 2011), 18.

21. Ronald Shouten and James Silver, *Almost a Psychopath: Do I (or Does Someone I Know) Have a Problem with Manipulation and Lack of Empathy?* (Center City, MN: Hazelden, 2012).

22. Pinker, *The Better Angels of Our Nature: Why Violence Has Declined*, 182.

23. Ibid., 695.

24. "From the factual knowledge that there is a universal human nature and the moral principle that no person has grounds for privileging his or her interests over others, we can deduce a great deal about how we ought to run our affairs," Pinker argues. "People are better off abjuring violence, if everyone else agrees to do so, and vesting authority in a disinterested third party. But since that third party will consist of human beings, not angels, their power must be checked by the power of other people, to force them to govern with the consent of the governed. They may not use violence against their citizens beyond the minimum necessary to prevent violence." Pinker, *The Better Angels of Our Nature: Why Violence has Declined*, 1183.

25. In 2005 Sheikh Al-Azhar, Ayatollah Sistani, and Sheikh Qaradawi, three prominent Islamic scholars (representing both Shia and Sunni perspectives), provided religious fatwas for the Amman Message. Signed by more than five hundred signatories from eighty-four countries, the message officially recognized thirteen Muslim sects, forbade *takfir*, and set forth preconditions for issuing fatwas. "The Amman Message," http://www.ammanmessage.com/, accessed November 29, 2014.

26. Scott Atran and Jeremy Ginges, "Religious and Sacred Imperatives in Human Conflict," *Science* 336 (2012): 885. Atran and his team are studying the impact of identify fusion (which occurs when individuals merge so strongly with the identity of a group that they begin to feel psychologically and viscerally at one with it, so much so that their personal identity collapses into a collective one), together with sacred values, which, according to their research, can generate a collective sense of invincibility and special destiny, as well as "devoted actors" willing to kill and die for a cause. Scott Atran, Hammad Sheikh, and Angel Gomez, "Devoted Actors Sacrifice for Close Commrade and Sacred Cause," *PNAS* 111, no. 50 (2014): 17702–3. While we agree that

sacred values may make conflicts more intractable, we have two issues with attributing ISIS's rise to sacred values. It is not only "belief
in gods and miracles" that are intensified when we are reminded of
death, but *any* value. Human beings are the only living beings forced
to live with the knowledge of their own demise, as far as we know; and
that knowledge creates existential anxiety. Ernest Becker, *The Denial
of Death* (New York: Simon & Schuster, 1997). Thomas Pyszczynski,
Sheldon Solomon, and Jeff Greenberg developed terror management
theory, which posits that all human behavior is motivated by uncon-
scious terror of death and the need to manage its attendant anxiety. We
do this through symbolic creations that make us feel that we will live
on, through our work (including terrorist acts), religious or political
affiliations, or culture. See Jeff Greenberg, et al., "Evidence for Terror
Management Theory II: The Effects of Mortality Salience on Reactions
to Those Who Threaten or Bolster the Cultural Worldview," *Journal
of Personality and Social Psychology* 58, no. 2 (February 1990): 308–18;
Brian Burke et al., "Two Decades of Terror Management Theory: A
Meta-Analysis of Mortality Salience Research," *Personality and Social
Psychology Review* 14, no. 2 (2010): 155–95.

27. We can't help but wonder why ISIS is focusing on those "sacred" values
that justify killing Shi'a, enslaving "polytheists," and pushing homo-
sexuals off tall buildings (although homosexual sex appears to be tolerated among ISIS's own fighters). Tim Arango, "A Boy in ISIS. A Suicide
Vest. A Hope to Live," *New York Times*, December 28, 2014, http://
www.nytimes.com/2014/12/27/world/middleeast/syria-isis-recruits
-teenagers-as-suicide-bombers.html.

28. There is enormous literature that attempts to address this question,
both at the level of individuals and mass movements. A good summary
is provided in James Waller, *Becoming Evil: How Ordinary People Commit Genocide and Mass Killing* (Oxford: Oxford University Press, 2007).
Stern argues in *Terror in the Name of God* that historical trauma and
collective humiliation are important explanatory factors; in the case
of ISIS's rise, both colonial rule and sectarian policies play a role. See
also the works of Vamik Volkan, discussed at the end of this chapter.

29. Rudolph J. Rummel, *Death by Government: Genocide and Mass Murder
Since 1900* (New Brunswick, NJ, and London: Transaction, 1994), 66,
as cited in Pinker, *The Better Angels of Our Nature,* 149.

30. Rummel, *Death by Government*, 66.

31. In England, public executions, often in the form of hangings, attracted enormous, unruly, often drunken crowds; and corpses were displayed on gibbets. Henry Fielding, writing in the mid-eighteenth century, complained that public execution lacked dignity and that hangings were staged more like carnivals than as solemn and edifying occasions. A member of Parliament named John Scott complained in 1773 that hangings carried out in public "degrade the man to the brute," "extinguish all compassion in the bosom of the punisher," and "harden the human heart." By the late eighteenth century, a movement arose to ban public executions. The last public execution occurred in England in 1868. Randall McGowen, "Civilizing Punishment: The End of the Public Execution in England," *Journal of British Studies* 33, no. 3 (July 1994): 257.

32. This is partly due to the United Nations General Assembly vote in 2007 in favor of a resolution for a non-binding moratorium on the death penalty, a measure that had failed in 1994 and 1999; Steven Pinker, *The Better Angels of Our Nature: Why Violence Has Declined*, 150.

    As of 2013, two-thirds of countries had either outlawed capital punishment or did not employ it in practice. For details, see "Death Sentences and Executions 2013," Amnesty International, http://www .amnesty.org/en/library/asset/ACT50/001/2014/en/652ac5b3-3979 -43e2-b1a1-6c4919e7a518/act500012014en.pdf.

33. "Guillotine," *Encyclopaedia Britannica*, online academic edition, 2014.

34. "Terrorism, n," *OED Online*, September 2014, http://www.oed.com /view/Entry/199608?redirectedFrom=terrorism, accessed September 11, 2014.

35. "Guillotine," *Encyclopaedia Britannica*.

36. "Horst Fisher Biography," Wollheim Memorial, accessed November 29, 2014.

37. Max Fisher, "Capital Punishment in China," *Atlantic*, September 22, 2011.

38. "Death Sentences and Executions 2013," Amnesty International.

39. Ester van Eijk, "Sharia and National Law in Saudi Arabia," in Jan Michiel Otto, ed., *Sharia Incorporated: A Comparative Overview of the Legal Systems of Twelve Muslim Countries in Past and Present*, edited by Jan Michiel Otto (Amsterdam: Leiden University Press, 2010).

40. "Saudi Arabia: Scheduled beheading reflects authorities' callous disregard to human rights," Amnesty International, August 22, 2014, accessed September 9, 2014.

41. Mahmoud Ahmed, "The Work of God," *Guardian*, June 5, 2003.

42. Video: "Israeli journalist Itai Anghel found himself face to face with captured ISIS fighters," TLV1, December 23, 2014, http://tlv1.fm/so-much-to-say/2014/12/23/israeli-journalist-itai-anghel-found-himself-face-to-face-with-captured-isis-fighters/.

43. In 2000, the UN adopted the Optional Protocol to the Convention on the Rights of the Child on the involvement of children in armed conflict, which established eighteen as the minimum age for any conscription or forced recruitment or direct participation in hostilities. Article 4 bans any recruitment or use of children under eighteen by nonstate actors, "Maybe We Live and Maybe We Die," Human Rights Watch, June 23, 2014.

44. Gail Sullivan, "Report: The Islamic State Puts Price Tags on Women, Literally, and Sells Them," *Washington Post,* October 3, 2014.

45. Leila Zerrougui, Report of the Secretary-General to the Security Council, A/68/878–S/2014/339, issued May 15, 2014.

46. "Maybe We Live and Maybe We Die," Human Rights Watch.

47. Al-Fares, "Frontline ISIS: How the Islamic State Is Brainwashing Children with Stone Age School Curriculum," September 1, 2014, http://www.ibtimes.co.uk/frontline-isis-how-islamic-state-brainwashing-children-stone-age-school-curriculum-1463474.

48. "The Islamic State: Grooming Children for Jihad," *Vice News* video, August 15, 2014.

49. Stewart Clegg, "Why Is Organization Theory So Ignorant? The Neglect of Total Institutions," *Journal of Management Inquiry* 15, no. 4 (2006): 427; Erving Goffman, *Asylums* (Harmondsworth, England: Penguin, 1961) as cited in ibid.

50. Miguel Pina e Cunha, Arménio Rego, and Stewart Clegg, "Obedience and Evil: From Milgram and Kampuchea to Normal Organizations," *Journal of Business Ethics* 97, no. 2 (2010): 291.

51. Mia Bloom and John G. Horgan, *Small Arms: Children and Terrorism* (Ithaca: Cornell University Press, forthcoming).

52. Omar Abdullah, "ISIS Teaches Children How to Behead in Training Camps," *Syria Deeply*, September 6, 2014.

53. "Maybe We Live and Maybe We Die," Human Rights Watch, 21.

54. Pina e Cunha, Rego, and Clegg, "Obedience and Evil," 291.

55. Kernberg, "Sanctioned Social Violence," 690.

56. "Maybe We Live and Maybe We Die," Human Rights Watch, 23.

57. "The Islamic State: Grooming Children for Jihad," *Vice News* video.

58. Ceylan Yeginsu, "ISIS Draws a Steady Stream of Recruits from Turkey," September 15, 2014.

59. "Maybe We Live and Maybe We Die," Human Rights Watch, 2014.

60. Kate Brannen, "Children of the Caliphate," *Foreign Policy,* October 27, 2014, http://www.foreignpolicy.com/articles/2014/10/24/children_of_the_caliphate_iraq_syria_child_soldiers.

61. Olivia Becker, "ISIS Is Radicalizing Kidnapped Kurdish Students," June 23, 2014; Salma Abdelaziz, "Syrian Radicals 'Brainwash' Kidnapped Kurdish School Children," *CNNWorld,* June 26, 2014.

62. "Maybe We Live and Maybe We Die," Human Rights Watch, June 23, 2014, 24.

63. Brannen, "Children of the Caliphate," 2.

64. The Rome Statute of the International Criminal Court, ratified in 2002, made the conscription or enlistment of children younger than fifteen into armed forces a war crime. "Rome Statute of the International Criminal Court," International Criminal Court, A/CONF.183/9, July 1, 2002, 8, http://www.icc-cpi.int/nr/rdonlyres/ea9aeff7-5752-4f84-be94-0a655eb30e16/0/rome_statute_english.pdf. That same year the United Nations adopted the Optional Protocol to the Convention on the Rights of the Child, which increased the age to eighteen in accordance with their definition of child.

65. R. Brett, "Contribution for Children and Political Violence," *WHO Global Report on Violence, Child Soldiering: Questions and Challenges for Health Professionals* (2001), 1, as cited in Patricia K. Kerig, Diana C. Bennett, Mamie Thompson, and Stephen P. Becker, "'Nothing Really Matters': Emotional Numbing as a Link between Trauma Exposure and Callousness in Delinquent Youth," *Journal of Traumatic Stress* 25, no. 3 (2012): 272.

66. "Child Recruitment," Office of the Special Representative of the Secretary-General for Children and Armed Conflict, https://childrenandarmedconflict.un.org/effects-of-conflict/six-grave-violations/child-soldiers/, accessed November 28, 2014.

67. Judith Lewis Herman, *Trauma and Recovery* (New York: Basic Books, 1992).

68. Bessel van der Kolk, "Developmental Trauma Disorder: Towards a Rational Diagnosis for Chronically Traumatized Children," *Psychiatric Annals* 35 (2005): 401–8, as cited in F. Klasen et al., "Posttraumatic Resilience in Former Ugandan Child Soldiers," *Child Development* 4 (2010): 1097.

69. Fionna Klasen, Gabriele Oettingen, Judith Daniels, and Hubertus Adam, "Multiple Trauma and Mental Health in Former Ugandan Child Soldiers," *Journal of Traumatic Stress* 23, no. 5 (2010): 573.

70. However, the study team points out that the finding of minimal violence may be an artifact of the sample; the group was living in a special-needs boarding school to help them readjust. Ibid., 578.

71. Theresa S. Betancourt, Ivelina I. Borisova, Marie de la Soudiére, and John Williamson, "Sierra Leone's Child Soldiers: War Exposures and Mental Health Problems by Gender," *Journal of Adolescent Health* no. 49, (2011): 21–28.

72. Theresa S. Betancourt, Elizabeth A. Newnham, Ryan McBain, and Robert T. Brennan, "Post-traumatic stress symptoms among former child soldiers in Sierra Leone: Follow-Up Study," *British Journal of Psychiatry* 203, no. 3 (2013): 196–202.

73. "Diagnostic criteria for PTSD include a history of exposure to a traumatic event that meets specific stipulations and symptoms from each of four symptom clusters: intrusion, avoidance, negative alterations in cognitions and mood, and alterations in arousal and reactivity. The sixth criterion concerns duration of symptoms; the seventh assesses functioning; and, the eighth criterion clarifies symptoms as not attributable to a substance or co-occurring medical condition." DSM-5 Diagnostic Criteria for PTSD, U.S. Department of Veterans Affairs, http://www.ptsd.va.gov/professional/PTSD-overview/dsm5_criteria_ptsd.asp.

74. Fiona Klasen et al., "Multiple Trauma and Mental Health in Former Ugandan Child Soldiers," 579.

75. Brett T. Litz, Nathan Stein, Eileen Delaney, Leslie Lebowitz, William P. Nash, Caroline Silva, Shira Maguen, "Moral Injury and Moral Repair in War Veterans: A Preliminary and Model and Intervention

Strategy," *Clinical Psychology Review* 29, no. 8 (2009): 695; Jonathan Shay, *Odysseus in America: Combat Trauma and the Trials of Homecoming* (New York: Scribner, 2010).

76. Examples from the Quran include, "Fight in the cause of God those who fight you, but do not transgress limits; for God loveth not transgressors" (2:190). "But if they cease, God is Oft Forgiving, Most Merciful. And fight them on until there is no more tumult or oppression, and there prevail justice and faith in God; but if they cease, let there be no hostility except to those who practice oppression" (2:192–193). There is also a well-cited Hadith from Abu Bakr al-Siddiq, the first caliph, who said to his army, "I advise you ten things: Do not kill women or children or an aged, infirm person. Do not cut down fruit-bearing trees. Do not destroy an inhabited place. Do not slaughter sheep or camels except for food. Do not burn bees and do not scatter them. Do not steal from the booty, and do not be cowardly." Malik's Muwatta', "Kitab al-Jihad," 21.3.10.

77. Casey T. Taft et al., "Risk Factors for Partner Violence Among a National Sample of Combat Veterans," *Journal of Consulting and Clinical Psychology* 73, no. 1 (2005): 151–59.

78. James Dao, "Drone Pilots Are Found to Get Stress Disorders Much as Those in Combat Do," *New York Times*, February 22, 2013, http://www.nytimes.com/2013/02/23/us/drone-pilots-found-to-get-stress-disorders-much-as-those-in-combat-do.html?_r=0; Rachel MacNair, *Perpetration-Induced Traumatic Stress: The Psychological Consequences of Killing* (Westport, CT: Greenwood Publishing Group, 2002). Rachel MacNair, "Psychological Reverberations for the Killers: Preliminary Historical Evidence for Perpetration-Induced Traumatic Stress," *Journal of Genocide Research* 3, no. 2 (2001): 273–82; S. Maguen et al., "Killing and Latent Classes of PTSD Symptoms in Iraq and Afghanistan Veterans," *Journal of Affective Disorders* 145, no. 3 (2013): 344–48; Angela Nickerson et al, "Accidental and Intentional Perpetration of Serious Injury or Death: Correlates and Relationship to Trauma Exposure," *Journal of Trauma: Injury, Infection, and Critical Care* 71, no. 6 (2011): 1821–28; B. Litz et al., "Moral Injury and Moral Repair in War Veterans: A Preliminary Model and Intervention Strategy," *Clinical Psychology Review* 29, no. 8 (2009): 695–706.

79. Roland Weierstall, Claudia Patricia Bueno Castellanos, Frank Neuner, and Thomas Elbert, "Relations Among Appetitive Aggression, Post-Traumatic Stress and Motives for Demobilization: A Study in Former Colombian Combatants," *Conflict and Health* 7, no. 1 (2013): 9; Roland Weierstall, Maggie Schauer, and Elbert Thomas, "An Appetite for Aggression," *Scientific American Mind* 24, no. 2 (2013): 46.

80. Steven Pinker refers to this site as the most comprehensive resource he could find: "Abolition of Slavery Timeline," Wikipedia, http://en.wikipedia.org/wiki/Abolition_of_slavery_timeline, accessed November 29, 2014.

81. "Report on the Protection of Civilians in Armed Conflict in Iraq: 6 July–10 September 2014," Office for the High Commissioner for Human Rights, September 2014, 15.

82. "Who, What, Why: Who are the Yazidis?" *BBC Magazine Monitor,* August 7, 2014, http://www.bbc.com/news/blogs-magazine-monitor-28686607.

83. Richard Spencer, "ISIL Carried out Massacres and Mass Sexual Enslavement of Yazidis, UN Confirms." *The Telegraph*, October 2014. http://www.telegraph.co.uk/news/worldnews/islamic-state/11160906/Isil-carried-out-massacres-and-mass-sexual-enslavement-of-Yazidis-UN-confirms.html; "Transgenerational Transmissions and Chosen Traumas: An Aspect of Large-Group Identity" *Group Analysis* 34, (March 2001): 79–97.

84. "The Revival of Slavery," *Dabiq,* no. 4 (September 2014): 14.

85. Ibid., 16.

86. "Su'al wa-Jawab fi al-Sabi wa Riqab: Questions and Answers on Taking Captives and Slaves," as translated by MEMRI Jihad and Terrorism Monitor, December 4, 2014, http://www.memrijttm.org/about-memri-jttm.html.

87. Ibid.

88. Ibid.

89. Ibid.

90. Ibid.

91. Vamik Volkan, "On Chosen Traumas," *Mind and Human Interaction* 3, 2013.

92. Email interview with Vamik Volkan, December 30, 2014.

93. Otto F. Kernberg, "Sanctioned Social Violence: A Psychoanalytic View, Part 1," 691.

CHAPTER 10. THE COMING FINAL BATTLE?

1. "Articles of Faith," ch. 3 in *The World's Muslims: Unity and Diversity*, Pew Research Center, August 9, 2012; http://www.pewforum.org/2012/08/09/the-worlds-muslims-unity-and-diversity-3-articles-of-faith/#end-times.

2. Miraiam Karouny, "Apocalyptic Prophecies Drive Both Sides to Syrian Battle for End of Time," Reuters, April 1, 2014.

3. Jean-Pierre Filiu, *Apocalypse in Islam* (Berkeley: University of California Press, 2011), 70.

4. David Cook, *Contemporary Muslim Apocalyptic Literature* (Syracuse, NY: Syracuse University Press, 2005), 173–74.

5. "The Five Letters to the African Corps," translated by the U.S. Defense Department's Harmony Program, AFGP-2002-600053, June 5, 2002, http://selectedwisdom.com/wp-content/uploads/2010/12/AFGP-2002-600053-trans-Meta.pdf; Vanguards of Khurasan, "'Vanguards of Khorasan' Editor Promotes al-Qaeda, Revolutions in First Appearance in as-Sahab Video," Global Terrorism Research Project, statement, released June 28, 2013, http://gtrp.haverford.edu/aqsi/aqsi-statement/775; Tara McKelvey, "New Militant Group Khorasan Creates Mystery and Fear," BBC News, September 2014, http://www.bbc.com/news/29334548.

6. Cook, *Contemporary Muslim Apocalyptic Literature*, 173.

7. William McCants, "The Foreign Policy Essay: The Sectarian Apocalypse," Brookings Institution, October 26, 2014. McCants is currently working on a book on the topic of this chapter, which we highly recommend.

8. Hassan Abbas, email communication, December 5, 2014.

9. Cook, *Contemporary Muslim Apocalyptic Literature*, 217; Filiu, *Apocalypse in Islam*, 5.

10. Cook, *Contemporary Muslim Apocalyptic Literature*, 8.

11. Ibid., 9.

12. Filiu, *Apocalypse in Islam*, 119, 130.

13. Ibid., 119, 139.

14. Karouny, "Apocalyptic Prophecies." A list and interpretation of signs that the end times are imminent can be found, for example, in "The Heart of Islam Is al-Sham and Its Covenant Is Ruling by Islam," Khilafah.com, http://www.khilafah.com/index.php/concepts/is lamic-culture/13692-the-heart-of-islam-is-al-sham-and-its-covenant -is-ruling-by-islam, last updated April 5, 2012; "The End of Times and the Signs of the Mahdi," End of Times, http://www.endoftimes .net/03signsofthemahdi05a.html; "Syria (Alsham—the Heartland of Islam)—Is a Sign of Great Coming," Islamicintrospection, http:// islamicintrospection.wordpress.com/2013/01/09/syriaalsham-the -heartland-of-islam-is-a-sign-of-great-coming/, last updated January 9, 2013.

15. David Kirkpatrick, "New Freedoms in Tunisia Drive Support for ISIS," *New York Times,* October 21, 2014, http://www.nytimes.com /2014/10/22/world/africa/new-freedoms-in-tunisia-drive-support-for -isis.html?_r=0.

16. Ibid.

17. Ibid.

18. Filiu, *Apocalypse,* 187.

19. Ibid.

20. Mustafa Setmariam Nasar, *Part I: The Roots, History and Experiences and Part II: The Call, Program and Method* (December 2004): 1518, as cited in Filiu, *Apocalypse,* 189.

21. David Cook, "Abu Musa'b Al-Suri and Abu Musa'b Al-Zarqawi: The Apocalyptic Theorist and the Apocalyptic Practitioner," unpublished ms.

22. Ibid.

23. William McCants, "ISIS Fantasies of an Apocalyptic Showdown in Northern Syria," Brookings Institution, October 3, 2014, http://www .brookings.edu/blogs/iran-at-saban/posts/2014/10/03-isis-apocalyptic -showdown-syria-mccants.

24. William McCants, "The Foreign Policy Essay.

25. Hannah Allam, "Peter Kassig's Friends Hope Unusual Islamic State Video Means He Fought His Beheading," McClatchy DC, November 16, 2014, http://www.mcclatchydc.com/2014/11/16/247033/is lamic-state-video-claims-beheading.html.

26. Mark Juergensmeyer, *Terror in the Mind of God: The Global Rise of Religious Violence* (Berkeley: University of California Press, 2000); Robert J. Lifton, *Destroying the World to Save It* (New York: Henry Holt, 1999).

27. The word *cult* is no longer used by scholars of religion. There has been academic debate over whether the use of the word *cult* is derogatory or whether this concern is just academic excessive political correctness. In 2007, though, a study looked at this question by surveying 2,500 people to look at the perspectives of nonacademics. The survey revealed "the remarkably negative view Nebraskans have of cults, [and] their general acceptance of new religious movements." Paul J. Olson, "The Public Perception of 'Cults' and 'New Religious Movements,'" *Journal for the Scientific Study of Religion* 45, no. 1 (2007): 97–106. Nonetheless, we will use the term cult here as it is commonly defined, for clarity.

28. John C. Danforth, "Final Report to the Attorney General Concerning the 1993 Confrontation at the Mt. Carmel Complex" [Redacted Version], November 8, 2000, http://upload.wikimedia.org/wikipedia /commons/8/85/Danforthreport-final.pdf; 6. Danforth concluded that the U.S. government agents were not responsible for setting the fire. The role of government agents in the siege is still contested.

29. Indeed, David Cook argues that one disadvantage to using apocalyptic teachings for terrorist groups is that it could lead followers to fatalism rather than action.

30. "Evidence Indicates Uganda Cult Held an Eerie Prelude to Fire," *New York Times,* March 26, 2000; "Cult in Uganda Poisoned Many, Police Say," *New York Times,* July 28, 2000.

31. Michael Barkun, *A Culture of Conspiracy: Apocalyptic Visions in Contemporary America* (Berkeley: University of California Press, 2003), 169, as cited in Filiu, *Apocalypse,* 193.

32. Filiu, *Apocalypse,* 76.

33. This paragraph summarizes Thomas Hegghammer and Stephane Lacroix, "The Meccan Rebellion," in *The Meccan Rebellion* (Bristol, England: Amal Press, 2011).

34. Alistair Crooke, "Middle East Time Bomb: The Aim of ISIS Is to Replace the Saud Family as the New Emirs of Arabia," *World Post,* November 2, 2014, http://www.huffingtonpost.com/alastair-crooke/isis -aim-saudi-arabia_b_5748744.html.

35. Hegghammer and Lacroix, "The Meccan Rebellion."

36. Lorne L. Dawson, "When Prophecy Fails and Faith Persists: A Theoretical Overview," *Nova Religio: The Journal of Alternative and Emergent Religions* 3, no. 1 (1999): 60–82.

37. Examples of millenarian cults that have survived the failure of prophecy include Jehovah's Witnesses, Lubavitcher Hasidim, and the Rouxists. Jehovah's Witnesses are a millenarian, evangelical Christian denomination who believe that the millennium of peace was spiritually laid in 1914, Lubavitcher Hasidim is an Orthodox Jewish movement whose living messiah passed away, and the Rouxists are a French messianic movement founded when their leader, George Roux, claimed to be the reincarnation of Christ. Dawson, "When Prophecy Fails and Faith Persists."

38. Ibid.

39. This concept is based primarily on 1 Thessalonians 4:17: "Then we which are alive and remain shall be caught up together with them in the clouds, to meet the Lord in the air: and so shall we ever be with the Lord." As interpreted by Christian fundamentalists, the rapture, which lifts the chosen few out of the mayhem of end-time destruction to meet the returning Messiah, is a reward for their steadfastness. Kerry Nobel, *Tabernable of Hate: Why They Bombed Oklahoma City* (Prescott, Ontario: Voyager, 1998), 120. The Darbyite movement of 1830s–1880s was the forerunner to modern Christian fundamentalism and supplied the theological basis for the rise of fundamentalism's emphasis on biblical literalism and inerrancy and the notion of "premillennial dispensationalism"—that Jesus will return prior to his millennial rule, and that mankind has entered the end time after receiving previous "dispensations" from God in the form of Adam's banishment from Eden, the Flood, and Christ's grace, to which man has failed to respond. Anglican John Nelson Darby's seven dispensations, although reflecting earlier thinking and sources regarding the rapture, premillenialism, and dispensationalism, was original in its focus on the rapture, which became a central feature in his prophetic system, and in his ideas pertaining to ingathering of the Jews and Israel. Charles B. Strozier, *On the Psychology of Fundamentalism in America* (Boston: Beacon Press, 1994), 183–84; James A. Aho, *The Politics of Righ-*

teousness: Idaho Christian Patriotism (Seattle: University of Washington Press, 1990), 53–54.

40. Will McCants, email communication to coauthor Jessica Stern, November 7, 2014.

41. Rosabeth Moss Kanter observed many of these commitment mechanisms in nineteenth-century utopias and communes. Rosabeth Moss Kanter, *Commitment and Community: Communes and Utopias in Sociological Perspective* (Cambridge, MA: Harvard University Press, 1972), 83–84.

42. Lifton defines totalistic groups as ideological organizations that strive to control all human behavior and thought. Lifton, *Destroying the World to Save It*.

43. Ibid., 5.

44. Ibid.

45. Jessica Stern, "Terrorist Motivations and Unconventional Weapons," in Peter Lavoy, Scott Sagan, and James Wirtz, eds., *Planning the Unthinkable* (Ithaca, NY: Cornell University Press, 2000).

46. Lifton, *Destroying the World to Save It*, 5.

CHAPTER 11. THE STATE OF TERROR

1. Reza Aslan, *No God but God* (New York: Random House, 2006), 263.

2. Karen Armstrong, *Islam—A Short History* (New York: Random House, 2002), 165.

3. Marwan Muasher, email communication, December 9, 2014.

4. Office of the Press Secretary, "Statement by the President on ISIL," The White House, September 10, 2014, http://www.whitehouse.gov /blog/2014/09/10/president-obama-we-will-degrade-and-ultimately -destroy-isil.

5. Geoff Earle, "Dempsey Hints at Ground Troups if US Attack on ISIS Fail," *New York Post*, September 16, 2014, http://nypost.com/2014/09/16 /army-general-to-congress-if-airstrikes-fail-us-should-deploy-ground -troops-in-iraq/.

6. Millenarianism involves the expectation of sweeping societal change, possibly as a result of the apocalypse.

7. Steven Pinker, email communication, September 13, 2014.

8. John D. Graham, and Jonathan Baert Wiener, eds., *Risk vs. Risk: Trade-offs in Protecting Health and the Environment* (Cambridge, MA: Harvard University Press, 1995), 234.

9. Anthony Patt and Richard Zeckhauser, "Behavioral Perceptions and Policies Toward the Environment," in Rajeev Gowda and Jeffrey C. Fox, eds., *Judgments, Decisions, and Public Policy* (Cambridge: Cambridge University Press, 2002), 256–302; Graham and Wiener, eds., *Risk vs. Risk*, 234.

10. Jessica Stern, "Dreaded Risks and the Control of Biological Weapons," *International Security* 27, no. 3 (2003): 89–123.

11. "Syrian Refugees," http://syrianrefugees.eu/, last updated October 2014; "Jordan: Vulnerable Refugees Forcibly Returned to Syria," Human Rights Watch, November 24, 2014, http://www.hrw.org /news/2014/11/23/jordan-vulnerable-refugees-forcibly-returned -syria.

12. Alessandra Masi, "Raqqa Civilians, Hit by New Assad Airstrikes, Tell Stories of ISIS Executions and Coalition Bombings," *International Business Times*, November 25, 2014, www.ibtimes.com/raqqa-civilians-hit -new-assad-airstrikes-tell-stories-isis-executions-coalition-1729295.

13. "Jabhat al-Nusra Eyes Idlib for Islamic Emirate," *Al-Monitor*, November 13, 2014, http://www.al-monitor.com/pulse/originals/2014/11 /jabhat-al-nusra-idlib-islamic-emirate.html.

14. Charles Lister, "In Syria, a Last Gasp Warning for U.S. Influence," Brookings Institution, December 5, 2014, http://www.brookings.edu /blogs/markaz/posts/2014/12/05-syria-united-states-losing-last-gasp -at-leverage.

15. Daniel Bolger, "The Truth About Wars," op-ed, *New York Times*, November 10, 2014, http://www.nytimes.com/2014/11/11/opinion/the -truth-about-the-wars-in-iraq-and-afghanistan.html?_r=0.

16. John Harwood, "An American General Explains How We Lost In Iraq And Afghanistan," NPR *On Point*, (radio), November 13, 2014, http:// onpoint.wbur.org/2014/11/13/lost-iraq-afghanistan-army-general. Former Senator James Webb and General Powell also warned against occupying Iraq, according to this show. In 2002, then Senator James Webb wrote an op-ed asking whether the American people were prepared to occupy Iraq for 30–50 years. James Webb, "Heading for Trouble: Do We Really Want to Occupy Iraq for the Next 30 Years?"

Washington Post, September 4, 2002, http://www.motherjones.com /mojo/2006/09/jim-webbs-2002-op-ed-against-invading-iraq.

17. Leslie Gelb, "Iraq Must Not Come Apart," *New York Times,* July 1, 2014, http://www.nytimes.com/2014/07/02/opinion/leslie-gelb-iraq-must -not-come-apart.html.

18. "David Petraeus: ISIS's Rise in Iraq Isn't a Surprise," *Frontline,* PBS, July 29, 2014.

19. Clint Watts, "The U.S. Can't Destroy ISIS, Only ISIS Can Destroy ISIS—The Unfortunate Merits of the 'Let Them Rot' Strategy," Foreign Policy Research Institute, September 2014, http://www.fpri.org /geopoliticus/2014/09/us-cant-destroy-isis-only-isis-can-destroy-isis -unfortunate-merits-let-them-rot-strategy.

20. Liz Sly, "The Islamic State is failing at being a state," *Washington Post,* December 24, 2014, http://www.washingtonpost.com/world /middle_east/the-islamic-state-is-failing-at-being-a-state/2014/12/24 /bfbf8962-8092-11e4-b936-f3afab0155a7_story.html; Kevin Sullivan and Karla Adam, "Hoping to create a new society, the Islamic State recruits entire families," *Washington Post,* December 24, 2014, http://www.washingtonpost.com/world/national-security/hoping -to-create-a-new-homeland-the-islamic-state-recruits-entire-families /2014/12/24/dbffceec-8917-11e4-8ff4-fb93129c9c8b_story.html

21. Karen Armstrong, *Islam* (New York: Random House, 2002), 165.

22. Allison Smith, Peter Suedfeld, Lucian Conway, and David Winter, "The Language of Violence: Distinguishing Terrorist from Nonterrorist Groups by Thematic Analysis," *Dynamics of Asymmetric Conflict: Pathways Toward Terrorism and Genocide* 1, no. 2 (2008) http://www .tandfonline.com/doi/full/10.1080/17467580802590449#.VHwBm TFvr6g.

23. "Suedfeld's Integrative Complexity Research," Peter Suedfeld's Home Page, last updated June 2004, http://www2.psych.ubc.ca/~psuedfeld /index2.html; L. Myyry, "Everday Value Conflicts and Integrative Complexity of Thought," *Scandinavian Journal of Psychology* 43, no. 5 (2002): 385–95.

24. Jose Liht, "Preventing Violent Extremism Though Value Complexity: Being Muslim Being British," *Journal of Strategic Security* 6, no. 4 (Winter 2013), http://scholarcommons.usf.edu/cgi/viewcontent.cgi ?article=1253&context=jss.

25. "Although the Disbelievers Dislike It," ISIS video, November 15, 2014.

26. "ADL Report Finds Right Wing Extremist Use Shortwave Radio to Target U.S. Audiences; Asks FCC to Investigate Possible Violation of Regulations," ADL, http://archive.adl.org/presrele/dirab_41/2655_41 .html.

27. "U.S. Designates Al-Manar as a Specially Designated Global Terrorist Entity[;] Television Station Is Arm of Hizballah Terrorist Network," U.S Department of the Treasury, press release, March 23, 2006, http://www.treasury.gov/press-center/press-releases/Pages/js4134.aspx.

28. Jillian York, "Terrorists on Twitter," Slate, June 25, 2014, http://www.slate.com/articles/technology/future_tense/2014/06/isis_twitter_suspended_how_attempts_to_silence_terrorists_online_could_backfire.html.

29. Colby Itkowitz, "State Department Trolls Islamic State Militants on Twitter," Washington Post, November 18, 2014, http://www.washingtonpost.com/blogs/in-the-loop/wp/2014/11/18/state-department-trolls-islamic-state-militants-on-twitter/.

30. "Islamic State Group 'Executes 700' in Syria," Al Jazeera, August 17, 2014, http://www.aljazeera.com/news/middleeast/2014/08/islamic-state-group-executes-700-syria-2014816123945662121.html; J.M. Berger, "For Global Jihadist Supporters, Islamic State's Massacre Wipes Out Any Sympathy Over U.S. Strikes," IntelWire, August 18, 2014, http://news.intelwire.com/2014/08/for-global-jihadist-supporters-islamic.html.

31. "Full Transcript of Bin Ladin's Speech," Al Jazeera, November 1, 2014.

32. Max Fisher, "6 Concrete Policy Ideas for Fixing America's Drone Dilemma," Washington Post, February 6, 2014, http://www.washingtonpost.com/blogs/worldviews/wp/2013/02/06/6-concrete-policy-ideas-for-fixing-americas-drone-dilemma/.

33. David Rothkopf, "Coming Clean, with Bloodstained Hands," Foreign Policy, December 9, 2014, http://foreignpolicy.com/2014/12/09/coming-clean-with-bloodstained-hands-senate-torture-report-cia-bush-administration-obama/.

34. George W. Bush, speech at the National Endowment for Democracy, Washington, DC, October 6, 2005, http://www.presidentialrhetoric.com/speeches/10.06.05.html.

35. For instance, in a 2007 study of public opinion in Egypt, Indonesia,

Morocco, and Pakistan, conducted by the Program on International Policy Attitudes at the University of Maryland, majorities in those countries believed that Washington's primary goal was to dominate the Middle East and weaken Islam and its people. Steven Kull, "Muslim Public Opinion on US Policy, Attacks on Civilians and Al Qaeda," Program on International Policy Attitudes, University of Maryland, April 24, 2007, http://www.worldpublicopinion.org/pipa/pdf/. . . /START_Apr07_rpt.pdf.

36. Thomas Carothers, "Promoting Democracy and Fighting Terror," *Foreign Affairs,* January/February 2003.

37. Alberto Abadie, "Poverty, Political Freedom, and the Roots of Terrorism," *American Economic Review* 96, no. 2 (2006): 50–56, http://www.jstor.org/stable/30034613, accessed January 30, 2014. Political scientist Erica Chenoweth found that post-1997, so-called anocracies, weak states between autocracies and democracies, have become most vulnerable to terrorism. However, she reports that terrorism in anocracies is most closely linked to the U.S. invasion of Iraq and Afghanistan. Removing Iraq, Aghanistan, and Pakistan from her data, Chenoweth shows that democracies still remain the most vulnerable to attacks. Erica Chenoweth, "Is Terrorism Still a Democratic Phenomenon?" *International Relations* 8, no. 32 (Winter 2012): 85–100. Burcu Savun and Brian J. Phillips report a similar idea, that it is not democracy itself but foreign policy that is a risk factor for vulnerability. They find that states involved in international politics, a policy often persued by democracies, are more likely to be targed for a terrorist attack than their less involved counterparts. Burcu Savun and Brian J. Phillips, "Democracy, Foreign Policy, and Terrorism," *Journal of Conflict Resolution* 53, no. 6 (2009): 878–904. Robert Pape found that suicide bombers almost always are deployed to fight a military occupation and that they almost always target democracies. Robert Anthony Pape, *Dying to Win: The Strategic Logic of Suicide Terrorism* (New York: Random House, 2005); Robert A. Pape and James K. Feldman, *Cutting the Fuse: The Explosion of Global Suicide Terrorism and How to Stop It* (Chicago: University of Chicago Press, 2010). Thomas Carothers believes that democracy promotion was tainted due to the Bush administration's association of democracy promotion with intervention and regime change, and their failure to put pressure on friendly authoritarian regimes under the pretense of

protecting economic and security interests. Thomas Carothers, "U.S. Democracy Promotion During and After Bush," Carnegie Endowment for International Peace, September 5, 2007.

38. Edward Mansfield and Jack Snyder, "Prone to Violence: The Paradox of Democratic Peace," *National Interest* no. 82 (Winter 2005): 39.

39. The term marries two closely connected ideas. It is liberal because it draws on the philosophical strain, beginning with the Greeks, that emphasizes individual liberty. It is constitutional because it rests on the tradition, beginning with the Romans, of the rule of law. Fareed Zakaria, "The Rise of Illberal Democracy," *Foreign Affairs*, November/December 1997, 2; Fareed Zakaria, *The Future of Freedom: Illiberal Democracy at Home and Abroad* (New York: Norton, 2007).

40. Zakaria, "The Rise of Illberal Democracy," 2; Zakaria, *The Future of Freedom*.

41. Marwan Muasher, *The Second Arab Awakening* (New Haven, CT: Yale University Press, 2014).

42. "Jordan's King: Fight on ISIS 'a Third World War," CBS News, December 5, 2014, http://www.cbsnews.com/news/jordan-king-abdullah-on-isis-middle-east-conflict/.

APPENDIX

1. Fred M. Donner, "Muhammad and the Caliphate: Political History of the Islamic Empire Up to the Mongol Conquest," in *The Oxford History of Islam*, ed. John L. Esposito (Oxford: Oxford University Press, 1999), 9.

2. Richard W Bulliet, *Islam: The View from the Edge* (New York: Columbia University Press, 1994), 5.

3. Ibid., 110.

4. Vali Nasr, *The Shia Revival* (New York: Norton, 2006), 49.

5. Ibid., 57.

6. Ibid., 70.

7. Nelly Lahoud, *The Jihadis' Path to Self-Destruction* (New York: Columbia University Press, 2010), 105.

8. Donner, "Muhammad and the Caliphate," 31.

9. Ibid., 31–2.

10. Albert Hourani, *Arabic Thought in the Liberal Age: 1798–1939* (Cambridge: Cambridge University Press, 2011), 10.

11. Drew Desilver, "World's Muslim population more widespread than you might think," *Pew Research Center,* June 7, 2013; http://www.pew research.org/fact-tank/2013/06/07/worlds-muslim-population-more -widespread-than-you-might-think/.

12. Donner, "Muhammad and the Caliphate," 7.

13. Ibid.

14. John L. Esposito, *Islam: The Straight Path* (Oxford: Oxford University Press, 1988), 90.

15. Mohammad Abu Rumman, *I Am A Salafi* (Amman: Friedrich Ebert Stiftung, 2014), 43.

16. Hillel Fradkin, "The History and Unwritten Future of Salafism," *Current Trends in Islamic Ideology,* 6 (2008): 13.

17. R. Scott Appleby, "Introduction," in *Spokesmen for the Despised: Fundamentalist Leaders of the Middle East,* ed. R. Scott Appleby (Chicago: University of Chicago Press, 1997), 3–4.

18. A number of taxonomies have been offered by scholars studying Salafism and this particular language—quietist, political, and jihadi— is adapted from multiple sources. One prominent example of this approach to Salafism can be found in Quintan Wiktorowicz, "Anatomy of the Salafi Movement," *Studies in Conflict & Terrorism* 29 (2006). A similar taxonomy that breaks Salafism into more than three factions has been offered by Mohammad Abu Rumman, *I Am a Salafi.*

19. Wiktorowicz, "Anatomy of the Salafi Movement," 218.

20. Ibid., 221–2.

21. Hasan Al-Banna, "Our Mission," in *Five Tracts of Hasan Al-Banna,* trans. Charles Wendell (Berkeley: University of California Press, 1988), 46.

22. Lahoud, *The Jihadis' Path to Self-Destruction,* 109.

23. Ibid.

24. Muslim Brotherhood members fled persecution in Egypt by migrating to a number of countries. Saudi Arabia is singled out principally because its oil wealth made it possible for it to proselytize far beyond its borders thus spreading both political and jihadi Salafism across the region. For an analysis of Salafism in Jordan that discusses a similar pattern of politicization, see Mohammad Abu Rumman, *I Am a Salafi.*

25. Nasr, *The Shia Revival,* 155.

26. Wiktorowicz, "Anatomy of the Salafi Movement," 225.

27. Ibid., 222.

28. Ibid.

29. Lawrence Wright, *The Looming Tower: Al-Qaeda and the Road to 9/11* (New York: Knopf, 2006), 79.

30. Wiktorowicz, "Anatomy of the Salafi Movement," 225.

31. Ibid., 225–7.

32. Khaled Abou El Fadl, *The Great Theft* (New York: Harper One, 2007), 79.

33. To further complicate the issue, "Wahhabism" is an "outsider's designation" with a controversial and somewhat derogatory connotation. Individuals that we might describe as practicing Wahhabism are actually more likely to describe themselves as "Salafi" or "Muwahhidun" (typically translated as "unitarian" and understood to be a reference to the "unity and uniqueness of God"). Christina Hellmich, "Creating the Ideology of Al Qaeda: From Hypocrites to Salafi-Jihadists," *Studies in Conflict and Terrorism* 31 (2008): 114; "Tawhid," in *The Oxford Dictionary of Islam,* ed. John L. Esposito (Oxford: Oxford University Press, 2003).

34. For an example of this position, see Ed Husain, "Saudis Must Stop Exporting Extremism: ISIS Atrocities Started With Saudi Support for Salafi Hate," *New York Times*, August 22, 2014, http://www.nytimes.com/2014/08/23/opinion/isis-atrocities-started-with-saudi-support-for-salafi-hate.html.

35. El Fadl, *The Great Theft*, 76.

36. "Takfir," in *The Oxford Dictionary of Islam,* ed. John L. Esposito.

37. Wiktorowicz, "Anatomy of the Salafi Movement," 232.

38. Quintan Wiktorowicz, "A Genealogy of Radical Islam," *Studies in Conflict & Terrorism* 28 (2005): 77.

39. Wiktorowicz, "Anatomy of the Salafi Movement," 233.

40. Ibid., 230.

41. "Jahiliyyah," in *The Oxford Dictionary of Islam,* ed. John L. Esposito.

42. Wiktorowicz, "A Genealogy of Radical Islam," 78.

43. Seyyed Vali Reza Nasr, *Mawdudi and the Making of Islamic Revivalism* (Oxford: Oxford University Press, 1996), 68.

44. El Fadl, *The Great Theft*, 83.

45. Ibid., 221.

46. Nasr, *Mawdudi and the Making of Islamic Revivalism*, 70.

47. Ibid., 74.

48. Lahoud, *The Jihadis' Path to Self-Destruction*, 115–7.

49. Mohammad Abd al-Salam Faraj, *The Neglected Duty: The Creed of Sadat's Assassins*, quoted in Quintan Wiktorowicz, "A Genealogy of Radical Islam," 79.

50. Lahoud, *The Jihadis' Path to Self-Destruction*, 123–4.

51. Abdulaziz H. Al-Fahad, "From Exclusivism to Accommodation: Doctrinal and Legal Evolution of Wahhabism," *New York University Law Review* 79, no. 2 (May 2004): 514; Jessica Stern, "Mind over Martyr: How to Deradicalize Islamic Extremists," *Foreign Affairs* 89, no. 1 (January/February 2010): 99.

52. Lahoud, *The Jihadis' Path to Self-Destruction*, 121.

53. Thomas Hegghammer, *Jihad in Saudi Arabia: Violence and Pan-Islamism since 1979* (Cambridge: Cambridge University Press, 2010), 7.

54. John L. Esposito, "Contemporary Islam: Reformation or Revolution?" in *The Oxford History of Islam*, ed. John L. Esposito, 645.

55. Fred Donner, *Muhammad and the Believers* (Cambridge, MA: Harvard University Press, 2010), 86.

56. Lahoud, *The Jihadis' Path to Self-Destruction*, xv.

57. David Bukay, "The Religious Foundations of Suicide Bombings," *Middle East Quarterly* 13, no. 4 (Fall 2006), http://www.meforum.org/1003/the-religious-foundations-of-suicide-bombings.

58. Joas Wagemakers, *A Quietist Jihadi: The Ideology and Influence of Abu Muhammad al-Maqdisi* (Cambridge: Cambridge University Press, 2012), 72–3.

59. Wiktorowicz, "A Genealogy of Radical Islam," 87.

60. Ibid., 89.

61. Ibid., 89–90.

62. Ibid., 91.

63. Nasr, *The Shia Revival*, 94, 96–8.

64. Sabrina Tavernise and Robert F. Worth, "Relentless Rebel Attacks Test Shiite Endurance," *New York Times*, September 19, 2005, http://www.nytimes.com/2005/09/19/international/middleeast/19shiites.html.

65. El Fadl, *The Great Theft*, 247–8.

66. Timothy Furnish, "Beheading in the Name of Islam," *Middle East Quarterly* 12, no. 2 (Spring 2005), http://www.meforum.org/713/beheading-in-the-name-of-islam.

67. Hellmich, "Creating the Ideology of Al Qaeda: From Hypocrites to Salafi-Jihadists," 114.

68. Assaf Moghadam, "Motives for Martyrdom: Al-Qaida, Salafi Jihad, and the Spread of Suicide Attack," *International Security* 33, no. 3 (2009): 59–60.

69. Wiktorowicz, "A Genealogy of Radical Islam," 93.

70. Mary Anne Weaver, "The Short, Violent Life of Abu Musab al-Zarqawi," *Atlantic,* June 8, 2006, http://www.theatlantic.com /magazine/archive/2006/07/the-short-violent-life-of-abu-musab-al -zarqawi/304983/.

71. Ayman Zawahiri, "Zawahiri's Letter to Zarqawi," trans. Center for Combating Terrorism at West Point, July 9, 2005, https://www.ctc .usma.edu/posts/zawahiris-letter-to-zarqawi-english-translation-2.

72. William McCants, "Militant Ideology Atlas," Combating Terrorism Center at West Point, November 1, 2006, https://www.ctc.usma .edu/posts/militant-ideology-atlas, 8; Joas Wagemakers, "Reclaiming Scholarly Authority: Abu Muhammad al-Maqdisi's Critique of Jihadi Practices," *Studies in Conflict & Terrorism* 34, no. 7 (July 2011): 526.

73. Lahoud, *The Jihadis' Path to Self-Destruction,* 243.

74. Wagemakers, *A Quietest Jihadi,* 92–3.

75. Ibid., 47.

76. Wagemakers, "Reclaiming Scholarly Authority," 525–6.

77. Wagemakers, *A Quietest Jihadi,* 74.

78. Ibid., 83.

79. Ibid.

80. Wagemakers, "Reclaiming Scholarly Authority," 524.

81. Patrick Cockburn, "ISIS Consolidates," *London Review of Books* 36, no. 16 (August 21, 2014), http://www.lrb.co.uk/v36/n16/patrick-cock burn/isis-consolidates.

82. Douglas A. Ollivant and Brian Fishman, "State of Jihad: The Reality of the Islamic War in Iraq and Syria," *War on the Rocks,* May 21, 2014, http://warontherocks.com/2014/05/state-of-jihad-the-reality-of-the -islamic-state-in-iraq-and-syria/.

83. Aaron Y. Zelin, "ISIS is Dead, Long Live the Islamic State," Washington Institute, June 30, 2014, http://www.washingtoninstitute.org /policy-analysis/view/isis-is-dead-long-live-the-islamic-state.

84. Ibid.

AFTERWORD

1. This section has been adapted from: J.M. Berger, "Barack Obama Still Misunderestimates ISIL," Politico, May 22, 2015, http://www.politico .com/magazine/story/2015/05/barack-obama-still-misunderesti mates-isil-118204.html#ixzz3bdzZJF3Y.

2. David Kilkullen, "How to Defeat Islamic State," *Australian*, May 22, 2015, http://www.theaustralian.com.au/news/ramadi-palmyra -show-west-needs-new-strategy-to-defeat-islamic-state/story-fnolg d60-1227365271699.

3. This section has been adapted from J.M. Berger, "The Middle East's Franz Ferdinand Moment," *Foreign Policy*, April 8, 2015, http:// foreignpolicy.com/2015/04/08/the-middle-easts-franz-ferdinand-moment-yemen-saudi-arabia-iran-isis/.

4. Michael R. Gordon and Rukmini Callimachi, "Kurdish Fighters Re-take Iraqi City of Sinjar From ISIS," *New York Times*, November 13, 2015.

5. "US and Russia Sign Deal to Avoid Syria Air Incidents," BBC News, October 20, 2015; "Syria Conflict: Russia Violation of Turkish Airspace 'No Accident,'" BBC News, October 6, 2015.

6. Allessandria Masi, "ISIS Supporters Announced Imminent Attack Days Before Tunisia Museum Shooting," *International Business Times*, March 18, 2015. http://www.ibtimes.com/isis-supporters-announced-imminent-attack-days-tunisia-museum-shooting-1851444; "ISIS Claims Responsibility for Tunis Barracks Attack," ANSAmed, May 26, 2015, http://www.ansamed.info/ansamed/en/news/sections/ generalnews/2015/05/26/isis-claims-responsibility-for-tunis-barracks-attack_e245a5bf-c966-4ab5-a315-d23ed86027d4.html; "Gun-man, 7 Other Soldiers Killed in Tunis Barracks Shooting," Associated Press, May 25, 2015, http://www.nytimes.com/aponline/2015/05/25/ world/middleeast/ap-ml-tunisia-shooting.html?_r=0.

7. Charles Lister, "An Internal Struggle: Al Qaeda's Syrian Affiliate Is Grappling with Its Identity," Huffington Post, May 31, 2015, http:// www.huffingtonpost.com/charles-lister/an-internal-struggle-al-q_b_7479730.html.

8. Aaron Y. Zelin, "The State of al Qaeda," ICSR Insight, April 13, 2015, http://www.washingtoninstitute.org/policy-analysis/view/the-state -of-al-qaeda.

9. Thomas Joscelyn, "Zawahiri Argues Islamic State's Caliphate Is Illegitimate in Newly Released Message," *Long War Journal,* September 9, 2015, http://www.longwarjournal.org/archives/2015/09/zawahiri-says-islamic-states-caliphate-is-illegitimate-in-newly-released-message.php.

10. Andrew Roth, "Russia Confirms Sinai Plane Crash was the Work of Terrorists," *Washington Post,* November 17, 2015, https://www.washingtonpost.com/world/russia-confirms-sinai-crash-was-the-work-of-terrorists/2015/11/17/496286f4-8d05-11e5-ae1f-af46b7df8483_story.html.

11. Based on analysis of Abu Omar's social network by J.M. Berger.

12. Paul Cruickshank, "Inside the ISIS Plot to Attack the Heart of Europe," CNN.com, February 13, 2015, http://www.cnn.com/2015/02/13/europe/europe-belgium-isis-plot/.

13. Don Melvin and Matthew Chance, "Russia Says Bomb Brought Down Jet in Sinai, offers $50 million reward," CNN.com, November 17, 2015, http://www.cnn.com/2015/11/17/middleeast/russian-metrojet-crash-bomb/index.html; "Beirut attacks: Suicide bombers kill dozens in Shia suburb," BBC News, November 12, 2015, http://www.bbc.com/news/world-middle-east-34795797.

14. Eyder Peralta, "The Paris Attacks: What We Know Right Now," November 16, 2015, http://www.npr.org/sections/thetwo-way/2015/11/15/456094470/the-paris-attacks-what-we-know-right-now; Paul Cruickshank, "Paris Attacks: 'Ringleader' Abdelhamid Abaaoud Killed in Raid," CNN.com, February 13, 2015, http://www.cnn.com/2015/02/13/europe/europe-belgium-isis-plot/.

15. *Dabiq,* no. 12 (November 2015).

16. Gabe Joselow, "Paris Attacks: Syria Refugees in France's 'Jungle' Fear Backlash," NBC News, November 19, 2015, http://www.nbcnews.com/storyline/paris-terror-attacks/paris-attacks-syria-refugees-frances-jungle-fear-backlash-n466076; "Refugees From War Aren't the Enemy," *New York Times,* November 18, 2015, http://www.nytimes.com/2015/11/19/opinion/refugees-from-war-arent-the-enemy.html?_r=0.

17. Richard Landes, *Heaven on Earth: The Varieties of the Millennial Experience* (Oxford: Oxford University Press, 2011), 15.

INDEX